endorsed for
Edexcel :::

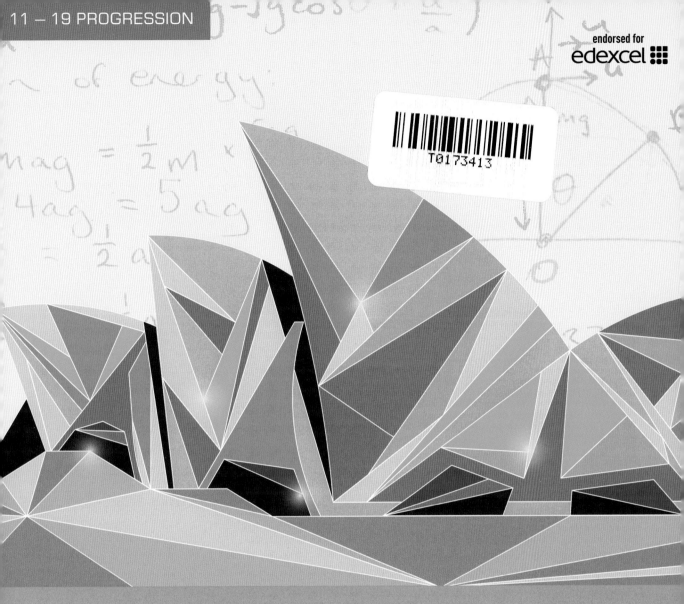

Edexcel AS and A level Further Mathematics

Further Mechanics 2

FM2

Series Editor: Harry Smith

Authors: Dave Berry, Keith Gallick, Susan Hooker, Michael Jennings, Jean Littlewood, Bronwen Moran, Su Nicholson, Laurence Pateman, Keith Pledger, Harry Smith, Jack Williams

Pearson

Published by Pearson Education Limited, 80 Strand, London WC2R 0RL.

www.pearsonschoolsandfecolleges.co.uk

Copies of official specifications for all Pearson qualifications may be found on the website:
qualifications.pearson.com

Text © Pearson Education Limited 2018
Edited by Tech-Set Ltd, Gateshead
Typeset by Tech-Set Ltd, Gateshead
Original illustrations © Pearson Education Limited 2018
Cover illustration Marcus@kja-artists

The rights of Dave Berry, Keith Gallick, Susan Hooker, Michael Jennings, Jean Littlewood, Bronwen
Moran, Su Nicholson, Laurence Pateman, Keith Pledger, Harry Smith and Jack Williams to be
identified as authors of this work have been asserted by them in accordance with the Copyright,
Designs and Patents Act 1988.

First published 2018

23
10 9 8 7 6 5 4 3

British Library Cataloguing in Publication Data
A catalogue record for this book is available from the British Library

ISBN 978 1292 18332 9

Printed in the UK by Bell and Bain Ltd, Glasgow

Acknowledgements
The authors and publisher would like to thank the following for their kind permission to
reproduce their photographs:

(Key: b-bottom; c-centre; l-left; r-right; t-top)

Alamy Stock Photo: Günter Lenz/ImageBROKER 1, 124l, Alvey & Towers Picture Library 36, 124c, G.
Lacz/Arco Images GmbH 146, 212l, Andy Fritzsche/EyeEm 171, 212r, **Shutterstock**: I T A L O 77, 124r

All other images © Pearson Education

A note from the publisher
In order to ensure that this resource offers high-quality support for the associated Pearson
qualification, it has been through a review process by the awarding body. This process confirms
that this resource fully covers the teaching and learning content of the specification or part
of a specification at which it is aimed. It also confirms that it demonstrates an appropriate
balance between the development of subject skills, knowledge and understanding, in addition
to preparation for assessment.

Endorsement does not cover any guidance on assessment activities or processes (e.g. practice
questions or advice on how to answer assessment questions), included in the resource nor does
it prescribe any particular approach to the teaching or delivery of a related course.

While the publishers have made every attempt to ensure that advice on the qualification
and its assessment is accurate, the official specification and associated assessment guidance
materials are the only authoritative source of information and should always be referred to for
definitive guidance.

Pearson examiners have not contributed to any sections in this resource relevant to
examination papers for which they have responsibility.

Examiners will not use endorsed resources as a source of material for any assessment set by
Pearson.

Endorsement of a resource does not mean that the resource is required to achieve this Pearson
qualification, nor does it mean that it is the only suitable material available to support the
qualification, and any resource lists produced by the awarding body shall include this and
other appropriate resources.

Pearson has robust editorial processes, including answer and fact checks, to ensure the
accuracy of the content in this publication, and every effort is made to ensure this publication
is free of errors. We are, however, only human, and occasionally errors do occur. Pearson is not
liable for any misunderstandings that arise as a result of errors in this publication, but it is
our priority to ensure that the content is accurate. If you spot an error, please do contact us at
resourcescorrections@pearson.com so we can make sure it is corrected.

Contents

Overarching themes

The following three overarching themes have been fully integrated throughout the Pearson Edexcel AS and A level Mathematics series, so they can be applied alongside your learning and practice.

1. Mathematical argument, language and proof

- Rigorous and consistent approach throughout
- Notation boxes explain key mathematical language and symbols
- Dedicated sections on mathematical proof explain key principles and strategies
- Opportunities to critique arguments and justify methods

2. Mathematical problem solving

- Hundreds of problem-solving questions, fully integrated into the main exercises
- Problem-solving boxes provide tips and strategies
- Structured and unstructured questions to build confidence
- Challenge boxes provide extra stretch

The Mathematical Problem-solving cycle

specify the problem

collect information

process and represent information

interpret results

3. Mathematical modelling

- Dedicated modelling sections in relevant topics provide plenty of practice where you need it
- Examples and exercises include qualitative questions that allow you to interpret answers in the context of the model
- Dedicated chapter in Statistics & Mechanics Year 1/AS explains the principles of modelling in mechanics

Finding your way around the book

Access an online digital edition using the code at the front of the book.

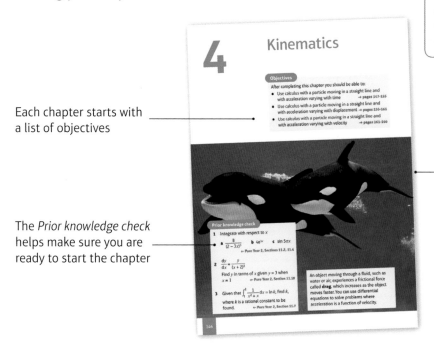

Each chapter starts with a list of objectives

The *Prior knowledge check* helps make sure you are ready to start the chapter

The real world applications of the maths you are about to learn are highlighted at the start of the chapter with links to relevant questions in the chapter

Exercise questions are carefully graded so they increase in difficulty and gradually bring you up to exam standard

Exercises are packed with exam-style questions to ensure you are ready for the exams

Challenge boxes give you a chance to tackle some more difficult questions

Exam-style questions are flagged with Ⓔ

Problem-solving questions are flagged with Ⓟ

A level content is clearly flagged with Ⓐ

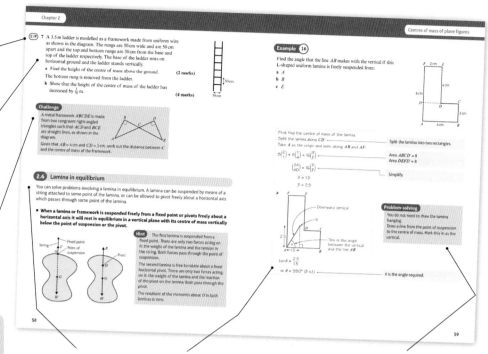

Each section begins with explanation and key learning points

Step-by-step worked examples focus on the key types of questions you'll need to tackle

Each chapter ends with a *Mixed exercise* and a *Summary of key points*

Problem-solving boxes provide hints, tips and strategies, and *Watch out* boxes highlight areas where students often lose marks in their exams

Every few chapters a *Review exercise* helps you consolidate your learning with lots of exam-style questions

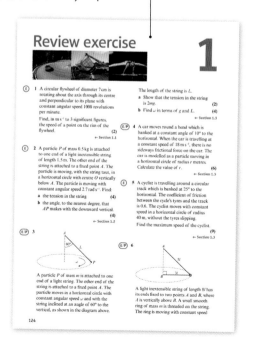

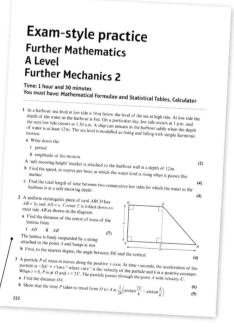

AS and A level practice papers at the back of the book help you prepare for the real thing.

v

Extra online content

Whenever you see an *Online* box, it means that there is extra online content available to support you.

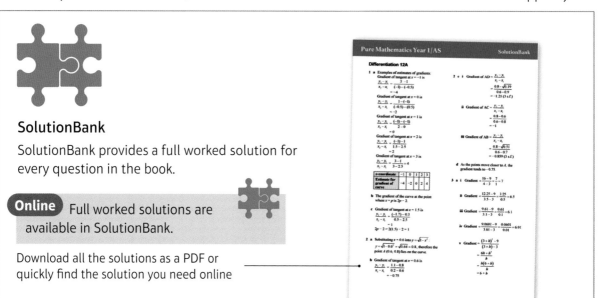

SolutionBank

SolutionBank provides a full worked solution for every question in the book.

Online Full worked solutions are available in SolutionBank.

Download all the solutions as a PDF or quickly find the solution you need online

Use of technology

Explore topics in more detail, visualise problems and consolidate your understanding using pre-made GeoGebra activities.

Online Find the point of intersection graphically using technology.

GeoGebra-powered interactives

Interact with the maths you are learning using GeoGebra's easy-to-use tools

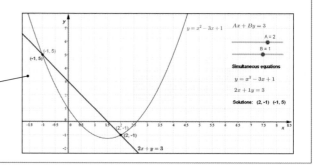

Access all the extra online content for free at:

www.pearsonschools.co.uk/fm1maths

You can also access the extra online content by scanning this QR code:

Circular motion

Objectives

After completing this chapter you should be able to:

* Understand and calculate angular speed of an object moving in a circle → **pages 2–4**
* Understand and calculate angular acceleration of an object moving on a circular path → **pages 5–10**
* Solve problems with objects moving in horizontal circles → **pages 11–18**
* Solve problems with objects moving in vertical circles → **pages 19–25**
* Solve problems when objects do not stay on a circular path → **pages 26–30**

Prior knowledge check

1 A smooth ring is threaded on a light inextensible string. The ends of the string are attached to a horizontal ceiling, and make angles of 30° and 60° with the ceiling respectively. The ring is held in equilibrium by a horizontal force of magnitude 8 N.

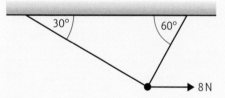

Find
   a the tension in the string   **b** the mass of the ring.
   ← **Statistics and Mechanics 2, Section 7.1**

A car travelling around a bend can be modelled as a particle on a circular path. Police use models such as this to determine likely speeds of cars following accidents. → **Exercise 1C, Q18**

2 A box of mass 4 kg is projected with speed 10 m s⁻¹ up the line of greatest slope of a rough plane, which is inclined at an angle of 20° to the horizontal. The coefficient of friction between the box and the plane is 0.15. Find:
   a the distance travelled by the box before it comes to instantaneous rest
   b the work done against friction as the box reaches instantaneous rest. ← **Further Mechanics 1, Section 2.3**

1.1 Angular speed

When an object is moving in a straight line, the speed, usually measured in m s⁻¹ or km h⁻¹, describes the rate at which distance is changing. For an object moving on a circular path, you can use the same method for measuring speed, but it is often simpler to measure the speed by considering the rate at which the radius is turning.

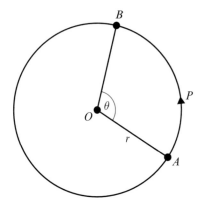

As the particle P moves from point A to point B on the circumference of a circle of radius r m, the radius of the circle turns through an angle θ radians.

The distance moved by P is $r\theta$ m, so if P is moving at v m s⁻¹

we know that $v = \dfrac{\mathrm{d}}{\mathrm{d}t}(r\theta) = r\dfrac{\mathrm{d}\theta}{\mathrm{d}t} = r \times \dot{\theta}$

Notation

$\dot{\theta}$ is the rate at which the radius is turning about O.

It is called the **angular speed of the particle** about O.

The angular speed of a particle is usually denoted by ω, and measured in rad s⁻¹.

- **If a particle is moving around a circle of radius r m with linear speed v m s⁻¹ and angular speed ω rad s⁻¹ then $v = r\omega$.**

Example 1

A particle moves in a circle of radius 4 m with speed 2 m s⁻¹. Calculate the angular speed.

Using $v = r\omega$, $2 = 4\omega$, so $\omega = 0.5$ rad s⁻¹

Example 2

Express an angular speed of 200 revolutions per minute in radians per second.

Each complete revolution is 2π radians, so 200 revolutions is 400π radians per minute.
Therefore the angular speed is
$\dfrac{400\pi}{60} = 20.9$ rad s⁻¹ (3 s.f.)

Watch out Sometimes an angular speed is described in terms of the number of revolutions completed in a given time.

Example 3

A particle moves round a circle in 10 seconds at a constant speed of $15 \, \text{m s}^{-1}$. Calculate the angular speed of the particle and the radius of the circle.

The particle rotates through an angle of 2π radians in 10 seconds, so $\omega = \dfrac{2\pi}{10} = 0.628 \, \text{rad s}^{-1}$ (3 s.f.)

Using $v = r\omega$, $r = \dfrac{v}{\omega} = \dfrac{15}{0.628} = 23.9 \, \text{m}$ (3 s.f.)

Exercise 1A

1 Express:

   a an angular speed of 5 revolutions per minute in rad s^{-1}

   b an angular speed of 120 revolutions per minute in rad s^{-1}

   c an angular speed of $4 \, \text{rad s}^{-1}$ in revolutions per minute

   d an angular speed of $3 \, \text{rad s}^{-1}$ in revolutions per hour.

2 Find the speed in m s^{-1} of a particle moving on a circular path of radius $20 \, \text{m}$ at:

   a $4 \, \text{rad s}^{-1}$

   b $40 \, \text{rev min}^{-1}$

3 A particle moves on a circular path of radius $25 \, \text{cm}$ at a constant speed of $2 \, \text{m s}^{-1}$. Find the angular speed of the particle:

   a in rad s^{-1}

   b in rev min^{-1}

4 Find the speed in m s^{-1} of a particle moving on a circular path of radius $80 \, \text{cm}$ at:

   a $2.5 \, \text{rad s}^{-1}$

   b $25 \, \text{rev min}^{-1}$

5 An athlete is running round a circular track of radius $50 \, \text{m}$ at $7 \, \text{m s}^{-1}$.

   a How long does it take the athlete to complete one circuit of the track?

   b Find the angular speed of the athlete in rad s^{-1}.

6 A disc of radius $12 \, \text{cm}$ rotates at a constant angular speed, completing one revolution every 10 seconds. Find:

   a the angular speed of the disc in rad s^{-1}

   b the speed of a particle on the outer rim of the disc in m s^{-1}

   c the speed of a particle at a point $8 \, \text{cm}$ from the centre of the disc in m s^{-1}.

7 A cyclist completes two circuits of a circular track in 45 seconds. Calculate:

 a his angular speed in rad s^{-1}

 b the radius of the track given that his speed is 40 km h^{-1}.

8 Anish and Bethany are on a fairground roundabout. Anish is 3 m from the centre and Bethany is 5 m from the centre. If the roundabout completes 10 revolutions per minute, calculate the speeds with which Anish and Bethany are moving.

9 A model train completes one circuit of a circular track of radius 1.5 m in 26 seconds. Calculate:

 a the angular speed of the train in rad s^{-1}

 b the linear speed of the train in m s^{-1}.

10 A train is moving at 150 km h^{-1} round a circular bend of radius 750 m. Calculate the angular speed of the train in rad s^{-1}.

(P) 11 The hour hand on a clock has radius 10 cm, and the minute hand has radius 15 cm. Calculate:

 a the angular speed of the end of each hand

 b the linear speed of the end of each hand.

12 The drum of a washing machine has diameter 50 cm. The drum spins at 1200 rev min^{-1}. Find the linear speed of a point on the drum.

13 A gramophone record rotates at 45 rev min^{-1}. Find:

 a the angular speed of the record in rad s^{-1}

 b the distance from the centre of a point moving at 12 cm s^{-1}.

(P) 14 The Earth completes one orbit of the sun in a year. Taking the orbit to be a circle of radius 1.5×10^{11} m, and a year to be 365 days, calculate the speed at which the Earth is moving.

(P) 15 A bead moves around a hoop of radius r m with angular velocity 1 rad s^{-1}. The bead moves at a speed greater than 5 m s^{-1}. Find the range of possible values for r.

Challenge

Two separate circular turntables, with different radii, are both mounted horizontally on a common vertical axis which acts as the centre of rotation for both. The smaller turntable, of radius 18 cm, is uppermost and rotates clockwise. The larger turntable has radius 20 cm and rotates anticlockwise. Both turntables have constant angular velocities, with magnitudes in the same ratio as their radii.

A blue dot is placed at a point on the circumference of the smaller turntable, and a red dot likewise on the larger one. Starting from the instant that the two dots are at their closest possible distance apart, it is known that 10 seconds later these dots are at their maximum distance apart for the first time. Find the exact angular velocity of the larger turntable.

1.2 Acceleration of an object moving on a horizontal circular path

When an object moves round a horizontal circular path at constant speed, the direction of the motion is changing. If the direction is changing, then, although the speed is constant, the velocity is not constant. If the velocity is changing then the object must have an acceleration.

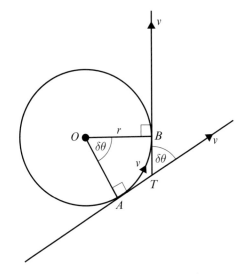

Suppose that the object is moving on a circular path of radius r at constant speed v.

Let the time taken to move from A to B be δt, and the angle AOB be $\delta\theta$.

At A, the velocity is v along the tangent AT. At B, the velocity is v along the tangent TB.

The velocity at B can be resolved into components:

$v\cos\delta\theta$ parallel to AT and

$v\sin\delta\theta$ perpendicular to AT.

We know that acceleration $= \dfrac{\text{change in velocity}}{\text{time}}$, so to find the acceleration of the object at the instant when it passes point A, we need to consider what happens to $\dfrac{v\cos\delta\theta - v}{\delta t}$ and $\dfrac{v\sin\delta\theta - 0}{\delta t}$ as $\delta t \to 0$.

These will be the components of the acceleration parallel to AT and perpendicular to AT respectively.

For a small angle $\delta\theta$ measured in radians, $\cos\delta\theta \approx 1$ and $\sin\delta\theta \approx \delta\theta$, so the acceleration parallel to AT is zero, and the acceleration perpendicular to AT is $v\dfrac{\delta\theta}{\delta t} = v\omega$.

Using $v = r\omega$, $v\omega$ can be written as $r\omega^2$ or $\dfrac{v^2}{r}$.

- **An object moving on a circular path with constant linear speed v and constant angular speed ω has acceleration $r\omega^2$ or $\dfrac{v^2}{r}$, towards the centre of the circle.**

Example 4

A particle is moving on a horizontal circular path of radius $20\,\text{cm}$ with constant angular speed $2\,\text{rad s}^{-1}$. Calculate the acceleration of the particle.

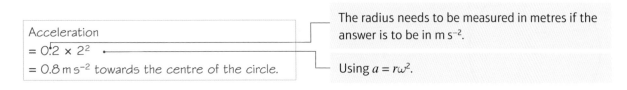

Acceleration
$= 0.2 \times 2^2$
$= 0.8\,\text{m s}^{-2}$ towards the centre of the circle.

The radius needs to be measured in metres if the answer is to be in m s^{-2}.

Using $a = r\omega^2$.

Example 5

A particle of mass $150\,\text{g}$ moves in a horizontal circle of radius $50\,\text{cm}$ at a constant speed of $4\,\text{m s}^{-1}$. Find the force towards the centre of the circle that must act on the particle.

Acceleration is given by:

$$a = \frac{v^2}{r} = \frac{4^2}{0.5} = 32\,\text{m s}^{-2}$$

Write down the formula for acceleration in terms of speed and radius.

$$F = ma = 0.15 \times 32 = 4.8\,\text{N}$$

Make sure lengths are in metres and masses are in kg before substituting.

Example 6

One end of a light inextensible string of length $20\,\text{cm}$ is attached to a particle P of mass $250\,\text{g}$. The other end of the string is attached to a fixed point O on a smooth horizontal table. P moves in a horizontal circle centre O at constant angular speed $3\,\text{rad s}^{-1}$. Find the tension in the string.

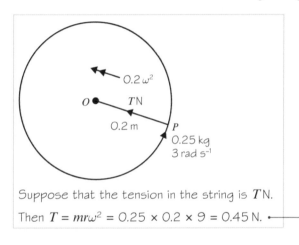

Online Explore circular motion of a particle attached to a light inextensible string using GeoGebra.

The force towards the centre of the circle is due to the tension in the string.

Suppose that the tension in the string is $T\,\text{N}$.

Then $T = mr\omega^2 = 0.25 \times 0.2 \times 9 = 0.45\,\text{N}$.

Use $F = ma$ with $a = r\omega^2$.

Example 7

A smooth wire is formed into a circle of radius $15\,\text{cm}$. A bead of mass $50\,\text{g}$ is threaded onto the wire. The wire is horizontal and the bead is made to move along it with a constant speed of $20\,\text{cm s}^{-1}$. Find the horizontal component of the force on the bead due to the wire.

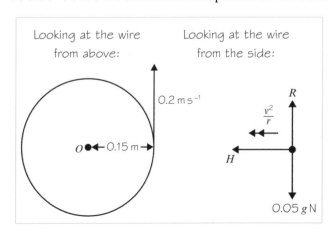

Looking at the wire from above:

Looking at the wire from the side:

Watch out If a question just says "speed" then it is referring to linear speed.

The forces acting on the bead are weight $0.05g\,\text{N}$, the normal reaction, R, and the horizontal force, H.

The force towards the centre of the circle is due to the horizontal component of the reaction of the wire on the bead.

Let the horizontal component of the force exerted on the bead by the wire be H.

$$H = \frac{mv^2}{r} = \frac{0.05 \times 0.2^2}{0.15} = 0.013 \, \text{N (2 s.f.)}$$ ◄——— Resolve towards the centre of the circle.

Example 8

A particle P of mass $10\,\text{g}$ rests on a rough horizontal disc at a distance $15\,\text{cm}$ from the centre. The disc rotates at constant angular speed of $1.2\,\text{rad s}^{-1}$, and the particle does not slip. Calculate the force due to the friction acting on the particle.

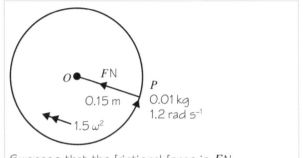

O $F\,\text{N}$ P
$0.15\,\text{m}$ $0.01\,\text{kg}$
$1.2\,\text{rad s}^{-1}$
$1.5\,\omega^2$

The force towards the centre of the circle is due to the friction between the particle and the disc. This is the force that is providing the angular acceleration of the particle.

Suppose that the frictional force is $F\,\text{N}$.
Then $F = mr\omega^2 = 0.01 \times 0.15 \times 1.2^2 = 0.00216\,\text{N}$. ◄——— Resolve towards the centre of the circle.

Example 9

A car of mass $M\,\text{kg}$ is travelling on a flat road round a bend which is an arc of a circle of radius $140\,\text{m}$. The greatest speed at which the car can travel round the bend without slipping is $45\,\text{km h}^{-1}$. Find the coefficient of friction between the tyres of the car and the road.

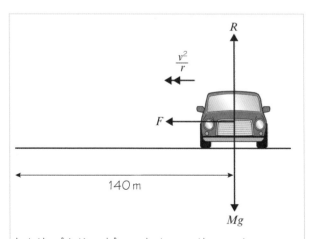

R
$\frac{v^2}{r}$
F
$140\,\text{m}$
Mg

The force towards the centre of the circle is due to the friction between the tyres of the car and the road.

Let the frictional force between the car tyres and the road be F, and the coefficient of friction be μ. The normal reaction between the car and the road is R.

Mark the forces on the diagram and resolve in the direction of the acceleration and perpendicular to it, i.e. horizontally and vertically.

$$R(\uparrow): R = Mg$$

$$R(\leftarrow): F = F_{max} = \frac{mv^2}{r}$$

As the car is about to slip at this speed, we know that $F = F_{max} = \mu R$.

Resolve towards the centre of the circle.

$$v = 45\ \text{km h}^{-1} = \frac{45 \times 1000}{3600} = 12.5\ \text{m s}^{-1}$$

Convert the speed from km h^{-1} to m s^{-1} so that the units are consistent.

$$\Rightarrow \mu Mg = \frac{M \times 12.5^2}{140}$$

$$\mu = \frac{12.5^2}{140 \times g} = 0.11\ (2\ \text{s.f.})$$

Problem-solving

You can cancel M from both sides of the equation. This tells you that the answer is independent of the mass of the car.

Exercise 1B

Whenever a numerical value of g is required take $g = 9.8\ \text{m s}^{-2}$.

1 A particle is moving on a horizontal circular path of radius 16 cm with a constant angular speed of 5 rad s^{-1}. Calculate the acceleration of the particle.

2 A particle is moving on a horizontal circular path of radius 0.3 m at a constant speed of 2.5 m s^{-1}. Calculate the acceleration of the particle.

3 A particle is moving on a horizontal circular path of radius 3 m. Given that the acceleration of the particle is 75 m s^{-2} towards the centre of the circle, find:
   a the angular speed of the particle
   b the linear speed of the particle.

4 A particle is moving on a horizontal circular path of diameter 1.2 m. Given that the acceleration of the particle is 100 m s^{-2} towards the centre of the circle, find:
   a the angular speed of the particle
   b the linear speed of the particle.

5 A car is travelling round a bend which is an arc of a circle of radius 90 m. The speed of the car is 50 km h^{-1}. Calculate its acceleration.

6 A car moving along a horizontal road which follows an arc of a circle of radius 75 m has an acceleration of 6 m s^{-2} directed towards the centre of the circle. Calculate the angular speed of the car.

7 One end of a light inextensible string of length 0.15 m is attached to a particle P of mass 300 g. The other end of the string is attached to a fixed point O on a smooth horizontal table. P moves in a horizontal circle centre O at constant angular speed 4 rad s^{-1}. Find the tension in the string.

8 One end of a light inextensible string of length 25 cm is attached to a particle P of mass 150 g. The other end of the string is attached to a fixed point O on a smooth horizontal table. P moves in a horizontal circle centre O at constant speed $9 \, \text{m s}^{-1}$. Find the tension in the string.

9 A smooth wire is formed into a circle of radius 0.12 m. A bead of mass 60 g is threaded onto the wire. The wire is horizontal and the bead is made to move along it with a constant speed of $3 \, \text{m s}^{-1}$. Find:

   a the vertical component of the force on the bead due to the wire

   b the horizontal component of the force on the bead due to the wire.

10 A particle P of mass 15 g rests on a rough horizontal disc at a distance 12 cm from the centre. The disc rotates at a constant angular speed of $2 \, \text{rad s}^{-1}$, and the particle does not slip. Calculate:

   a the linear speed of the particle

   b the force due to the friction acting on the particle.

11 A particle P rests on a rough horizontal disc at a distance 20 cm from the centre. When the disc rotates at constant angular speed of $1.2 \, \text{rad s}^{-1}$, the particle is just about to slip. Calculate the value of the coefficient of friction between the particle and the disc.

12 A particle P of mass 0.3 kg rests on a rough horizontal disc at a distance 0.25 m from the centre of the disc. The coefficient of friction between the particle and the disc is 0.25.
Given that P is on the point of slipping, find the angular speed of the disc.

13 A car is travelling round a bend on a flat road which is an arc of a circle of radius 80 m. The greatest speed at which the car can travel round the bend without slipping is $40 \, \text{km h}^{-1}$. Find the coefficient of friction between the tyres of the car and the road.

14 A car is travelling round a bend on a flat road which is an arc of a circle of radius 60 m. The coefficient of friction between the tyres of the car and the road is $\frac{1}{3}$. Find the greatest angular speed at which the car can travel round the bend without slipping.

(P) 15 A centrifuge consists of a vertical hollow cylinder of radius 20 cm rotating about a vertical axis through its centre at $90 \, \text{rev s}^{-1}$.

   a Calculate the magnitude of the normal reaction between the cylinder and a particle of mass 5 g on the inner surface of the cylinder.

   b Given that the particle remains at the same height on the cylinder, calculate the least possible coefficient of friction between the particle and the cylinder.

(E/P) 16 A fairground ride consists of a vertical hollow cylinder of diameter 5 m which rotates about a vertical axis through its centre. When the ride is rotating at $W \, \text{rad s}^{-1}$ the floor of the cylinder opens. The people on the ride remain, without slipping, in contact with the inner surface of the cylinder.

   a Given that the coefficient of friction between a person and the inner surface of the cylinder is $\frac{2}{3}$, find the minimum value for W. **(5 marks)**

   b State, with a reason, whether this would be a safe speed at which to operate the ride. **(1 mark)**

(E) **17** Two particles P and Q, both of mass 80 g, are attached to the ends of a light inextensible string of length 30 cm. Particle P is on a smooth horizontal table, the string passes through a small smooth hole in the centre of the table, and particle Q hangs freely below the table at the other end of the string. P is moving on a circular path about the centre of the table at constant linear speed. Find the linear speed at which P must move if Q is in equilibrium 10 cm below the table.

(4 marks)

(E) **18** A car travels travels around a bend on a flat road. The car is modelled as a particle travelling at a constant speed of v m s^{-1} along a path which is an arc of a circle of radius R m. Given that the car does not slip,

  a find the minimum value for the coefficient of friction between the car and the road, giving your answer in terms of R and g. **(4 marks)**

  b Describe one weakness of the model. **(1 mark)**

(A)(P) **19** One end of a light extensible string of natural length 0.3 m and modulus of elasticity 10 N is attached to a particle P of mass 250 g. The other end of the string is attached to a fixed point O on a smooth horizontal table. P moves in a horizontal circle centre O at constant angular speed 3 rad s^{-1}. Find the radius of the circle.

(E/P) **20** A particle P of mass 4 kg rests on a rough horizontal disc, centre O, which is rotating at ω rad s^{-1}. The coefficient of friction between the particle and the disc is 0.3. The particle is attached to O by means of a light elastic string of natural length 1.5 m and modulus of elasticity 12 N. The distance OP is 2 m. Given that the particle does not slide across the surface of the disc, find the maximum possible value of ω. **(7 marks)**

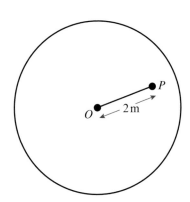

Challenge

A particle is moving in the horizontal x-y plane. Its x- and y-coordinates at time t seconds are given by the parametric equations

$$x = pt, \ y = qt^2, \ t \geqslant 0$$

where t is the time in seconds, and p and q are positive constants.

a Sketch the path of P and write its equation in the form $y = f(x)$.

b Find the acceleration of the particle and its speed, v m s^{-1}, at the origin.

c Find the equation of the lower half of a circle with centre $(0, R)$ and radius R, giving your answer in the form $y = g(x)$.

d By comparing second derivatives, find, in terms of p and q, the value of R for which this circle most closely matches the path of P at the origin.

A second particle Q moves around this circle with linear speed v m s^{-1}.

e Find the acceleration of Q.

f Comment on your answer.

1.3 Three-dimensional problems with objects moving in horizontal circles

In this section you will find out how the method of resolving forces can be used to solve a problem about an object moving in a horizontal circle.

Example 10

A particle of mass 2 kg is attached to one end of a light inextensible string of length 50 cm. The other end of the string is attached to a fixed point A. The particle moves with constant angular speed in a horizontal circle of radius 40 cm. The centre of the circle is vertically below A. Calculate the tension in the string and the angular speed of the particle.

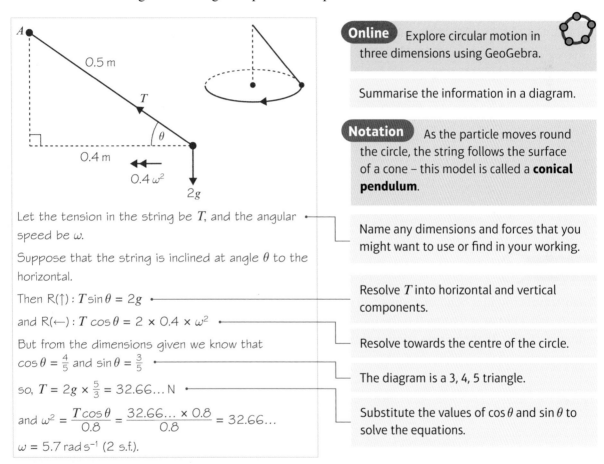

Online Explore circular motion in three dimensions using GeoGebra.

Summarise the information in a diagram.

Notation As the particle moves round the circle, the string follows the surface of a cone – this model is called a **conical pendulum**.

Let the tension in the string be T, and the angular speed be ω.

Name any dimensions and forces that you might want to use or find in your working.

Suppose that the string is inclined at angle θ to the horizontal.

Then R(↑): $T \sin \theta = 2g$

Resolve T into horizontal and vertical components.

and R(←): $T \cos \theta = 2 \times 0.4 \times \omega^2$

Resolve towards the centre of the circle.

But from the dimensions given we know that $\cos \theta = \frac{4}{5}$ and $\sin \theta = \frac{3}{5}$

The diagram is a 3, 4, 5 triangle.

so, $T = 2g \times \frac{5}{3} = 32.66\ldots$ N

and $\omega^2 = \dfrac{T \cos \theta}{0.8} = \dfrac{32.66\ldots \times 0.8}{0.8} = 32.66\ldots$

Substitute the values of $\cos \theta$ and $\sin \theta$ to solve the equations.

$\omega = 5.7 \text{ rad s}^{-1}$ (2 s.f.).

Example 11

A particle of mass m is attached to one end of a light inextensible string of length l. The other end of the string is attached to a fixed point A. The particle moves with constant angular speed in a horizontal circle. The string is taut and the angle between the string and the vertical is θ. The centre of the circle is vertically below A. Find the angular speed of the particle.

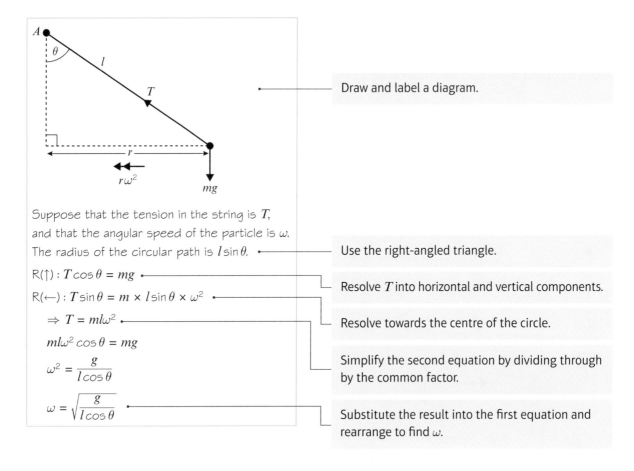

Suppose that the tension in the string is T, and that the angular speed of the particle is ω. The radius of the circular path is $l\sin\theta$. — Use the right-angled triangle.

$R(\uparrow): T\cos\theta = mg$ — Resolve T into horizontal and vertical components.

$R(\leftarrow): T\sin\theta = m \times l\sin\theta \times \omega^2$ — Resolve towards the centre of the circle.

$\Rightarrow T = ml\omega^2$

$ml\omega^2\cos\theta = mg$ — Simplify the second equation by dividing through by the common factor.

$\omega^2 = \dfrac{g}{l\cos\theta}$

$\omega = \sqrt{\dfrac{g}{l\cos\theta}}$ — Substitute the result into the first equation and rearrange to find ω.

Draw and label a diagram.

Example 12

A car travels round a bend of radius 500 m on a flat road which is banked at an angle θ to the horizontal. The car is assumed to be moving at constant speed in a horizontal circle and there is no tendency to slip. If there is no frictional force acting on the car when it is travelling at 90 km h^{-1}, find the value of θ.

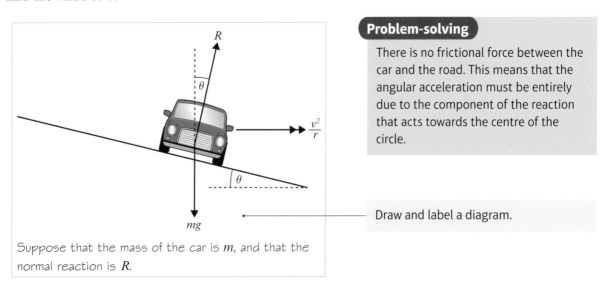

Problem-solving

There is no frictional force between the car and the road. This means that the angular acceleration must be entirely due to the component of the reaction that acts towards the centre of the circle.

Draw and label a diagram.

Suppose that the mass of the car is m, and that the normal reaction is R.

$$90\,\text{km}\,\text{h}^{-1} = \frac{90 \times 1000}{3600} = 25\,\text{m}\,\text{s}^{-1}$$

$R(\uparrow): R\cos\theta = mg$ — Resolve the normal reaction into vertical and horizontal components.

$R(\leftarrow): R\sin\theta = \dfrac{m \times 25^2}{500}$ — Resolve towards the centre of the circle.

$\Rightarrow \tan\theta = \dfrac{25^2}{500 \times g} = 0.128\ldots, \theta = 7.3° \text{ (2 s.f.)}$ — Divide the second equation by the first.

Example 13

The diagram shows a particle P of mass m attached by two strings to fixed points A and B, where A is vertically above B. The strings are both taut and P is moving in a horizontal circle with constant angular speed $2\sqrt{3g}$ rad s^{-1}.

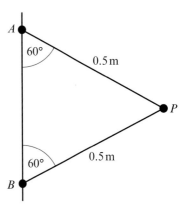

Both strings are 0.5 m in length and inclined at 60° to the vertical.

a Calculate the tensions in the two strings.

The strings will break if the tension in them exceeds $8mg$ N.

The angular speed of the particle is increased until the strings break.

b State which string will break first.

c Find the maximum angular speed of the particle before the string breaks.

a

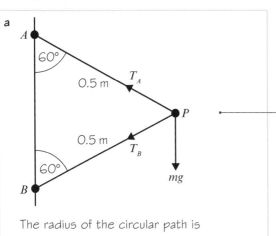

Copy the diagram and show all the forces.

The radius of the circular path is

$0.5\cos 30° = \dfrac{\sqrt{3}}{4}$ m — This is an equilateral triangle.

$R(\uparrow): T_A\cos 60 = T_B\cos 60 + mg$

$\therefore T_A - T_B = 2mg$ (1)

Resolve both tensions into their horizontal and vertical components.

$R(\leftarrow): T_A\cos 30 + T_B\cos 30 = mr\omega^2$

$\therefore \dfrac{\sqrt{3}}{2}(T_A + T_B) = m \times \dfrac{\sqrt{3}}{4} \times 4 \times 3g$

Resolve towards the centre of the circle.

$T_A + T_B = 6mg$ (2)

Simplify and solve the pair of simultaneous equations (1) and (2).

$\Rightarrow T_A = 4mg$ N and $T_B = 2mg$ N

b The upper string will always have greater tension so will break first.

c Let maximum angular speed be ω_{max}.

At this speed, $T_A = 8\,mg$, so from (1),

$T_B = 6\,mg$.

So $\dfrac{\sqrt{3}}{2}(8mg + 6mg) = m\dfrac{\sqrt{3}}{4}\omega_{max}^2$

$28g = \omega_{max}^2$

$\omega_{max} = \sqrt{28g} = 17\text{ rad s}^{-1}$ (2 s.f.)

The string will snap if $T_A > 8mg$.

Problem-solving

Don't repeat work from part **a** when answering part **c**. All of the working in part **a** up to the point where ω is substituted still applies, so use $T_A - T_B = 2mg$ and $T_A\cos 30° + T_B\cos 30° = mr\omega^2$

Example 14

An aircraft of mass 2 tonnes flies at 500 km h^{-1} on a path which follows a horizontal circular arc in order to change course from due north to due east. The aircraft turns in the clockwise direction from due north to due east. It takes 40 seconds to change course, with the aircraft banked at an angle α to the horizontal. Calculate the value of α and the magnitude of the lift force perpendicular to the surface of the aircraft's wings.

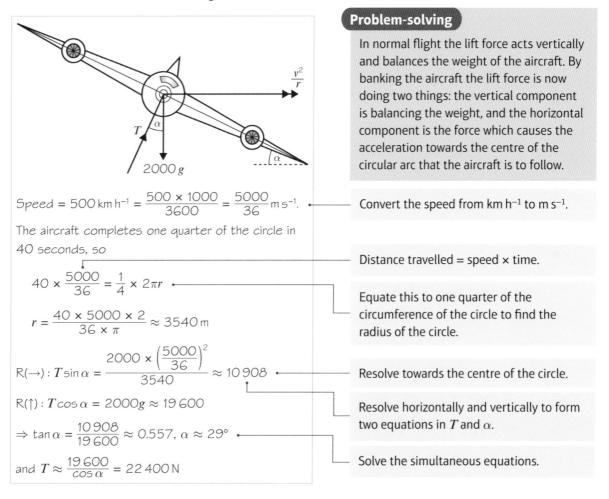

Problem-solving

In normal flight the lift force acts vertically and balances the weight of the aircraft. By banking the aircraft the lift force is now doing two things: the vertical component is balancing the weight, and the horizontal component is the force which causes the acceleration towards the centre of the circular arc that the aircraft is to follow.

$\text{Speed} = 500\text{ km h}^{-1} = \dfrac{500 \times 1000}{3600} = \dfrac{5000}{36}\text{ m s}^{-1}.$

Convert the speed from km h^{-1} to m s^{-1}.

The aircraft completes one quarter of the circle in 40 seconds, so

$40 \times \dfrac{5000}{36} = \dfrac{1}{4} \times 2\pi r$

Distance travelled = speed × time.

$r = \dfrac{40 \times 5000 \times 2}{36 \times \pi} \approx 3540\text{ m}$

Equate this to one quarter of the circumference of the circle to find the radius of the circle.

$R(\rightarrow): T\sin\alpha = \dfrac{2000 \times \left(\dfrac{5000}{36}\right)^2}{3540} \approx 10\,908$

Resolve towards the centre of the circle.

$R(\uparrow): T\cos\alpha = 2000g \approx 19\,600$

$\Rightarrow \tan\alpha = \dfrac{10\,908}{19\,600} \approx 0.557, \alpha \approx 29°$

Resolve horizontally and vertically to form two equations in T and α.

and $T \approx \dfrac{19\,600}{\cos\alpha} = 22\,400\text{ N}$

Solve the simultaneous equations.

Example 15

In this question use $g = 9.81 \text{ m s}^{-2}$.

A hollow right circular cone is fixed with its axis of symmetry vertical and its vertex V pointing downwards. A particle, P of mass 25 g moves in a horizontal circle with centre C, and radius 0.27 m, on the rough inner surface of the cone. P remains in contact as it moves with constant angular speed, ω, and does not slip. The angle between VP and the vertical is θ, such that $\tan \theta = 0.45$.

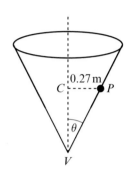

The coefficient of friction between the particle and the cone is 0.15.

Find the greatest possible value of ω.

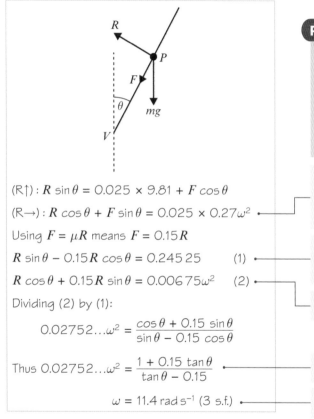

$(R\uparrow): R \sin \theta = 0.025 \times 9.81 + F \cos \theta$

$(R\rightarrow): R \cos \theta + F \sin \theta = 0.025 \times 0.27\omega^2$

Using $F = \mu R$ means $F = 0.15R$

$R \sin \theta - 0.15R \cos \theta = 0.24525 \quad (1)$

$R \cos \theta + 0.15R \sin \theta = 0.00675\omega^2 \quad (2)$

Dividing (2) by (1):

$\quad 0.02752...\omega^2 = \dfrac{\cos \theta + 0.15 \sin \theta}{\sin \theta - 0.15 \cos \theta}$

Thus $0.02752...\omega^2 = \dfrac{1 + 0.15 \tan \theta}{\tan \theta - 0.15}$

$\quad\quad\quad \omega = 11.4 \text{ rad s}^{-1} \text{ (3 s.f.)}$

Problem-solving

Begin by drawing a cross-section showing the three forces acting on the particle whilst in motion. You are looking for the **greatest** possible value of ω, so the particle is on the point of slipping **up** the side of the cone. This means that the frictional force acts down towards V.

The angular acceleration is due to the horizontal components of the normal reaction and the frictional force.

Resolve vertically and horizontally to form two simultaneous equations.

Substitute to eliminate F.

Divide the numerator and denominator by $\cos \theta$.

Use the fact that $\tan \theta = 0.45$

Exercise 1C

Whenever a numerical value of g is required take $g = 9.8 \text{ m s}^{-2}$.

1 A particle of mass 1.5 kg is attached to one end of a light inextensible string of length 60 cm. The other end of the string is attached to a fixed point A. The particle moves with constant angular speed in a horizontal circle of radius 36 cm. The centre of the circle is vertically below A. Calculate the tension in the string and the angular speed of the particle.

2 A particle of mass 750 g is attached to one end of a light inextensible string of length 0.7 m. The other end of the string is attached to a fixed point A. The particle moves with constant angular speed in a horizontal circle whose centre is 0.5 m vertically below A. Calculate the tension in the string and the angular speed of the particle.

3 A particle of mass 1.2 kg is attached to one end of a light inextensible string of length 2 m. The other end of the string is attached to a fixed point A. The particle moves in a horizontal circle with constant angular speed. The centre of the circle is vertically below A. The particle takes 2 seconds to complete one revolution. Calculate the tension in the string and the angle between the string and the vertical, to the nearest degree.

4 A conical pendulum consists of a light inextensible string AB of length 1 m, fixed at A and carrying a small ball of mass 6 kg at B. The particle moves in a horizontal circle, with centre vertically below A, at constant angular speed 3.5 rad s^{-1}. Find the tension in the string and the radius of the circle.

(E/P) 5 A conical pendulum consists of a light inextensible string AB of length l, fixed at A and carrying a small ball of mass m at B. The particle moves in a horizontal circle, with centre vertically below A, at constant angular speed ω. Find, in terms of m, l and ω, the tension in the string. **(5 marks)**

(E/P) 6 A conical pendulum consists of a light inextensible string AB fixed at A and carrying a small ball of mass m at B. With the string taut the particle moves in a horizontal circle at constant angular speed ω. The centre of the circle is at distance x vertically below A. Show that $\omega^2 x = g$. **(5 marks)**

(P) 7 A hemispherical bowl of radius r cm is resting in a fixed position with its rim horizontal. A small marble of mass m is moving in a horizontal circle around the smooth inside surface of the bowl. The plane of the circle is 3 cm below the plane of the rim of the bowl. Find the angular speed of the marble.

Problem-solving

The normal reaction of the bowl on the marble will act towards the centre of the sphere.

(P) 8 A hemispherical bowl of radius 15 cm is resting in a fixed position with its rim horizontal. A particle P of mass m is moving at 14 rad s^{-1} in a horizontal circle around the smooth inside surface of the bowl. Find the distance d of the plane of the circle below the plane of the rim of the bowl.

9 A cone is fixed with its base horizontal and its vertex 4 m below the centre of the base. The base has a diameter of 8 m. A particle moves around the smooth inside of the cone a vertical distance 1 m below the base on a horizontal circle. Find the angular and linear speed of the particle.

(E/P) **10** A particle P is moving in a horizonal circle, with centre C and radius r. P is in contact with the rough inside surface of a hollow right circular cone. The cone is fixed with its axis of symmetry vertical and its vertex V pointing downwards. The radius at the top of the cone is 6 m and the cone has a perpendicular height of 2 m.

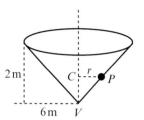

When $r = 0.1$ m, the maximum constant angular speed at which the particle can move, without slipping from its path, is $14\sqrt{5}$ rad s^{-1}.

Find the maximum angular speed without slipping for $r = 0.3$ m. **(10 marks)**

11 A car travels round a bend of radius 750 m on a road which is banked at an angle θ to the horizontal. The car is assumed to be moving at constant speed in a horizontal circle and there is no tendency to slip. If there is no frictional force acting on the car when it is travelling at 126 km h^{-1}, find the value of θ.

12 A car travels round a bend of radius 300 m on a road which is banked at an angle of 10° to the horizontal. The car is assumed to be moving at constant speed in a horizontal circle and there is no tendency to slip. Given that the road is smooth, find the speed of the car.

13 A cyclist rides round a circular track of diameter 50 m. The track is banked at 20° to the horizontal. There is no force due to friction and there is no tendency to slip. By modelling the cyclist and bicycle as a particle of mass 75 kg, find the speed at which the cyclist is moving.

(E) **14** A bend in the road is modelled as a horizontal circular arc of radius r. The surface of the bend is banked at an angle α to the horizontal, and the friction between the tyres and the road is modelled as being negligible. When a vehicle is driven round the bend there is no tendency to slip.

    a Show that according to this model, the speed of the vehicle is given by $\sqrt{rg\tan\alpha}$. **(5 marks)**

    b Suggest, with reasons, which modelling assumption is likely to give rise to the greatest inaccuracy in this calculation. **(1 mark)**

15 A girl rides her cycle round a circular track of diameter 60 m. The track is banked at 15° to the horizontal. The coefficient of friction between the track and the tyres of the cycle is 0.25. Modelling the girl and her cycle as a particle of mass 60 kg moving in a horizontal circle, find the minimum speed at which she can travel without slipping.

(E) **16** A van is moving on a horizontal circular bend in the road of radius 75 m. The bend is banked at $\arctan\frac{1}{3}$ to the horizontal. The maximum speed at which the van can be driven round the bend without slipping is 90 km h^{-1}. Calculate the coefficient of friction between the road surface and the tyres of the van. **(4 marks)**

17 A car moves on a horizontal circular path round a banked bend in a race track. The radius of the path is 100 m. The coefficient of friction between the car tyres and the track is 0.3. The maximum speed at which the car can be driven round the bend without slipping is 144 km h^{-1}. Find the angle at which the track is banked, to the nearest degree.

E/P **18** A bend in a race track is banked at 30°. A car will follow a horizontal circular path of radius 70 m round the bend. The coefficient of friction between the car tyres and the track surface is 0.4. Find the maximum and minimum speeds at which the car can be driven round the bend without slipping. **(10 marks)**

E/P **19** An aircraft of mass 2 tonnes flies at 400 km h^{-1} on a path which follows a horizontal circular arc in order to change course from a bearing of 060° to a bearing of 015°. It takes 25 seconds to change course, with the aircraft banked at α° to the horizontal.

    a Calculate the two possible values of α, to the nearest degree and the corresponding values of the magnitude of the lift force perpendicular to the surface of the aircraft's wings. **(4 marks)**

    b Without further calculation, state how your answers will change if the aircraft wishes to complete its turn in a shorter time. **(3 marks)**

E/P **20** A particle of mass m is attached to one end of a light, inextensible string of length l. The other end of the string is attached to a point vertically above the vertex of a smooth cone. The cone is fixed with its axis vertical, as shown in the diagram. The semi-vertical angle of the cone is θ, and the string makes a constant angle of θ with the horizontal, where $\dfrac{\pi}{4} < \theta < \dfrac{\pi}{2}$

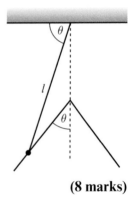

Given that the particle moves in a horizontal circle with angular speed ω, show that the tension in the string is given by $\frac{1}{2}m(\omega^2 l + g \operatorname{cosec} \theta)$ **(8 marks)**

A
E/P **21** A light elastic string AB has natural length 2 m and modulus of elasticity 30 N. The end A is attached to a fixed point. A particle of mass 750 g is attached to the end B. The particle is moving in a horizontal circle below A with the string inclined at 40° to the vertical. Find the angular speed of the particle. **(7 marks)**

1.4 Objects moving in vertical circles

When an object moves in a vertical circle it gains height as it follows its circular path. If it gains height then it must gain gravitational potential energy. Therefore, using the work–energy principle it follows that it must lose kinetic energy, and its speed will not be constant.

You can use vectors to understand motion in a vertical circle.

If O is the centre of the circle of radius r and P is the particle, we can set up coordinate axes in the plane of the circle with the x-axis horizontal, and the y-axis vertical.

Let the unit vectors $\mathbf{i}$ and $\mathbf{j}$ be parallel to the x-axis and y-axis respectively.

At time t the angle between the radius OP and the x-axis is θ and the position vector of P is $\mathbf{r}$.

$$\mathbf{r} = (r\cos\theta)\,\mathbf{i} + (r\sin\theta)\,\mathbf{j}$$

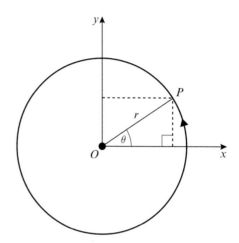

By differentiating this with respect to time we obtain the velocity vector

$$\mathbf{v} = \frac{\mathrm{d}}{\mathrm{d}t}(\mathbf{r}) = (-r\sin\theta)\dot{\theta}\mathbf{i} + (r\cos\theta)\dot{\theta}\mathbf{j} = r\dot{\theta}(-\sin\theta\,\mathbf{i} + \cos\theta\,\mathbf{j})$$

Looking at the directions of $\mathbf{r}$ and $\mathbf{v}$, we find that the lines representing them have gradients $\dfrac{r\sin\theta}{r\cos\theta}$ and $\left(-\dfrac{r\cos\theta}{r\sin\theta}\right)$ respectively.

But $\dfrac{r\sin\theta}{r\cos\theta} \times \left(-\dfrac{r\cos\theta}{r\sin\theta}\right) = -1$, so these two vectors are perpendicular. Alternatively, using the scalar product we see that the vectors are perpendicular since $(\cos\theta\,\mathbf{i} + \sin\theta\,\mathbf{j}) \cdot (-\sin\theta\,\mathbf{i} + \cos\theta\,\mathbf{j}) = 0$.

This means that the acceleration has two components, one of magnitude $r\dot{\theta}^2$ directed towards the centre of the circle, and one of magnitude $r\ddot{\theta}$ directed along the tangent to the circle.

Using $\dot{\theta} = \omega$ gives:

- **For motion in a vertical circle of radius r, the components of the acceleration are $r\omega^2$ or $\dfrac{v^2}{r}$ towards the centre of the circle and $r\ddot{\theta} = \dot{v}$ along the tangent.**

The force directed towards the centre of the circle is perpendicular to the direction of motion of the particle, so it does no work. If the only other force acting on the particle is gravity, then it follows (using the work–energy principle) that the sum of the kinetic energy and the potential energy of the particle will be constant. You will use this fact to solve problems about motion in a vertical circle.

> **Links** The work-energy principle states that the change in the total energy of a particle is equal to the work done on the particle. This means that where the only force acting on a particle is gravity, the sum of its kinetic and gravitational potential energies remains constant.
> ← **Further Mechanics 1, Section 2.3**

Example 16

A particle of mass 0.4 kg is attached to one end A of a light rod AB of length 0.3 m. The rod is free to rotate in a vertical plane about B. The particle is held at rest with AB horizontal. The particle is released. Calculate:

a the speed of the particle as it passes through the lowest point of the path,

b the tension in the rod at this point.

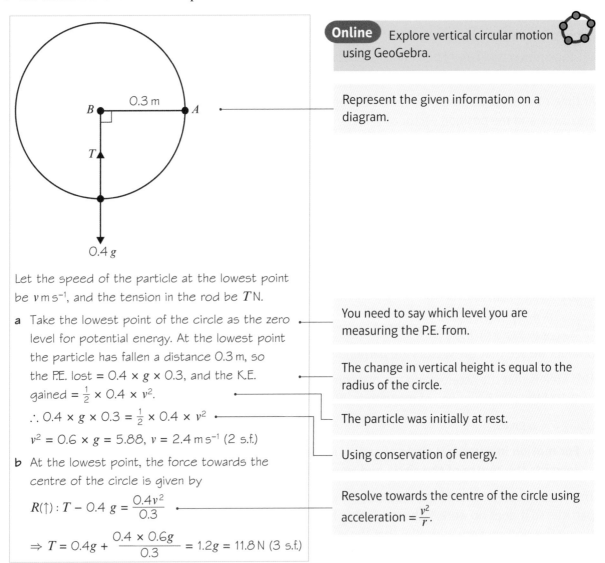

Online Explore vertical circular motion using GeoGebra.

Represent the given information on a diagram.

Let the speed of the particle at the lowest point be v m s⁻¹, and the tension in the rod be T N.

a Take the lowest point of the circle as the zero level for potential energy. At the lowest point the particle has fallen a distance 0.3 m, so the P.E. lost = 0.4 × g × 0.3, and the K.E. gained = $\frac{1}{2}$ × 0.4 × v^2.

$\therefore$ 0.4 × g × 0.3 = $\frac{1}{2}$ × 0.4 × v^2

v^2 = 0.6 × g = 5.88, v = 2.4 m s⁻¹ (2 s.f.)

b At the lowest point, the force towards the centre of the circle is given by

$R(\uparrow): T - 0.4\,g = \dfrac{0.4v^2}{0.3}$

$\Rightarrow T = 0.4g + \dfrac{0.4 \times 0.6g}{0.3}$ = 1.2g = 11.8 N (3 s.f.)

You need to say which level you are measuring the P.E. from.

The change in vertical height is equal to the radius of the circle.

The particle was initially at rest.

Using conservation of energy.

Resolve towards the centre of the circle using acceleration = $\dfrac{v^2}{r}$.

Questions about motion in a vertical circle will often ask you to consider whether or not an object will perform complete circles. The next two examples illustrate the importance of considering how the circular motion occurs.

Example 17

A

A particle of mass 0.4 kg is attached to one end A of a light rod AB of length 0.3 m. The rod is free to rotate in a vertical plane about B. The rod is hanging vertically with A below B when the particle is set in motion with a horizontal speed of u m s^{-1}. Find:

a an expression for the speed of the particle when the rod is at an angle θ to the downward vertical through B

b the minimum value of u for which the particle will perform a complete circle.

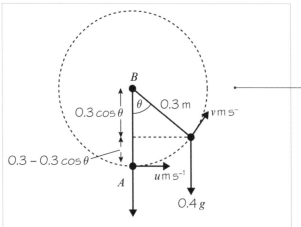

Represent the given information on a diagram.

a Take the lowest point of the circle as the zero level for potential energy.

You need to say which level you are measuring the P.E. from.

At the lowest level the particle has

$\quad$ K.E. $= \frac{1}{2} \times 0.4 \times u^2 = 0.2u^2$

$\quad$ P.E. $= 0$

When the rod is at angle θ to the vertical the particle has

$\quad$ K.E. $= \frac{1}{2} \times 0.4 \times v^2 = 0.2v^2$

$\quad$ P.E. $= 0.4 \times g \times 0.3(1 - \cos\theta)$

$\therefore 0.2u^2 = 0.2v^2 + 0.12g(1 - \cos\theta)$

$\quad v = \sqrt{u^2 - 0.6g(1 - \cos\theta)}$

Conservation of energy means that the total energy at each point will be equal.

b If the particle is to reach to top of the circle then we require $v > 0$ when $\theta = 180°$.

$\Rightarrow u^2 - 0.6g(1 - \cos 180°) > 0$

$\qquad u^2 > 0.6g \times 2$

$\qquad u > \sqrt{1.2g}$

$\cos 180° = -1$

Problem-solving

Note that if $u = \sqrt{1.2g}$ then the speed of the particle at the top of the circle would be zero. In this case the rod would be in thrust, with the force in the rod balancing the weight of the particle.

Example 18

A

A particle A of mass $0.4\,\text{kg}$ is attached to one end of a light inextensible string of length $0.3\,\text{m}$. The other end of the string is attached to a fixed point B. The particle is hanging in equilibrium when it is set in motion with a horizontal speed of $u\,\text{m s}^{-1}$. Find:

a an expression for the tension in the string, in terms of u, when it is at an angle θ to the downward vertical through B

b the minimum value of u for which the particle will perform a complete circle.

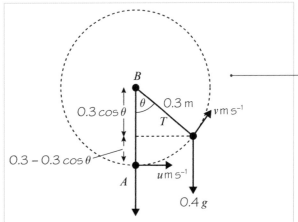

Represent the given information on a diagram.

a Take the lowest point of the circle as the zero level for potential energy.

You need to say which level you are measuring the P.E. from.

At the lowest level the particle has

$$\text{K.E.} = \tfrac{1}{2} \times 0.4 \times u^2 = 0.2u^2$$
$$\text{P.E.} = 0$$

When the string is at angle θ to the vertical the particle has

$$\text{K.E.} = \tfrac{1}{2} \times 0.4 \times v^2$$
$$\text{P.E.} = 0.4 \times g \times 0.3(1 - \cos\theta)$$
$$\therefore 0.2u^2 = 0.2v^2 + 0.12g(1 - \cos\theta)$$

Conservation of energy means that the total energy at each point will be equal.

Resolving towards the centre of the circle:

$$R(\nwarrow): T - 0.4g\cos\theta = \frac{mv^2}{r} = \frac{0.4v^2}{0.3}$$

Use $a = \dfrac{v^2}{r}$

$$T = 0.4g\cos\theta + \tfrac{4}{3}(u^2 - 0.6g + 0.6g\cos\theta)$$

Express v^2 in terms of u^2

$$= 1.2g\cos\theta + \frac{4u^2}{3} - 0.8g$$

b If the particle is to reach to top of the circle then we require $T > 0$ when $\theta = 180°$.

$$\Rightarrow -1.2g + \frac{4u^2}{3} - 0.8g > 0$$
$$\frac{4u^2}{3} > 2g$$
$$u^2 > \frac{6g}{4}$$
$$u > \sqrt{\frac{3g}{2}}$$

Problem-solving

In the previous example the rod could be in thrust, and could support the particle. In this example the string must remain taut for the particle to perform a complete circle. The condition for the string to remain taut is that the tension on the string remains positive.

A Examples 17 and 18 above illustrate the difference between particles attached to strings and rods. You can use these conditions to determine whether particles moving in a vertical circle perform complete circles.

- **A particle attached to the end of a light rod will perform complete vertical circles if it has speed > 0 at the top of the circle.**
- **A small bead threaded on to a smooth circular wire will perform complete vertical circles if it has speed > 0 at the top of the circle.**
- **A particle attached to a light inextensible string will perform complete vertical circles if the tension in the string > 0 at the top of the circle. This means that the speed of the particle when it reaches the top of the circle must be large enough to keep the string taut at the top of the circle.**

Exercise 1D

Whenever a numerical value of g is required take $g = 9.8\,\text{m s}^{-2}$.

1 A particle of mass $0.6\,\text{kg}$ is attached to end A of a light rod AB of length $0.5\,\text{m}$. The rod is free to rotate in a vertical plane about B. The particle is held at rest with AB horizontal. The particle is released. Calculate:
   a the speed of the particle as it passes through the lowest point of the path
   b the tension in the rod at this point.

2 A particle of mass $0.4\,\text{kg}$ is attached to end A of a light rod AB of length $0.3\,\text{m}$. The rod is free to rotate in a vertical plane about B. The particle is held at rest with A vertically above B. The rod is slightly displaced so that the particle moves in a vertical circle. Calculate:
   a the speed of the particle as it passes through the lowest point of the path
   b the tension in the rod at this point.

3 A particle of mass $0.4\,\text{kg}$ is attached to end A of a light rod AB of length $0.3\,\text{m}$. The rod is free to rotate in a vertical plane about B. The particle is held at rest with AB at $60°$ to the upward vertical. The particle is released. Calculate:
   a the speed of the particle as it passes through the lowest point of the path
   b the tension in the rod at this point.

4 A particle of mass $0.6\,\text{kg}$ is attached to end A of a light rod AB of length $0.5\,\text{m}$. The rod is free to rotate in a vertical plane about B. The particle is held at rest with AB at $60°$ to the upward vertical. The particle is released. Calculate:
   a the speed of the particle as it passes through the point where AB is horizontal
   b the tension in the rod at this point.

5 A smooth bead of mass $0.5\,\text{kg}$ is threaded onto a circular wire ring of radius $0.7\,\text{m}$ that lies in a vertical plane. The bead is at the lowest point on the ring when it is projected horizontally with speed $10\,\text{m s}^{-1}$. Calculate:
   a the speed of the bead when it reaches the highest point on the ring
   b the reaction of the ring on the bead at this point.

A
P

6 A particle of mass 0.5 kg moves around the interior of a sphere of radius 0.7 m. The particle moves in a circle in the vertical plane containing the centre of the sphere. The line joining the centre of the sphere to the particle makes an angle of θ with the vertical. The particle is resting on the bottom of the sphere when it is projected horizontally with speed u m s^{-1}. Find

    a an expression for the speed of the particle in terms of u and θ

    b the restriction on u if the particle is to reach the highest point of the sphere.

E/P **7** A particle A of mass 1.5 kg is attached to one end of a light inextensible string of length 2 m. The other end of the string is attached to a fixed point B. The particle is hanging in equilibrium when it is set in motion with a horizontal speed of u m s^{-1}. Find:

    a an expression for the tension in the string when it is at an angle θ to the downward vertical through B **(3 marks)**

    b the minimum value of u for which the particle will perform a complete circle.   **(3 marks)**

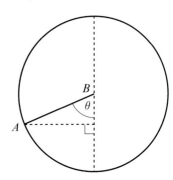

E/P **8** A small bead of mass 50 g is threaded on a smooth circular wire of radius 75 cm which is fixed in a vertical plane. The bead is at rest at the lowest point of the wire when it is hit with an impulse of I N s horizontally causing it to start to move round the wire. Find the value of I if:

    a the bead just reaches the top of the circle   **(4 marks)**

    b the bead just reaches the point where the radius from the bead to the centre of the circle makes an angle of $\arctan\frac{3}{4}$ with the upward vertical and then starts to slide back to its original position.   **(3 marks)**

E/P **9** A particle of mass 50 g is attached to one end of a light inextensible string of length 75 cm. The other end of the string is attached to a fixed point. The particle is hanging at rest when it is hit with an impulse of I N s horizontally causing it to start to move in a vertical circle. Find the value of I if:

    a the particle just reaches the top of the circle   **(4 marks)**

    b the string goes slack at the instant when the particle reaches the point where the string makes an angle of $\arctan\frac{3}{4}$ with the upward vertical.   **(3 marks)**

    c Describe the subsequent motion in part **b** qualitatively.   **(1 mark)**

10 A particle of mass 0.8 kg is attached to end A of a light rod AB of length 2 m. The end B is attached to a fixed point so that the rod is free to rotate in a vertical circle with its centre at B. The rod is held in a horizontal position and then released. Calculate the speed of the particle and the tension in the rod when

    a the particle is at the lowest point of the circle

    b the rod makes an angle of $\arctan\frac{3}{4}$ with the downward vertical through B.

11 A particle of mass 500 g describes complete vertical circles on the end of a light inextensible string of length 1.5 m. Given that the speed of the particle is 8 m s^{-1} at the highest point, find:

    a the speed of the particle when the string is horizontal

    b the magnitude of the tangential acceleration when the string is horizontal

    c the tension in the string when the particle is at the lowest point of the circle.

A **12** A light rod AB of length 1 m has a particle of mass 4 kg
E/P   attached at A. End B is pivoted to a fixed point so that
     AB is free to rotate in a vertical plane. When the rod is
     vertical with A below B the speed of the particle is 6.5 m s⁻¹.
     Find the angle between AB and the vertical at the instant
     when the tension in the rod is zero, and calculate the speed
     of the particle at that instant. **(7 marks)**

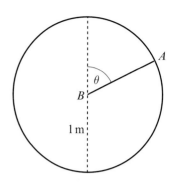

P **13** A particle P of mass m kg is attached to one end of a light rod of length r m which is free to
     rotate in a vertical plane about its other end. The particle describes complete vertical circles.
     Given that the tension at the lowest point of P's path is three times the tension at the highest
     point, find the speed of P at the lowest point on its path.

P **14** A particle P of mass m kg is attached to one end of a light inextensible string of length r m.
     The other end of the string is attached to a fixed point O, and P describes complete vertical
     circles about O. Given that the speed of the particle at the lowest point is one-and-a-half times
     the speed of the particle at the highest point, find:

     a the speed of the particle at the highest point

     b the tension in the string when the particle is at the highest point.

E/P **15** A light inelastic string of length r has one end attached to a fixed point O. A particle P of mass
     m kg is attached to the other end. P is held with OP horizontal and the string taut. P is then
     projected vertically downwards with speed $\sqrt{gr}$.

     a Find, in terms of θ, m and g, the tension in the string when OP makes an angle θ with the
       horizontal. **(4 marks)**

     b Given that the string will break when the tension in the string is $2mg$ N, find, to 3 significant
       figures the angle between the string and the horizontal when the string breaks. **(3 marks)**

E/P **16** The diagram shows the cross-section of an industrial roller.
     The roller is modelled as a cylinder of radius 4 m. The cylinder
     is oriented with its long axis horizontal, and is free to spin
     about this axis.
     A handle of mass 0.4 kg is attached to the outer surface of
     the cylinder at a point S, which is 3.8 m vertically above O.
     The cylinder is held in place by this handle, then released
     from rest. The handle is modelled as a particle, P.
     In the subsequent motion, OP moves in part of a vertical
     circle, making an angle θ above the horizontal,

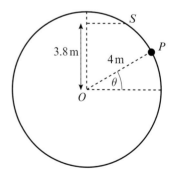

     a show that the linear speed of the handle at any point in its motion is given by
       $\sqrt{7.6g - 8g \sin \theta}$ **(5 marks)**

     b According to the model, state the height of the handle above O at the point where the
       cylinder next comes to rest. **(1 mark)**

     c State, with a reason, how this answer is likely to differ in reality. **(1 mark)**

1.5 Objects not constrained on a circular path

A In some models, for example a bead threaded on a ring or a particle attached to the end of a light rod, the object has to stay on the circular path. If the initial speed is not sufficient for the object to reach the top of the circular path then it will fall back and oscillate about the lowest point of the path. Other particles may not be constrained to stay on a circular path, for example a particle moving on the convex surface of a sphere.

- **If an object is not constrained to stay on its circular path then as soon as the contact force associated with the circular path becomes zero the object can be treated as a projectile moving freely under gravity.**

Example 19

A particle P of mass m is attached to one end of a light inextensible string of length l. The other end of the string is attached to a fixed point O. The particle is hanging in equilibrium at point A, directly below O, when it is set in motion with a horizontal speed $2\sqrt{gl}$. When OP has turned through an angle θ and the string is still taut, the tension in the string is T. Find:

a an expression for T

b the height of P above A at the instant when the string goes slack

c the maximum height above A reached by P before it starts to fall to the ground again.

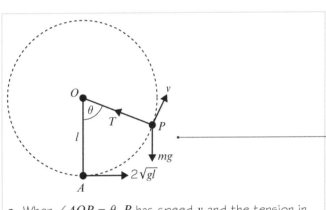

Online Explore motion of a particle not constrained on a circular path using GeoGebra.

Draw and label a diagram.

a When $\angle AOP = \theta$, P has speed v and the tension in the string is T.

Let A be the zero level for P.E.

At A, P has P.E. = 0 and K.E. = $\frac{1}{2} \times m \times u^2 = \frac{1}{2}m \times 4gl$

When $\angle AOP = \theta$, P has P.E. = $mgl(1 - \cos \theta)$ and

Find the total of P.E. + K.E. at both levels.

$$\text{K.E.} = \frac{1}{2}mv^2$$

$$\therefore 2mgl = mgl(1 - \cos \theta) + \frac{1}{2}mv^2$$

$$v^2 = 2gl(1 + \cos \theta)$$

Energy is conserved.

Resolving parallel to OP:

$$R(\nwarrow): T - mg \cos \theta = \frac{mv^2}{l} = \frac{m \times 2gl(1 + \cos \theta)}{l}$$

Using the equation for circular motion.

$$\Rightarrow T = 2mg + 2mg \cos \theta + mg \cos \theta$$

$$= 2mg + 3mg \cos \theta$$

b When $T = 0$, $\cos\theta = -\frac{2}{3}$, so the height of P above A

is $l(1 - \cos\theta) = \dfrac{5l}{3}$

> String slack, so $T = 0$.
>
> Substitute for $\cos\theta$.

c From the energy equation, we know that when the

string becomes slack $v^2 = 2gl(1 + \cos\theta) = \dfrac{2gl}{3}$

At this point the horizontal component of the velocity

is $v\cos(180 - \theta) = \dfrac{2}{3}\sqrt{\dfrac{2gl}{3}}$

If the additional height before the particle begins to

fall is h, then

$$mgh + \tfrac{1}{2} \times m \times \tfrac{4}{9} \times \dfrac{2gl}{3} = \tfrac{1}{2} \times m \times v^2 = \tfrac{1}{2} \times m \times \dfrac{2gl}{3},$$

$$h + \dfrac{4l}{27} = \tfrac{1}{3} \Rightarrow h = \dfrac{5l}{27}$$

$\therefore$ total height above original level $= \dfrac{5l}{27} + \dfrac{5l}{3} = \dfrac{50l}{27}$

> **Problem-solving**
>
> P is now moving freely under gravity. The horizontal component of the velocity will not change.
> At the maximum height the vertical component of the velocity is zero.
>
> Conservation of energy.
>
> **Watch out** The particle is not necessarily above A when at its maximum height.

Example 20

A smooth hemisphere with radius 5 m and centre O is resting in a fixed position on a horizontal plane with its flat face in contact with the plane. A particle P of mass 4 kg is slightly disturbed from rest at the highest point of the hemisphere.

When OP has turned through an angle θ and the particle is still on the surface of the hemisphere the normal reaction of the sphere on the particle is R. Find:

a an expression for R

b the angle between OP and the upward vertical when the particle leaves the surface of the hemisphere

c the distance of the particle from the centre of the hemisphere when it hits the ground.

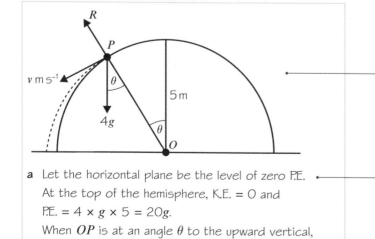

> Draw and label a diagram.

a Let the horizontal plane be the level of zero P.E.

At the top of the hemisphere, K.E. = 0 and

P.E. $= 4 \times g \times 5 = 20g$.

When OP is at an angle θ to the upward vertical,

K.E. $= \tfrac{1}{2}mv^2 = 2v^2$

P.E. $= 4 \times g \times 5\cos\theta = 20g\cos\theta$

> Choose a zero level for P.E.
>
> Find the total of P.E. + K.E. at both points.

A

$\therefore 20g = 2v^2 + 20g\cos\theta$ •————————— Energy is conserved.

$v^2 = 10g(1 - \cos\theta)$

Resolving parallel to PO:

$R(\searrow) : 4g\cos\theta - R = \dfrac{mv^2}{r} = \dfrac{4 \times 10g(1 - \cos\theta)}{5}$ •————— Use the equation for circular motion and substitute for v^2.

$\qquad\qquad = 8g(1 - \cos\theta)$

so $R = 4g\cos\theta - 8g + 8g\cos\theta = 12g\cos\theta - 8g$

b The particle leaves the hemisphere when $R = 0$.

This is when $\cos\theta = \dfrac{2}{3}$

$\theta = \arccos\dfrac{2}{3} = 48.2°$ (3 s.f.)

Problem-solving

The particle leaves the hemisphere when there is no contact force.

c When the particle leaves the hemisphere: •—————

vertical distance $OP = 5\cos\theta = \dfrac{10}{3}$

horizontal distance $OP = 5\sin\theta = \dfrac{5\sqrt{5}}{3}$

and $v^2 = 10g\left(1 - \dfrac{2}{3}\right) = \dfrac{10g}{3}$

The particle is now a projectile with initial velocity $\sqrt{\dfrac{10g}{3}}$ at an angle $\arccos\dfrac{2}{3}$ below the horizontal.

initial vertical speed $= v\sin\theta = \sqrt{\dfrac{10g}{3}} \times \dfrac{\sqrt{5}}{3}$, so

$\dfrac{10}{3} = \sqrt{\dfrac{50g}{27}}t + \dfrac{1}{2}gt^2$

$3\sqrt{3}gt^2 + 2\sqrt{50g}\,t - 20\sqrt{3} = 0$ •————— Using $s = ut + \dfrac{1}{2}at^2$ and solving the quadratic equation for t.

$t = 0.4976\ldots$

Horizontal distance travelled in this time

$= v\cos\theta \times t = \sqrt{\dfrac{10g}{3}} \times \dfrac{2}{3} \times 0.4976\ldots = 1.896\ldots$ •————— No horizontal acceleration.

Total distance from $O = \dfrac{5\sqrt{5}}{3} + 1.896\ldots = 5.6\,\text{m}$ (2 s.f.) •————— Add the two horizontal distances.

Exercise (**1E**)

Whenever a numerical value of g is required take $g = 9.8\,\text{m s}^{-2}$.

(**P**) **1** A particle P of mass m is attached to one end of a light inextensible string of length l. The other end of the string is attached to a fixed point O. The particle is hanging in equilibrium at a point A, directly below O, when it is set in motion with a horizontal speed $\sqrt{3gl}$. When OP has turned through an angle θ and the string is still taut, the tension in the string is T. Find:

a an expression for T

b the height of P above A at the instant when the string goes slack

c the maximum height above A reached by P before it starts to fall to the ground again.

A **2** A smooth solid hemisphere with radius 6 m and centre O is resting in a fixed position on a
P horizontal plane with its flat face in contact with the plane. A particle P of mass 3 kg is slightly
disturbed from rest at the highest point of the hemisphere.

When OP has turned through an angle θ and the particle is still on the surface of the hemisphere
the normal reaction of the sphere on the particle is R. Find:

a an expression for R

b the angle, to the nearest degree, between OP and the upward vertical when the particle leaves
the surface of the hemisphere

c the distance of the particle from the centre of the hemisphere when it hits the ground.

P **3** A smooth solid hemisphere is fixed with its plane face on a horizontal table and its curved
surface uppermost. The plane face of the hemisphere has centre O and radius r. The point A is
the highest point on the hemisphere. A particle P is placed on the hemisphere at A. It is then
given an initial horizontal speed u, where $u^2 = \dfrac{rg}{4}$. When OP makes an angle θ with OA, and
while P remains on the hemisphere, the speed of P is v. Find:

a an expression for v^2

b the value of $\cos\theta$ when P leaves the hemisphere

c the value of v when P leaves the hemisphere.

After leaving the hemisphere P strikes the table at B, find:

d the speed of P at B

e the angle, to the nearest degree, at which P strikes the table.

P **4** A smooth sphere with centre O and radius 2 m is fixed to a horizontal surface. A particle P of
mass 3 kg is slightly disturbed from rest at the highest point of the sphere and starts to slide
down the surface of the sphere. Find:

a the angle, to the nearest degree, between OP and the upward vertical at the instant when P
leaves the surface of the sphere

b the magnitude and direction, to the nearest degree, of the velocity of the particle as it hits the
horizontal surface.

E/P **5** A particle of mass m is projected with speed v from the top of the outside of a smooth sphere of
radius a. In the subsequent motion the particle slides down the surface of the sphere and leaves
the surface of the sphere with speed $\dfrac{\sqrt{3ga}}{2}$. Find:

a the vertical distance travelled by the particle before it loses contact with the surface of the
sphere **(4 marks)**

b v **(4 marks)**

c the magnitude and direction, to the nearest degree, of the velocity of the particle when it is at
the same horizontal level as the centre of the sphere. **(5 marks)**

6 A smooth hemisphere with centre O and radius 50 cm is fixed with its plane face in contact with a horizontal surface. A particle P is released from rest at point A on the sphere, where OA is inclined at 10° to the upward vertical. The particle leaves the sphere at point B.

    a Find the angle, to the nearest degree, between OB and the upward vertical.

    b Describe the subsequent motion qualitatively.

 7 A smooth laundry chute is built in two sections, PQ and QR. Each section is in the shape of an arc of a circle. PQ has radius 5 m and subtends an angle of 70° at its centre, A. QR has radius 7 m and subtends an angle of 40° at its centre, B. The points A, Q and B are in a vertical straight line. The laundry bags are collected in a large bin $\frac{1}{2}$ m below R. To test the chute, a beanbag of mass 2 kg is released from rest at P.

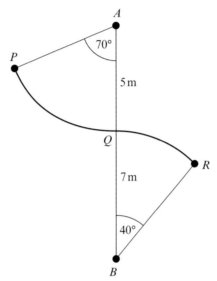

The beanbag is modelled as a particle and the laundry chute is modelled as being smooth.

    a Calculate the speed with which the beanbag reaches the laundry bin.   **(2 marks)**

    b Show that the beanbag loses contact with the chute before it reaches R.   **(5 marks)**

    In practice, laundry bags do remain in contact with the chute throughout.

    c State a possible refinement to the model which could account for this discrepancy.   **(1 mark)**

 8 Part of a hollow spherical shell, centre O and radius a, is removed to form a smooth bowl with a plane circular rim. The bowl is fixed with the rim uppermost and horizontal. The centre of the circular rim is $\frac{4a}{3}$ vertically above the lowest point of the bowl. A marble is placed inside the bowl and projected horizontally from the lowest point of the bowl with speed u.

    a Find the minimum value of u for which the marble will leave the bowl and not fall back in to it.   **(10 marks)**

    In reality the marble is subject to frictional forces from the surface of the bowl and air resistance.

    b State how this will affect your answer to part **a**.   **(1 mark)**

Mixed exercise 1

 1 A particle of mass m moves with constant speed u in a horizontal circle of radius $\frac{3a}{2}$ on the inside of a fixed smooth hollow sphere of radius $2a$. Show that $9ag = 2\sqrt{7}u^2$.

2 A particle P of mass m is attached to one end of a light inextensible string of length $3a$. The other end of the string is attached to a fixed point A which is a vertical distance a above a smooth horizontal table. The particle moves on the table in a circle whose centre O is vertically below A, as shown in the diagram. The string is taut and the speed of P is $2\sqrt{ag}$. Find:

  a the tension in the string

  b the normal reaction of the table on P.

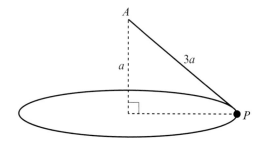

3 A light inextensible string of length $25l$ has its ends fixed to two points A and B, where A is vertically above B. A small smooth ring of mass m is threaded on the string. The ring is moving with constant speed in a horizontal circle with centre B and radius $12l$, as shown in the diagram. Find:

  a the tension in the string

  b the speed of the ring.

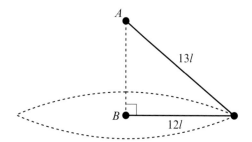

4 A car moves round a bend which is banked at a constant angle of $12°$ to the horizontal. When the car is travelling at a constant speed of $15\,\text{m s}^{-1}$ there is no sideways frictional force on the car. The car is modelled as a particle moving in a horizontal circle of radius r metres. Calculate the value of r.

5 A particle P of mass m is attached to the ends of two light inextensible strings AP and BP each of length l. The ends A and B are attached to fixed points, with A vertically above B and $AB = l$, as shown in the diagram. The particle P moves in a horizontal circle with constant angular speed ω. The centre of the circle is the midpoint of AB and both strings remain taut.

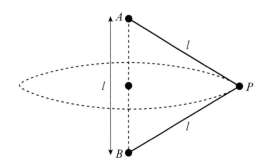

  a Show that the tension in AP is $\frac{m}{2}(2g + l\omega^2)$. **(3 marks)**

  b Find, in terms of m, l, ω and g, an expression for the tension in BP. **(2 marks)**

  c Deduce that $\omega^2 > \frac{2g}{l}$. **(1 mark)**

(P) 6 A particle P of mass m is attached to one end of a light string of length l. The other end of the string is attached to a fixed point A. The particle moves in a horizontal circle with constant angular speed ω and with the string inclined at an angle of 45° to the vertical, as shown in the diagram.

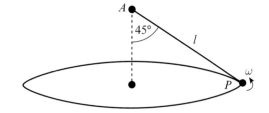

    a Show that the tension in the string is $\sqrt{2}mg$.

    b Find ω in terms of g and l.

7 A particle P of mass 0.6 kg is attached to one end of a light inextensible string of length 1.2 m. The other end of the string is attached to a fixed point A. The particle is moving, with the string taut, in a horizontal circle with centre O vertically below A. The particle is moving with constant angular speed 3 rad s^{-1}. Find:

    a the tension in the string

    b the angle, to the nearest degree, between AP and the downward vertical.

(E) 8 A particle P of mass m moves on the smooth inner surface of a spherical bowl of internal radius r. The particle moves with constant angular speed in a horizontal circle, which is at a depth $\frac{r}{4}$ below the centre of the bowl. Find:

    a the normal reaction of the bowl on P **(2 marks)**

    b the time it takes P to complete three revolutions of its circular path.   **(4 marks)**

(E/P) 9 A bend of a race track is modelled as an arc of a horizontal circle of radius 100 m. The track is not banked at the bend. The maximum speed at which a motorcycle can be ridden round the bend without slipping sideways is 21 m s^{-1}. The motorcycle and its rider are modelled as particles.

    a Show that the coefficient of friction between the motorcycle and the track is 0.45.   **(6 marks)**

    The bend is now reconstructed so that the track is banked at an angle α to the horizontal. The maximum speed at which the motorcycle can now be ridden round the bend without slipping sideways is 28 m s^{-1}. The radius of the bend and the coefficient of friction between the motorcycle and the track are unchanged.

    b Find the value of $\tan \alpha$.   **(8 marks)**

(A)
(E/P) 10 A light rod rests on the surface of a sphere of radius r, as shown in the diagram. The rod is attached to a point vertically above the centre of the sphere, a distance r from the top of the sphere. A particle, P, of mass m is attached to the rod at the point where the rod meets the sphere. The rod pivots freely such that the particle completes horizontal circles on the smooth outer surface of the sphere with angular speed ω.

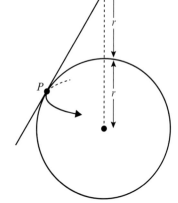

    a Find the tension in the rod above the particle, giving your answer in terms of m, g, ω and r.   **(8 marks)**

Given that the rod remains on the surface of the sphere,

b show that the time taken for the particle to make one complete revolution is at least $\pi\sqrt{\dfrac{6r}{g}}$.

(3 marks)

c Without further calculation, state how your answer to part **b** would change if the particle was moved:

  i up the rod towards the pivot

  ii down the rod away from the pivot.

(2 marks)

E/P **11** A rough disc rotates in a horizontal plane with constant angular velocity ω about a fixed vertical axis. A particle P of mass m lies on the disc at a distance $\frac{3}{5}a$ from the axis. The coefficient of friction between P and the disc is $\frac{3}{7}$. Given that P remains at rest relative to the disc,

a prove that $\omega^2 \leqslant \dfrac{5g}{7a}$

(7 marks)

The particle is now connected to the axis by a horizontal light elastic string of natural length $\dfrac{a}{2}$ and modulus of elasticity $\dfrac{5mg}{2}$. The disc again rotates with constant angular velocity ω about the axis and P remains at rest relative to the disc at a distance $\frac{3}{5}a$ from the axis.

b Find the range of possible values of ω^2.

(8 marks)

12 A particle P of mass m is attached to one end of a light inextensible string of length a. The other end of the string is fixed at a point O. The particle is held with the string taut and OP horizontal. It is then projected vertically downwards with speed u, where $u^2 = \frac{4}{3}ga$. When OP has turned through an angle θ and the string is still taut, the speed of P is v and the tension in the string is T, as shown in the diagram. Find:

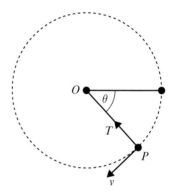

a an expression for v^2 in terms of a, g and θ

b an expression for T in terms of m, g and θ

c the value of θ when the string becomes slack to the nearest degree.

d Explain why P would not complete a vertical circle if the string were replaced by a light rod.

13 A particle P of mass $0.4\,\text{kg}$ is attached to one end of a light inelastic string of length $1\,\text{m}$. The other end of the string is fixed at point O. P is hanging in equilibrium below O when it is projected horizontally with speed $u\,\text{m s}^{-1}$. When OP is horizontal it meets a small smooth peg at Q, where $OQ = 0.8\,\text{m}$. Calculate the minimum value of u if P is to describe a complete circle about Q.

E/P **14** A smooth solid hemisphere is fixed with its plane face on a horizontal table and its curved surface uppermost. The plane face of the hemisphere has centre O and radius a. The point A is the highest point on the hemisphere. A particle P is placed on the hemisphere at A.

It is then given an initial horizontal speed u, where $u^2 = \dfrac{ag}{2}$. When OP makes an angle θ with OA, and while P remains on the hemisphere, the speed of P is v.

A
  a Find an expression for v^2. **(2 marks)**

  b Show that P is still on the hemisphere when $\theta = \arccos 0.9$. **(2 marks)**

  c Find the value of:

    i $\cos\theta$ when P leaves the hemisphere

    ii v when P leaves the hemisphere. **(3 marks)**

After leaving the hemisphere P strikes the table at B, find:

  d the speed of P at B **(2 marks)**

  e the angle, to the nearest degree, at which P strikes the table. **(3 marks)**

(E/P) 15 Part of a hollow spherical shell, centre O and radius r, is removed to form a bowl with a plane circular rim. The bowl is fixed with the circular rim uppermost and horizontal. The point C is the lowest point of the bowl. The point B is on the rim of the bowl and OB is at an angle α to the upward vertical as shown in the diagram. Angle α satisfies $\tan\alpha = \frac{4}{3}$. A smooth small marble of mass m is placed inside the bowl at C and given an initial horizontal speed u. The direction of motion of the marble lies in the vertical plane COB. The marble stays in contact with the bowl until it reaches B. When the marble reaches B it has speed v.

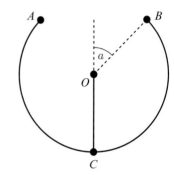

  a Find an expression for v^2. **(4 marks)**

  b If $u^2 = 4gr$, find the normal reaction of the bowl on the marble as the marble reaches B. **(3 marks)**

  c Find the least possible value of u for the marble to reach B. **(3 marks)**

The point A is the other point of the rim of the bowl lying in the vertical plane COB.

  d Find the value of u which will enable the marble to leave the bowl at B and meet it again at A. **(4 marks)**

(E/P) 16 A particle is at the highest point A on the outer surface of a fixed smooth hemisphere of radius a and centre O. The hemisphere is fixed to a horizontal surface with the plane face in contact with the surface. The particle is projected horizontally from A with speed u, where $u < \sqrt{ag}$. The particle leaves the sphere at the point B, where OB makes an angle θ with the upward vertical, as shown in the diagram.

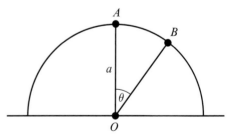

  a Find an expression for $\cos\theta$ in terms of u, g and a. **(3 marks)**

The particle strikes the horizontal surface with speed $\sqrt{\dfrac{5ag}{2}}$.

  b Find the value of θ, to the nearest degree. **(4 marks)**

A Challenge

The diagram shows the curve with equation $y = f(x)$, $x > 0$, where f is a strictly increasing function.

The curve is rotated through 2π radians about the y-axis to form a smooth surface of revolution, which is oriented with the y-axis pointing vertically upwards. A particle is placed on the inside of the surface and completes horizontal circles at a fixed vertical height, with angular speed ω.

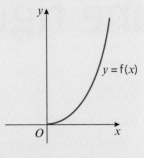

Problem-solving

f is a strictly increasing function, so $f'(x) > 0$ for all $x > 0$. This means that the particle will be able to complete circles at any position on the inside of the surface, and that the height of the particle will be uniquely determined by its horizontal distance from the y-axis.

a In the case where $f(x) = x^2$, show that ω is independent of the vertical height of the particle, and that $\omega = \sqrt{2g}$.

b Conversely, show that if ω is independent of the height of the particle, then $f(x)$ must be of the form $px^2 + q$, where p and q are constants.

Summary of key points

1 If a particle is moving around a circle of radius r m with linear speed v m s^{-1} and angular speed ω rad s^{-1} then $v = r\omega$.

2 An object moving on a circular path with constant linear speed v and constant angular speed ω has acceleration $r\omega^2$ or $\dfrac{v^2}{r}$, towards the centre of the circle.

3 For motion in a vertical circle of radius r, the components of the acceleration are $r\omega^2$ or $\dfrac{v^2}{r}$ towards the centre of the circle and $r\ddot{\theta} = \dot{v}$ along the tangent.

4 A particle attached to the end of a light rod will perform complete vertical circles if it has speed > 0 at the top of the circle.

5 A small bead threaded on to a smooth circular wire will perform complete vertical circles if it has speed > 0 at the top of the circle.

6 A particle attached to a light inextensible string will perform complete vertical circles if the tension in the string > 0 at the top of the circle. This means that the speed of the particle when it reaches the top of the circle must be large enough to keep the string taut at the top of the circle.

7 If an object is not constrained to stay on its circular path then as soon as the contact force associated with the circular path becomes zero the object can be treated as a projectile moving freely under gravity.

2 Centres of mass of plane figures

Objectives

After completing this chapter you should be able to:

The centre of mass of large vehicles must be calculated, tested and sometimes adjusted, so that the vehicle does not topple over easily.

Prior knowledge check

1 Work out the values of x and y:

$$8\begin{pmatrix}x\\y\end{pmatrix} = 4\begin{pmatrix}3\\1\end{pmatrix} + 6\begin{pmatrix}2\\2\end{pmatrix}$$

2 A uniform plank AB of length 6 m and mass 16 kg lies on the edge of a table. A mass of 4 kg is attached to one end of the plank at B, causing the plank to be on the point of tilting.

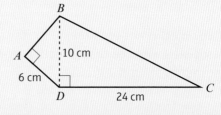

Find the distance AC.

← **Statistics and Mechanics 2, Chapter 4**

3 Find the area of quadrilateral $ABCD$.

B

A

10 cm

6 cm

D

24 cm

C

2.1 Centre of mass of a set of particles on a straight line

You can find the **centre of mass** of a set of particles arranged along a straight line by considering moments. You will use the fact that $\sum m_i x_i = \bar{x} \sum m_i$

Example 1

A system of 3 particles, with masses $2\,\text{kg}$, $5\,\text{kg}$ and $3\,\text{kg}$ are placed along the x-axis at the points $(3, 0)$, $(4, 0)$, ..., $(6, 0)$ respectively. Find the centre of mass of the system.

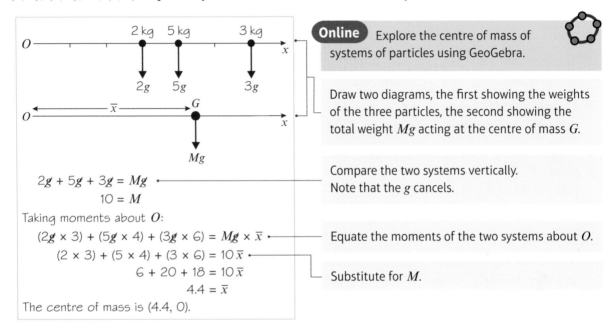

Online Explore the centre of mass of systems of particles using GeoGebra.

Draw two diagrams, the first showing the weights of the three particles, the second showing the total weight Mg acting at the centre of mass G.

$2g + 5g + 3g = Mg$ — Compare the two systems vertically. Note that the g cancels.

$10 = M$

Taking moments about O:

$(2g \times 3) + (5g \times 4) + (3g \times 6) = Mg \times \bar{x}$ — Equate the moments of the two systems about O.

$(2 \times 3) + (5 \times 4) + (3 \times 6) = 10\bar{x}$

$6 + 20 + 18 = 10\bar{x}$ — Substitute for M.

$4.4 = \bar{x}$

The centre of mass is $(4.4, 0)$.

Example 2

A system of n particles, with masses $m_1, m_2, \ldots, m_n$ are placed along the x-axis at the points $(x_1, 0)$, $(x_2, 0)$, ..., $(x_n, 0)$ respectively. Find the centre of mass of the system.

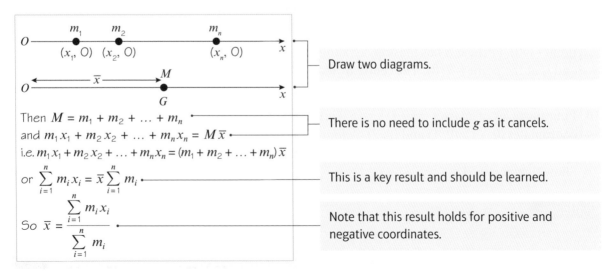

Draw two diagrams.

Then $M = m_1 + m_2 + \ldots + m_n$ — There is no need to include g as it cancels.

and $m_1 x_1 + m_2 x_2 + \ldots + m_n x_n = M\bar{x}$

i.e. $m_1 x_1 + m_2 x_2 + \ldots + m_n x_n = (m_1 + m_2 + \ldots + m_n)\bar{x}$

or $\displaystyle\sum_{i=1}^{n} m_i x_i = \bar{x} \sum_{i=1}^{n} m_i$ — This is a key result and should be learned.

So $\bar{x} = \dfrac{\displaystyle\sum_{i=1}^{n} m_i x_i}{\displaystyle\sum_{i=1}^{n} m_i}$ — Note that this result holds for positive and negative coordinates.

■ If a system of n particles with masses $m_1, m_2, ..., m_n$ are placed along the x-axis at the points $(x_1, 0), (x_2, 0), ..., (x_n, 0)$ respectively, then:

$$\sum_{i=1}^{n} m_i x_i = \bar{x} \sum_{i=1}^{n} m_i$$

where $(\bar{x}, 0)$ is the position of the centre of mass of the system.

Note This result could also be used for a system of particles placed along the y-axis:

$$\sum_{i=1}^{n} m_i y_i = \bar{y} \sum_{i=1}^{n} m_i$$

Exercise 2A

1 Find the position of the centre of mass of four particles of masses 1 kg, 4 kg, 3 kg and 2 kg placed on the x-axis at the points $(6, 0)$, $(3, 0)$, $(2, 0)$ and $(4, 0)$ respectively.

2 Three masses 1 kg, 2 kg and 3 kg, are placed at the points with coordinates $(0, 2)$, $(0, 5)$ and $(0, 1)$ respectively. Find the coordinates of G, the centre of mass of the three masses.

3 Three particles of masses 2 kg, 3 kg and 5 kg, are placed at the points $(-1, 0)$, $(-4, 0)$ and $(5, 0)$ respectively. Find the coordinates of the centre of mass of the three particles.

4 A light rod PQ of length 4 m has particles of masses 1 kg, 2 kg and 3 kg attached to it at the points P, Q and R respectively, where $PR = 2$ m. The centre of mass of the loaded rod is at the point G. Find the distance PG.

5 Three particles of masses 5 kg, 3 kg and m kg lie on the y-axis at the points $(0, 4)$, $(0, 2)$ and $(0, 5)$ respectively. The centre of mass of the system is at the point $(0, 4)$. Find the value of m.

(P) 6 A light rod PQ of length 2 m has particles of masses 0.4 kg and 0.6 kg fixed to it at the points P and R respectively, where $PR = 0.5$ m. Find the mass of the particle which must be fixed at Q so that the centre of mass of the loaded rod is at its midpoint.

(P) 7 The centre of mass of four particles of masses $2m$, $3m$, $7m$ and $8m$, which are positioned at the points $(0, a)$, $(0, 2)$, $(0, -1)$ and $(0, 1)$ respectively, is the point G. Given that the coordinates of G are $(0, 1)$, find the value of a.

(P) 8 Particles of masses 3 kg, 2 kg and 1 kg lie on the y-axis at the points with coordinates $(0, -2)$, $(0, 7)$ and $(0, 4)$ respectively. Another particle of mass 6 kg is added to the system so that the centre of mass of all four particles is at the origin. Find the position of this particle.

(E/P) 9 Three particles A, B and C are placed along the x-axis. Particle A has mass 5 kg and is at the point $(2, 0)$. Particle B has mass m_1 kg and is at the point $(3, 0)$ and particle C has mass m_2 kg and is at the point $(-2, 0)$. The centre of mass of the three particles is at the point $G(1, 0)$. Given that the total mass of the three particles is 10 kg, find the values of m_1 and m_2. **(3 marks)**

(E/P) 10 Four particles of masses $(m - 1)$ kg, $(5 - m)$ kg and m kg lie on the y-axis at the points with coordinates $(0, -1)$, $(0, 1)$ and $(0, 2)$. A fourth particle of mass $(m + 1)$ kg is added at the point $(0, 0)$ so that the centre of mass of all four particles is at the point $(0, 1)$.
Show that $m = 0.5$ kg. **(3 marks)**

Challenge

Three particles, of masses 1 kg, 2 kg and 3 kg respectively, lie on the x-axis at points P, Q and R with $PQ:QR = 2:3$. The centre of mass of the particles is at G. Show that the ratio of the lengths $PQ:PG$ is $12:19$.

2.2 Centre of mass of a set of particles arranged in a plane

You can use $\sum m_i x_i = \bar{x} \sum m_i$ and $\sum m_i y_i = \bar{y} \sum m_i$ to find the centre of mass of a set of point masses arranged in a plane by considering the x-coordinate and y-coordinate of the centre of mass separately.

Example ③

Find the coordinates of the centre of mass of the following system of particles:

  2 kg at (1, 2); 3 kg at (3, 1); 5 kg at (4, 3)

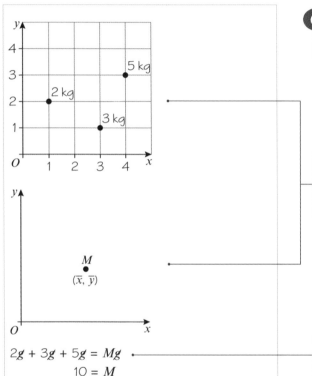

Online Explore the centre of mass of particles arranged in a plane using GeoGebra.

Draw two diagrams, the first showing the three particles, the second showing the total mass M placed at the centre of mass $(\bar{x}, \bar{y})$.

$2g + 3g + 5g = Mg$

$10 = M$

Equate the total weights.
Note that g cancels.

Method 1

Taking moments about the y-axis:

$(2g \times 1) + (3g \times 3) + (5g \times 4) = Mg\bar{x}$

$(2 \times 1) + (3 \times 3) + (5 \times 4) = (2 + 3 + 5)\bar{x}$

$2 + 9 + 20 = 10\bar{x}$

$3.1 = \bar{x}$

Equate the moments of the systems about the y-axis.

Substitute for M.

Taking moments about the x-axis:

$(2g \times 2) + (3g \times 1) + (5g \times 3) = (2 + 3 + 5)g\bar{y}$

$2.2 = \bar{y}$

The centre of mass is (3.1, 2.2).

Equate the moments of the systems about the x-axis.

Method 2

$$2\binom{1}{2} + 3\binom{3}{1} + 5\binom{4}{3} = (2 + 3 + 5)\binom{\bar{x}}{\bar{y}}$$

$$\binom{2}{4} + \binom{9}{3} + \binom{20}{15} = \binom{10x}{10y}$$

$$\binom{31}{22} = \binom{10\bar{x}}{10\bar{y}}$$

$$\binom{3.1}{2.2} = \binom{\bar{x}}{\bar{y}}$$

The centre of mass is at (3.1, 2.2).

Problem-solving

You can reduce the working by using position vectors.

The top line is
$$\sum m_i x_i = \bar{x}\sum m_i$$
and the bottom line is
$$\sum m_i y_i = \bar{y}\sum m_i$$

Divide both sides by 10.

- **If a system consists of n particles: mass m_1 with position vector $\mathbf{r}_1$, mass m_2 with position vector $\mathbf{r}_2$, ..., mass m_n with position vector $\mathbf{r}_n$, then**
$$\sum m_i \mathbf{r}_i = \bar{\mathbf{r}}\sum m_i$$
where $\bar{\mathbf{r}}$ is the position vector of the centre of mass of the system.

Notation The position vector of a point can be written in terms of $\mathbf{i}$ and $\mathbf{j}$ or as a column vector. For example, the position vector of the point (3, 4) is $3\mathbf{i} + 4\mathbf{j}$ or $\binom{3}{4}$.

Example 4

Find the coordinates of the centre of mass of the following system of particles:

4 kg at (−1, 3); 2 kg at (−2, −4); 8 kg at (4, 0); 6 kg at (1, −3)

$$4\binom{-1}{3} + 2\binom{-2}{-4} + 8\binom{4}{0} + 6\binom{1}{-3} = (4+2+8+6)\binom{\bar{x}}{\bar{y}}$$

$$\binom{-4}{12} + \binom{-4}{-8} + 3\binom{2}{0} + \binom{6}{-18} = 20\binom{\bar{x}}{\bar{y}}$$

$$\binom{30}{-14} = 20\binom{\bar{x}}{\bar{y}}$$

$$\binom{1.5}{-0.7} = \binom{\bar{x}}{\bar{y}}$$

Centre of mass is (1.5, −0.7)

The result applies with positive or negative coordinates.

Simplify the LHS.

- **If a question does not specify axes or coordinates you will need to choose your own axes and origin.**

Example 5

A light rectangular plate $ABCD$ has $AB = 20$ cm and $AD = 50$ cm. Particles of masses 2 kg, 3 kg, 5 kg and 5 kg are attached to the plate at the points A, B, C and D respectively.
Find the distance of the centre of mass of the loaded plate from:

a AD **b** AB

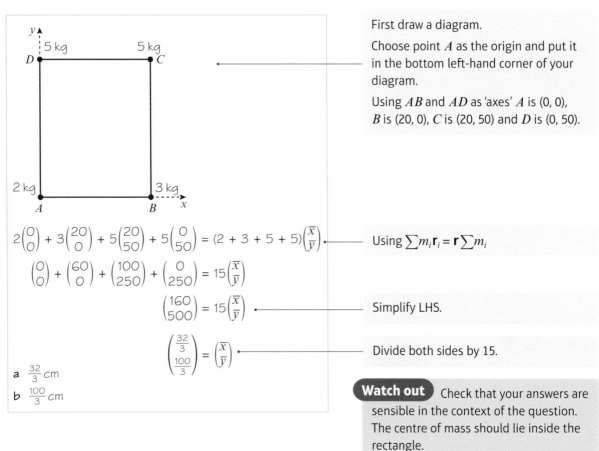

First draw a diagram.

Choose point A as the origin and put it in the bottom left-hand corner of your diagram.

Using AB and AD as 'axes' A is $(0, 0)$, B is $(20, 0)$, C is $(20, 50)$ and D is $(0, 50)$.

$$2\begin{pmatrix}0\\0\end{pmatrix} + 3\begin{pmatrix}20\\0\end{pmatrix} + 5\begin{pmatrix}20\\50\end{pmatrix} + 5\begin{pmatrix}0\\50\end{pmatrix} = (2 + 3 + 5 + 5)\begin{pmatrix}\bar{x}\\\bar{y}\end{pmatrix}$$

Using $\sum m_i \mathbf{r}_i = \mathbf{r} \sum m_i$

$$\begin{pmatrix}0\\0\end{pmatrix} + \begin{pmatrix}60\\0\end{pmatrix} + \begin{pmatrix}100\\250\end{pmatrix} + \begin{pmatrix}0\\250\end{pmatrix} = 15\begin{pmatrix}\bar{x}\\\bar{y}\end{pmatrix}$$

$$\begin{pmatrix}160\\500\end{pmatrix} = 15\begin{pmatrix}\bar{x}\\\bar{y}\end{pmatrix}$$

Simplify LHS.

$$\begin{pmatrix}\frac{32}{3}\\\frac{100}{3}\end{pmatrix} = \begin{pmatrix}\bar{x}\\\bar{y}\end{pmatrix}$$

Divide both sides by 15.

a $\frac{32}{3}$ cm

b $\frac{100}{3}$ cm

Watch out Check that your answers are sensible in the context of the question. The centre of mass should lie inside the rectangle.

Example 6

Particles of masses 4 kg, 3 kg, 2 kg and 1 kg are placed at the points (x, y), $(3, 2)$, $(1, -5)$ and $(6, 0)$ respectively. Given that the centre of mass of the four particles is at the point $(2.5, -2)$, find the values of x and y.

$$4\begin{pmatrix}x\\y\end{pmatrix} + 3\begin{pmatrix}3\\2\end{pmatrix} + 2\begin{pmatrix}1\\-5\end{pmatrix} + 1\begin{pmatrix}6\\0\end{pmatrix} = (4 + 3 + 2 + 1)\begin{pmatrix}2.5\\-2\end{pmatrix}$$

Using $\sum m_i \mathbf{r}_i = \mathbf{r} \sum m_i$

$$\begin{pmatrix}4x\\4y\end{pmatrix} + \begin{pmatrix}9\\6\end{pmatrix} + \begin{pmatrix}2\\-10\end{pmatrix} + \begin{pmatrix}6\\0\end{pmatrix} = \begin{pmatrix}25\\-20\end{pmatrix}$$

$$\begin{pmatrix}4x + 17\\4y - 4\end{pmatrix} = \begin{pmatrix}25\\-20\end{pmatrix}$$

$$4x + 17 = 25$$
$$4y - 4 = -20$$

Equate the **i** and **j** components.

$x = 2, y = -4$

Solve the two equations for x and y.

Example 7

Three particles of masses 2 kg, 1 kg and m kg are situated at the points $(-1, 3)$, $(2, 9)$ and $(2, -1)$ respectively. Given that the centre of mass of the three particles is at the point $(1, \bar{y})$, find:

a the value of m

b the value of $\bar{y}$.

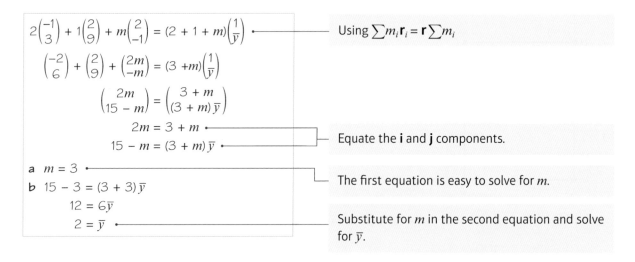

$$2\binom{-1}{3} + 1\binom{2}{9} + m\binom{2}{-1} = (2 + 1 + m)\binom{1}{\bar{y}}$$

Using $\sum m_i \mathbf{r}_i = \mathbf{r} \sum m_i$

$$\binom{-2}{6} + \binom{2}{9} + \binom{2m}{-m} = (3 + m)\binom{1}{\bar{y}}$$

$$\binom{2m}{15 - m} = \binom{3 + m}{(3 + m)\bar{y}}$$

$$2m = 3 + m$$

$$15 - m = (3 + m)\bar{y}$$

Equate the **i** and **j** components.

a $m = 3$

b $15 - 3 = (3 + 3)\bar{y}$

The first equation is easy to solve for m.

$$12 = 6\bar{y}$$

$$2 = \bar{y}$$

Substitute for m in the second equation and solve for $\bar{y}$.

Exercise 2B

1 Two particles of equal mass are placed at the points $(1, -3)$ and $(5, 7)$. Find the centre of mass of the particles.

2 Four particles of equal mass are situated at the points $(2, 0)$, $(-1, 3)$, $(2, -4)$ and $(-1, -2)$. Find the coordinates of the centre of mass of the particles.

3 A system of three particles consists of 10 kg placed at $(2, 3)$, 15 kg placed at $(4, 2)$ and 25 kg placed at $(6, 6)$. Find the coordinates of the centre of mass of the system.

4 Find the position vector of the centre of mass of three particles of masses 0.5 kg, 1.5 kg and 2 kg which are situated at the points with position vectors $(6\mathbf{i} - 3\mathbf{j})$, $(2\mathbf{i} + 5\mathbf{j})$ and $(3\mathbf{i} + 2\mathbf{j})$ respectively.

5 Particles of masses m, $2m$, $5m$ and $2m$ are situated at $(-1, -1)$, $(3, 2)$, $(4, -2)$ and $(-2, 5)$ respectively. Find the coordinates of the centre of mass of the particles.

6 A light rectangular metal plate $PQRS$ has $PQ = 4$ cm and $PS = 2$ cm. Particles of masses 3 kg, 5 kg, 1 kg and 7 kg are attached respectively to the corners P, Q, R and S of the plate. Find the distance of the centre of mass of the loaded plate from:

    a the side PQ **b** the side PS.

(P) **7** Three particles of masses 1 kg, 2 kg and 3 kg are positioned at the points $(1, 0)$, $(4, 3)$ and (p, q) respectively. Given that the centre of mass of the particles is at the point $(2, 0)$, find the values of p and q.

(P) **8** A system consists of three particles with masses $3m$, $4m$ and $5m$. The particles are situated at the points with coordinates $(-3, -4)$, $(0.5, 4)$ and $(0, -5)$ respectively. Find the coordinates of the position of a fourth particle of mass $7m$, given that the centre of mass of all four particles is at the origin.

(E) **9** A light rectangular piece of card $ABCD$ has $AB = 8$ cm and $AD = 6$ cm. Four particles of masses 300 g, 200 g, 600 g and 100 g are fixed to the rectangle at the midpoints of the sides AB, BC, CD and DE respectively. Find the distance of the centre of mass of the loaded rectangle from the sides AB and AD. **(4 marks)**

(E/P) **10** A light rectangular piece of card $ABCD$ has $AB = 8$ cm and $AD = 6$ cm. Three particles of masses 3 g, 2 g and 2 g are attached to the rectangle at the points A, B and C respectively.

   a Find the mass of a particle which must be placed at the point D for the centre of mass of the whole system of four particles to lie 3 cm from the line AB. **(2 marks)**

   b With this fourth particle in place, find the distance of the centre of mass of the system from the side AD. **(4 marks)**

> **Challenge**
>
> A light triangular piece of card ABC has sides $AB = 6$ cm, $AC = 5$ cm and $BC = 5$ cm. Three particles of masses m kg, 0.2 kg and 0.2 kg are fixed to the triangle at the midpoints of the sides AB, BC and AC respectively. The point P lies at the intersection of the lines joining each vertex of the triangle with the midpoint of the opposite side. Given that the centre of mass of the whole system lies at P, find the value of m.

2.3 Centres of mass of standard uniform plane laminas

You can find the positions of the centres of mass of standard uniform plane laminas, including a rectangle, a triangle and a semicircle.

- **An object which has one dimension (its thickness) very small compared with the other two (its length and width) is modelled as a lamina. This means that it is regarded as being two-dimensional with area but no volume.**

> **Hint** For example, a sheet of paper or a piece of card could be modelled as a lamina.

A lamina is **uniform** if its mass is evenly spread throughout its area.

- **If a uniform lamina has an axis of symmetry then its centre of mass must lie on the axis of symmetry. If the lamina has more than one axis of symmetry then it follows that the centre of mass must be at the point of intersection of the axes of symmetry.**

Uniform circular disc

Since every diameter of the disc is a line of symmetry the centre of mass of the disc is at their intersection. This is the centre of the disc.

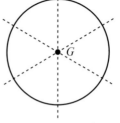

Uniform rectangular lamina

A uniform rectangular lamina has two lines of symmetry, each one joining the midpoints of a pair of opposite sides. The centre of mass is at the point where the two lines meet.

Uniform triangular lamina

A uniform triangular lamina only has axes of symmetry if it is either equilateral or isosceles.

A uniform equilateral triangle has three axes of symmetry, each one joining a vertex to the midpoint of the opposite side. These three lines are called the **medians** of the triangle.

- **The centre of mass of a uniform triangular lamina is at the intersection of the medians. This point is called the centroid of the triangle.**

Note that the medians are not axes of symmetry of the triangle unless the triangle is equilateral (in which case all three medians are axes of symmetry) or isosceles (in which case one median is also an axis of symmetry).

Hint It can be proved that the centroid G (and therefore the centre of mass) of any triangle is two-thirds of the way down each median from each vertex:

where A' is the midpoint of BC, B' is the midpoint of CA and C' is the midpoint of AB:

i.e. $\dfrac{AG}{GA'} = \dfrac{BG}{GB'} = \dfrac{CG}{GC'} = \dfrac{2}{1}$

- **If the coordinates of the three vertices of a uniform triangular lamina are (x_1, y_1), (x_2, y_2) and (x_3, y_3) then the coordinates of the centre of mass are given by taking the average (mean) of the coordinates of the vertices:**

  G is the point $\left(\dfrac{x_1 + x_2 + x_3}{3}, \dfrac{y_1 + y_2 + y_3}{3}\right)$

Hint This is the two-dimensional version of a similar result for a uniform rod: if the ends of the rod are (x_1, y_1) and (x_2, y_2) then its centre of mass is its midpoint, $\left(\dfrac{x_1 + x_2}{2}, \dfrac{y_1 + y_2}{2}\right)$.

Uniform sector of a circle

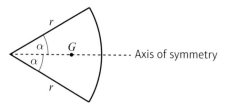

A uniform sector of a circle of radius r and centre angle 2α, where α is measured in radians, has its centre of mass on the axis of symmetry at a distance $\dfrac{2r\sin\alpha}{3\alpha}$ from the centre.

Links Although this result is given in the formulae booklet, A-level students may need to prove it using calculus.

→ **Section 3.1**

Example 8

A uniform triangular lamina has vertices $A(1, 4)$, $B(3, 2)$ and $C(5, 3)$. Find the coordinates of its centre of mass.

Online Explore centres of mass of standard uniform plane laminas using GeoGebra.

G is the point $\left(\dfrac{1+3+5}{3}\right), \left(\dfrac{4+2+3}{3}\right) = (3, 3)$

Find the mean of the vertices of the triangle. This is the **centroid** of the triangle.

Example 9

Find the centre of mass of the uniform triangular lamina shown:

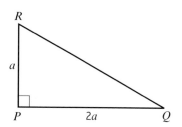

Here we need to choose our own axes and origin.

Taking P as the origin and PQ and PR as axes:

P is $(0, 0)$; Q is $(2a, 0)$; R is $(0, a)$

Write down the coordinates of each of the three vertices.

G is the point

$\left(\dfrac{0+2a+0}{3}, \dfrac{0+0+a}{3}\right) = \left(\dfrac{2a}{3}, \dfrac{a}{3}\right)$

Find the mean of the three vertices.

The centre of mass is $\dfrac{2a}{3}$ from PR and $\dfrac{a}{3}$ from PQ.

Watch out When you choose your own axes you must not leave your answer in coordinate form.

Example 10

The diagram shows a uniform semicircular lamina of radius 6 cm with centre O.
Find the centre of mass of the lamina.

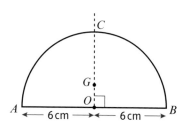

The centre of mass must lie on the line through O which is perpendicular to AB. •——— This is the axis of symmetry of the lamina.

Let $OG = \bar{y}$. Then:

$$\bar{y} = \frac{2 \times 6 \times \sin\frac{\pi}{2}}{\frac{3\pi}{2}}$$ •——— We use the result for a sector which is in the formula booklet with $r = 6$ and $\alpha = \frac{\pi}{2}$

$$y = \frac{12 \times 1}{\frac{3\pi}{2}}$$ •

$$\sin\frac{\pi}{2} = 1$$
You must give the angle in radians for this formula.

$$y = 12 \times \frac{2}{3\pi}$$

$$y = \frac{8}{\pi}$$ •——— Simplify.

The centre of mass of the lamina is on the line OC at a distance $\frac{8}{\pi}$ cm from O.

Exercise 2C

1 Find the centre of mass of a uniform triangular lamina whose vertices are:

a (1, 2), (2, 6) and (3, 1)   **b** (−1, 4), (3, 5) and (7, 3)

c (−3, 2), (4, 0) and (0, 1)   **d** (a, a), (3a, 2a) and (4a, 6a)

2 Find the position of the centre of mass of a uniform semicircular lamina of radius 4 cm and centre O.

(P) **3** The centre of mass of a uniform triangular lamina ABC is at the point (2, a). Given that A is the point (4, 3), B is the point (b, 1) and C is the point (−1, 5), find the values of a and b.

4 Find the position of the centre of mass of the following uniform triangular laminas:

a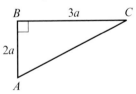

b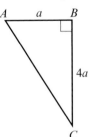

c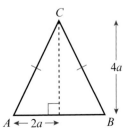

d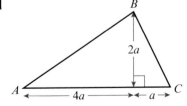

P **5** A uniform triangular lamina is isosceles and has the line $y = 4$ as its axis of symmetry. One of the vertices of the triangle is the point (2, 1). Given that the x-coordinate of the centre of mass of the lamina is -3, find the coordinates of the other two vertices.

P **6** A uniform rectangular lamina $ABCD$ is positioned such that AB lies on the line $y = 2x + 1$. Given that A is at the point (0, 1) and C is at the point (6, 7), find:

  a the coordinates of the points B and D

  b the coordinates of the centre of mass of the lamina.

E/P **7** A uniform triangular lamina ABC has coordinates $A(2, 1)$, $B(4, 1)$ and $C(x, y)$. The centre of mass of the lamina lies on the line $x = 3$. Given that the triangle ABC has an area of $4\,\text{cm}^2$, work out:

  a the possible values of x and y **(3 marks)**

  b the possible coordinates of the centre of mass.   **(4 marks)**

E/P **8** The diagram below shows an equilateral triangle ABC where AC is 4 cm.

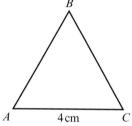

Show that the centre of mass lies $\dfrac{4\sqrt{3}}{3}$ cm from B.   **(3 marks)**

2.4   Centre of mass of a composite lamina

A **composite uniform lamina** consists of two or more standard uniform laminas joined together. You can find the centre of mass of a composite lamina by considering each part of the lamina as a particle positioned at its centre of mass. The masses of each part of the lamina will be proportional to their areas.

Example 11

A uniform lamina consists of a rectangle $PQRS$ joined to an isosceles triangle QRT, as shown in the diagram.

Find the distance of the centre of mass of the lamina from:

a PQ

b PS

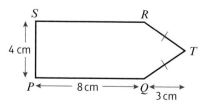

Let the mass per unit area be m kg per cm². •———

Split the lamina along the line QR:

Area of $PQRS = 8 \times 4 = 32$ cm²

So, mass of $PQRS = 32m$

Similarly, mass of $QRT = \frac{1}{2} \times 4 \times 3 \times m$ •———
$$= 6m$$

So, total mass of the lamina $= 32m + 6m = 38m$

Take P as the origin, and axes along PQ and PS. •———

The centre of mass of $PQRS$ is at the point $(4, 2)$. •———

The coordinates of Q are $(8, 0)$.

The coordinates of R are $(8, 4)$.

The coordinates of T are $(11, 2)$.

The centre of mass of $\triangle QRT$ will be

$$\left(\frac{8 + 8 + 11}{3}, \frac{0 + 4 + 2}{3} \right)$$ •———

$$= (9, 2)$$ •———

Replace the lamina by two particles:

$32m$ placed at $(4, 2)$ •———

and $6m$ placed at $(9, 2)$ •———

$$32m \binom{4}{2} + 6m \binom{9}{2} = 38m \binom{\bar{x}}{\bar{y}}$$ •———

$$\binom{128}{64} + \binom{54}{12} = 38 \binom{\bar{x}}{\bar{y}}$$ •———

$$\binom{182}{76} = 38 \binom{\bar{x}}{\bar{y}}$$

$$\frac{91}{19} = \bar{x}$$

$$2 = \bar{y}$$

a Distance from PQ is 2 cm.

b Distance from PS is $\frac{91}{19}$ cm.

Since the lamina is uniform, the mass per unit area will be a constant.

You must always split the lamina up into standard shapes.
G_1 is the centre of mass of the rectangle.
G_2 is the centre of mass of the triangle.

Area of $\triangle$ is $\frac{1}{2} \times$ base $\times$ height.

It's usually a good idea to take the origin at the bottom left-hand corner of your diagram.

This is the centre of the rectangle, G_1.

Take the mean of the coordinates of the vertices of the triangle.

This is G_2.

This is the key idea behind the method.

$\binom{\bar{x}}{\bar{y}}$ is the position vector of the centre of mass.

Cancel the ms and simplify.

Problem-solving

Note that you could have got the answer to part **a** using the fact that the lamina has an axis of symmetry. You should always use this as it will considerably reduce the amount of working required.

■ **The centre of mass of a uniform plane lamina or framework will always lie on an axis of symmetry.**

Example 12

The diagram shows a uniform lamina.

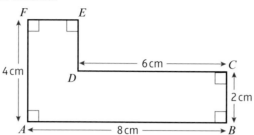

Find the distance of the centre of mass of the lamina from:

a *AF* **b** *AB*

> **Hint** You can find the centre of mass in three different ways.

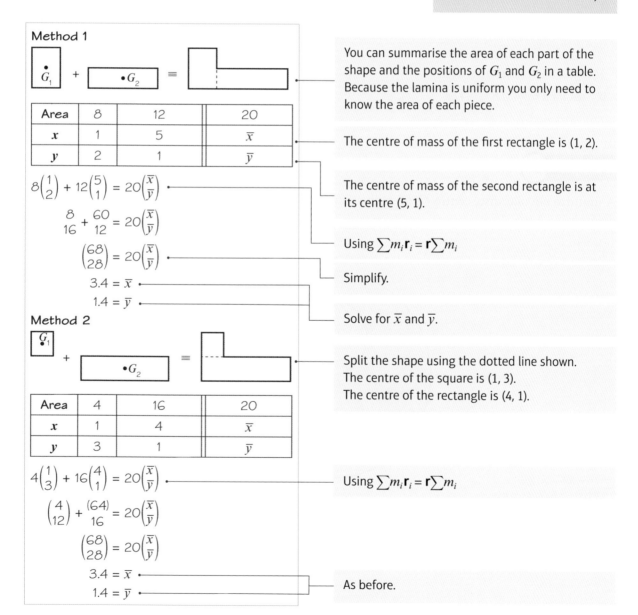

Method 1

Area	8	12	20
x	1	5	$\bar{x}$
y	2	1	$\bar{y}$

$$8\binom{1}{2} + 12\binom{5}{1} = 20\binom{\bar{x}}{\bar{y}}$$

$$\binom{8}{16} + \binom{60}{12} = 20\binom{\bar{x}}{\bar{y}}$$

$$\binom{68}{28} = 20\binom{\bar{x}}{\bar{y}}$$

$$3.4 = \bar{x}$$

$$1.4 = \bar{y}$$

You can summarise the area of each part of the shape and the positions of G_1 and G_2 in a table. Because the lamina is uniform you only need to know the area of each piece.

The centre of mass of the first rectangle is (1, 2).

The centre of mass of the second rectangle is at its centre (5, 1).

Using $\sum m_i \mathbf{r}_i = \mathbf{r}\sum m_i$

Simplify.

Solve for $\bar{x}$ and $\bar{y}$.

Method 2

Area	4	16	20
x	1	4	$\bar{x}$
y	3	1	$\bar{y}$

$$4\binom{1}{3} + 16\binom{4}{1} = 20\binom{\bar{x}}{\bar{y}}$$

$$\binom{4}{12} + \binom{(64)}{16} = 20\binom{\bar{x}}{\bar{y}}$$

$$\binom{68}{28} = 20\binom{\bar{x}}{\bar{y}}$$

$$3.4 = \bar{x}$$

$$1.4 = \bar{y}$$

Split the shape using the dotted line shown.
The centre of the square is (1, 3).
The centre of the rectangle is (4, 1).

Using $\sum m_i \mathbf{r}_i = \mathbf{r}\sum m_i$

As before.

Method 3

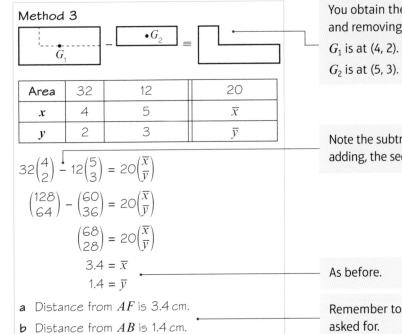

Area	32	12		20
x	4	5		$\bar{x}$
y	2	3		$\bar{y}$

$$32\binom{4}{2} - 12\binom{5}{3} = 20\binom{\bar{x}}{\bar{y}}$$

$$\binom{128}{64} - \binom{60}{36} = 20\binom{\bar{x}}{\bar{y}}$$

$$\binom{68}{28} = 20\binom{\bar{x}}{\bar{y}}$$

$$3.4 = \bar{x}$$
$$1.4 = \bar{y}$$

a Distance from AF is 3.4 cm.

b Distance from AB is 1.4 cm.

You obtain the lamina by starting with a rectangle and removing another rectangle.

G_1 is at (4, 2).

G_2 is at (5, 3).

Note the subtraction, since you are removing, not adding, the second rectangle.

As before.

Remember to give your answers in the form asked for.

Example 13

A uniform circular disc, centre O, of radius 5 cm has two circular holes cut in it, as shown in the diagram.

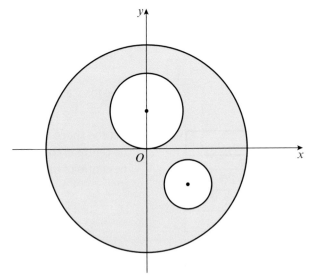

The larger hole has radius 2 cm and the smaller hole has radius 1 cm. The coordinates of the centres of the holes are (0, 2) and (2, −2) respectively. Find the coordinates of the centre of mass of the remaining lamina.

Area	π × 5²	π × 2²	π × 1²	π(5² − 2² − 1²)
x	0	0	2	$\bar{x}$
y	0	2	−2	$\bar{y}$

Problem-solving

When you remove part of a lamina, you can deal with the removed sections by adding sections with negative mass.

Setting out the key information in a table helps to clarify your working.

$$\pi5^2\binom{0}{0} - \pi2^2\binom{0}{2} - \pi1^2\binom{2}{-2} = \pi(5^2 - 2^2 - 1^2)\binom{\bar{x}}{\bar{y}}$$

Note the subtraction signs for each removed section.

$$\binom{0}{0} - \binom{0}{8} - \binom{2}{-2} = 20\binom{\bar{x}}{\bar{y}}$$

Cancel the πs and simplify.

$$\binom{-2}{-6} = 20\binom{\bar{x}}{\bar{y}}$$

Collect terms.

$$\binom{-0.1}{-0.3} = \binom{\bar{x}}{\bar{y}}$$

Solve for $\bar{x}$ and $\bar{y}$.

The coordinates of the centre of mass of the lamina are (−0.1, −0.3).

Exercise **2D**

1 The following diagrams show uniform plane figures. Each one is drawn on a grid of unit squares. Find, in each case, the coordinates of the centre of mass.

a
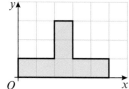

b
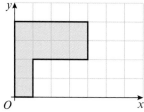

c
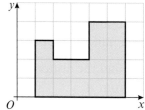

d
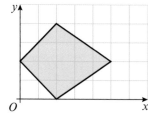

e
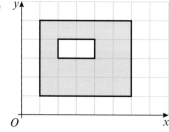

f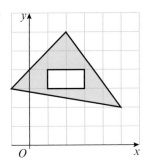

g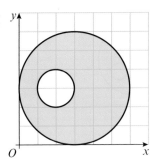

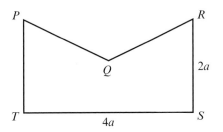

(E) **2** The uniform lamina *PQRST* is formed by removing the triangle *PQR* from the rectangle *PRST* with centre *Q*. The rectangle has sides of length 4*a* and 2*a*. Find the distance of the centre of mass of *PQRST* from *Q*.

(4 marks)

(E) **3** The uniform lamina shown in the diagram is formed by removing a square *DEFG* of side length *a* from the equilateral triangle *ABC* of side length 5*a*. The centre of mass of *DEFG* lies on the perpendicular bisector of *AC*. Given that *AC* and *DG* are parallel and a distance *a* apart, work out the distance of the centre of mass of the whole lamina from *B*.   **(6 marks)**

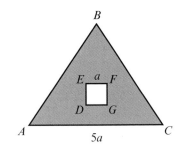

(E/P) **4** The diagram shows a metal template in the shape of a right-angled triangle *ABC*. The template is modelled as a uniform lamina.

a Find the distance of the centre of mass of the lamina from *A*.   **(4 marks)**

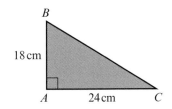

The mass of the template is 15 kg. A particle of mass 5 kg is attached to vertex *C* of the template.

b Find the position of the centre of mass of the template with the attached particle.   **(3 marks)**

> **Problem-solving**
>
> When you attach a particle to a lamina or framework, you can work out the new centre of mass by considering the lamina as a single particle whose weight acts at its centre of mass.

E/P **5** The diagram shows a uniform lamina formed from two rectangles. All the angles are right angles.

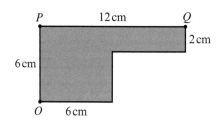

  a Find the position of the centre of mass of the lamina. **(4 marks)**

The mass per unit area of the lamina is $30\,\text{g}\,\text{cm}^{-2}$. Two particles of masses $200\,\text{g}$ and $500\,\text{g}$ are attached to points P and Q on the lamina respectively.

  b Find the new centre of mass of the lamina, giving any lengths correct to 3 significant figures. **(3 marks)**

E/P **6** The diagram shows a uniform piece of card $ABCDEFGH$ where $AB = 6\,\text{cm}$, $BC = 10\,\text{cm}$, $CD = 8\,\text{cm}$, $DE = 2\,\text{cm}$, $EF = 4\,\text{cm}$, $FG = 6\,\text{cm}$, $GH = 2\,\text{cm}$, $HA = 2\,\text{cm}$.

Assume that A is the origin and AH lies on the x-axis.

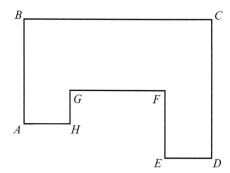

  a Show that the centre of mass lies at the point $\left(\frac{69}{13}, \frac{41}{13}\right)$. **(7 marks)**

The template needs to be changed so that the centre of mass lies at the point $\left(\frac{69}{13}, \frac{41}{13}\right)$. To achieve this, two squares of side length $3\,\text{cm}$ are cut out of the card. Their midpoints are $\left(a, \frac{41}{13}\right)$ and $\left(5, \frac{41}{13}\right)$.

  b Explain how you can use symmetry to determine the value of a. **(2 marks)**

  c Find the value of a. **(5 marks)**

E **7** A uniform circular disc has centre O and radius $3a$. The lines AB and CD are perpendicular diameters of the disc. A circular hole of radius x is made in the disc, with the centre of the hole at the point E on AB and the edge of the hole touching O to form the lamina shown on the right. Given that the centre of mass of the lamina lies on the line AB a distance of $\frac{23}{8}a$ from the point B, find the value of x in terms of a. **(6 marks)**

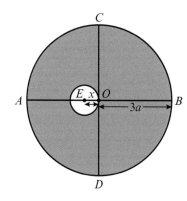

Challenge

A regular hexagon $ABCDEF$ of side length x has the triangle DEF removed to leave the irregular pentagon $ABCDF$ as shown in the diagram below.

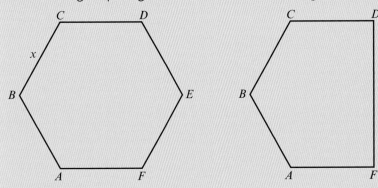

Given that the centre of mass of the hexagon is at the point M and that the centre of mass of the pentagon is at the point N, show that the length of MN is $\frac{2}{15}x$.

2.5 Centre of mass of a framework

You can find the centre of mass of a framework by using the centre of mass of each rod or wire which makes up the framework.

Links The centre of mass of a uniform straight rod is located at its midpoint.
← **Statistics and Mechanics 2, Section 4.4**

Uniform circular arc

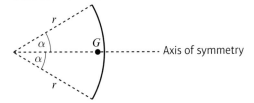

Hint This result can be found in the formula booklet.

A uniform circular arc of radius r and centre angle 2α, where α is measured in radians, has its centre of mass on the axis of symmetry at a distance $\dfrac{r \sin \alpha}{\alpha}$ from the centre.

- **A framework consists of a number of rods joined together or a number of pieces of wire joined together.**

Provided that you can identify the position of the centre of mass of each of the rods or pieces of wire that make up a framework you can find the position of the centre of mass of the whole framework.

Example 14

A framework consists of a uniform length of wire which has been bent into the shape of a letter L, as shown.

Find the distance of the centre of mass of the framework from AB and AF.

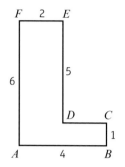

 Online Explore centres of mass of a framework using GeoGebra.

Since the wire is uniform, the mass of each edge will be proportional to its length. The centre of mass of each edge will be at its midpoint.

Taking A as the origin and axes along AB and AF:

Each term on the LHS consists of the length of an edge multiplied by the position vector of the midpoint of the edge.

$$4\binom{2}{0} + 1\binom{4}{0.5} + 2\binom{3}{1} + 5\binom{2}{3.5} + 2\binom{1}{6} + 6\binom{0}{3} = 20\binom{\bar{x}}{\bar{y}}$$

$$\binom{8}{0} + \binom{4}{0.5} + \binom{6}{2} + \binom{10}{17.5} + \binom{2}{12} + \binom{0}{18} = 20\binom{\bar{x}}{\bar{y}}$$

$$\binom{30}{50} = 20\binom{\bar{x}}{\bar{y}}$$

Simplify and collect terms.

$$\binom{1.5}{2.5} = \binom{\bar{x}}{\bar{y}}$$

This is $\bar{x}$.

Distance from AF is 1.5 cm.

Distance from AB is 2.5 cm.

This is $\bar{y}$.

Example 15

Find the position of the centre of mass of a framework constructed from a uniform piece of wire bent into the shape shown:

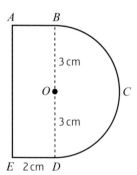

where the wire BCD is a semicircle, centre O, of radius 3 cm and wire $BAED$ forms three sides of a rectangle $ABDE$.

Take O as the origin and axes along OC and OB.

$$3\pi\begin{pmatrix}\frac{6}{\pi}\\0\end{pmatrix} + 2\begin{pmatrix}-1\\-3\end{pmatrix} + 2\begin{pmatrix}-1\\3\end{pmatrix} + 6\begin{pmatrix}-2\\0\end{pmatrix} = (10 + 3\pi)\begin{pmatrix}\bar{x}\\\bar{y}\end{pmatrix}$$

$$\begin{pmatrix}18\\0\end{pmatrix} + \begin{pmatrix}-2\\-6\end{pmatrix} + \begin{pmatrix}-2\\6\end{pmatrix} + \begin{pmatrix}-12\\0\end{pmatrix} = (10 + 3\pi)\begin{pmatrix}\bar{x}\\\bar{y}\end{pmatrix}$$

$$\begin{pmatrix}2\\0\end{pmatrix} = (10 + 3\pi)\begin{pmatrix}\bar{x}\\\bar{y}\end{pmatrix}$$

$$\bar{x} = \frac{2}{(10 + 3\pi)}$$

$$\bar{y} = 0$$

Use the result for the centre of mass of a uniform circular arc with $\alpha = \frac{\pi}{2}$ and $r = 3$ to find the centre of mass of the arc BCD. 3π is the length of the arc.

Simplify.

Collect terms.

You could have used the symmetry to deduce this without any working.

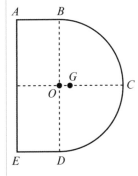

G is the centre of mass of the framework, on the axis of symmetry, a distance $\frac{2}{10 + 3\pi}$ cm to the right of O.

Exercise 2E

1 By regarding the shapes shown below as uniform plane wire frameworks, find the coordinates of the centre of mass of each shape.

a
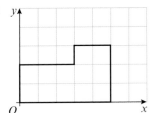

b
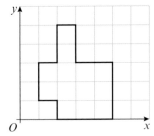

c
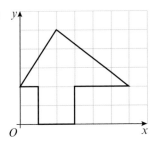

d
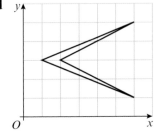

E/P **2** Find the position of the centre of mass of the framework shown in the diagram, which is formed by bending a uniform piece of wire of total length $(12 + 2\pi)$ cm to form a sector of a circle, centre O, radius 6 cm. **(4 marks)**

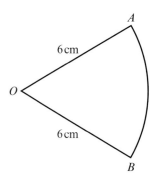

E **3** A framework consists of a uniform length of wire which has been bent into the shape of a letter T, as shown.

Find the distance of the centre of mass of the framework from AB. **(6 marks)**

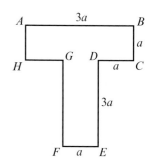

E/P **4** A uniform framework is constructed by bending wire into the shape of a semicircle and a diameter. The semicircle has radius 15 cm.

a Find the distance of the centre of mass of the framework from AB. **(4 marks)**

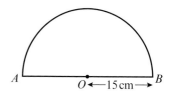

The metal used to form the framework has a mass of 8 grams per cm. Two identical particles of masses 100 g are attached to the framework, at A and B.

b Find:
  i the total mass of the loaded framework
  ii the distance of the centre of mass of the loaded framework from AB. **(4 marks)**

E/P **5** The diagram shows a triangular framework formed from uniform wire. Particles of masses 10 kg, 20 kg and 30 kg are attached to vertices A, B and C respectively. The mass of the unloaded framework is 15 kg.

Find the position of the centre of mass of the loaded framework. **(8 marks)**

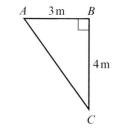

E/P **6** A uniform length of wire is bent to form the shape shown in the diagram.

ACB is a semicircle of radius 3 cm, centre O.
ADO and BEO are both semicircles of radius 1.5 cm.
Find the position of the centre of mass of the framework. **(6 marks)**

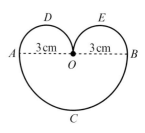

(E/P) **7** A 3.5 m ladder is modelled as a framework made from uniform wire as shown in the diagram. The rungs are 50 cm wide and are 50 cm apart and the top and bottom rungs are 50 cm from the base and top of the ladder respectively. The base of the ladder rests on horizontal ground and the ladder stands vertically.

    a Find the height of the centre of mass above the ground. **(2 marks)**

    The bottom rung is removed from the ladder.

    b Show that the height of the centre of mass of the ladder has increased by $\frac{5}{76}$ m. **(4 marks)**

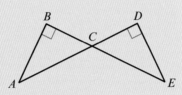

50 cm

50 cm

Challenge

A metal framework $ABCDE$ is made from two congruent right-angled triangles such that ACD and BCE are straight lines, as shown in the diagram.

Given that $AB = 4$ cm and $CD = 3$ cm, work out the distance between C and the centre of mass of the framework.

2.6 Lamina in equilibrium

You can solve problems involving a lamina in equilibrium. A lamina can be suspended by means of a string attached to some point of the lamina, or can be allowed to pivot freely about a horizontal axis which passes through some point of the lamina.

■ **When a lamina or framework is suspended freely from a fixed point or pivots freely about a horizontal axis it will rest in equilibrium in a vertical plane with its centre of mass vertically below the point of suspension or the pivot.**

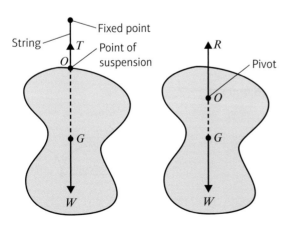

String — Fixed point
T — Point of suspension
O
G
W

R
Pivot
O
G
W

Hint The first lamina is suspended from a fixed point. There are only two forces acting on it: the weight of the lamina and the tension in the string. Both forces pass through the point of suspension.

The second lamina is free to rotate about a fixed horizontal pivot. There are only two forces acting on it: the weight of the lamina and the reaction of the pivot on the lamina. Both pass through the pivot.

The resultant of the moments about O in both laminas is zero.

Example 16

Find the angle that the line AB makes with the vertical if this L-shaped uniform lamina is freely suspended from:

a A

b B

c E

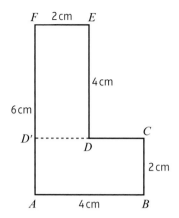

First find the centre of mass of the lamina.

Split the lamina along CD'. ———————————— Split the lamina into two rectangles.

Take A as the origin and axes along AB and AF:

$$8\binom{2}{1} + 8\binom{1}{4} = 16\binom{\bar{x}}{\bar{y}}$$ ———————————— Area $ABCD' = 8$
Area $DEFD' = 8$

$$\binom{24}{40} = 16\binom{\bar{x}}{\bar{y}}$$ ———————————— Simplify.

$$\bar{x} = 1.5$$

$$\bar{y} = 2.5$$

a

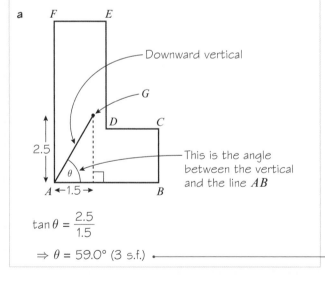

Downward vertical

This is the angle between the vertical and the line AB

$$\tan\theta = \frac{2.5}{1.5}$$

$$\Rightarrow \theta = 59.0° \text{ (3 s.f.)}$$ ———————————— θ is the angle required.

Problem-solving

You do not need to draw the lamina hanging.
Draw a line from the point of suspension to the centre of mass. Mark this in as the vertical.

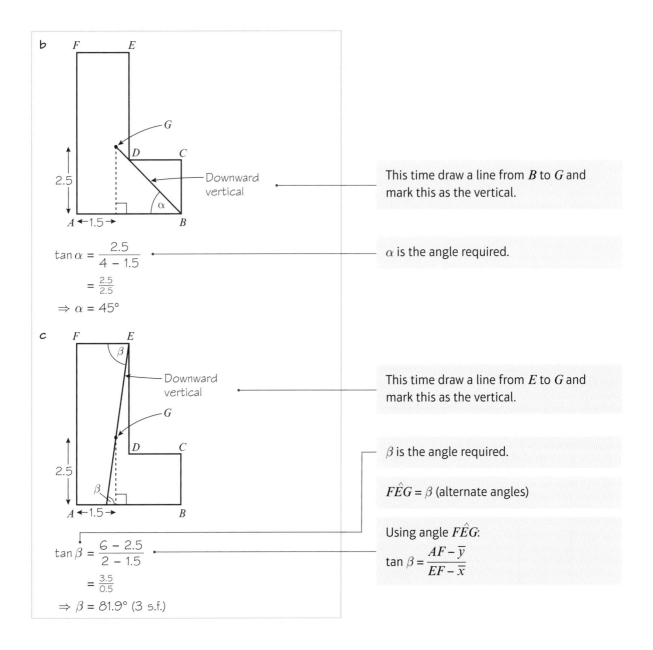

b

$$\tan \alpha = \frac{2.5}{4 - 1.5}$$

$$= \frac{2.5}{2.5}$$

$$\Rightarrow \alpha = 45°$$

This time draw a line from B to G and mark this as the vertical.

α is the angle required.

c

$$\tan \beta = \frac{6 - 2.5}{2 - 1.5}$$

$$= \frac{3.5}{0.5}$$

$$\Rightarrow \beta = 81.9° \text{ (3 s.f.)}$$

This time draw a line from E to G and mark this as the vertical.

β is the angle required.

$F\hat{E}G = \beta$ (alternate angles)

Using angle $F\hat{E}G$:

$$\tan \beta = \frac{AF - \bar{y}}{EF - \bar{x}}$$

Example 17

The L-shaped lamina in Example 16 has mass M kg. Find the angle that FE makes with the vertical when a mass of $\frac{1}{10}M$ kg is attached to B and the lamina is freely suspended from F.

$$M\begin{pmatrix} 1.5 \\ 2.5 \end{pmatrix} + \frac{1}{10}M\begin{pmatrix} 4 \\ 0 \end{pmatrix} = \frac{11}{10}M\begin{pmatrix} \bar{x} \\ \bar{y} \end{pmatrix}$$

$$\begin{pmatrix} 1.9 \\ 2.5 \end{pmatrix} = \frac{11}{10}\begin{pmatrix} \bar{x} \\ \bar{y} \end{pmatrix}$$

$$\begin{pmatrix} \bar{x} \\ \bar{y} \end{pmatrix} = \begin{pmatrix} \frac{19}{11} \\ \frac{25}{11} \end{pmatrix}$$

Problem-solving

Consider the lamina as a single particle whose weight acts at its centre of mass.

Recall from Example 16 that the centre of mass of the lamina is at (1.5, 2.5).

The centre of mass of the added mass is at (4, 0).

So the centre of mass of the loaded lamina is at the point $\left(\frac{19}{11}, \frac{25}{11}\right)$.

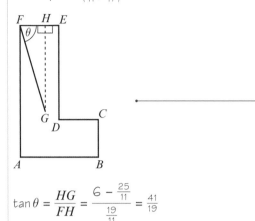

Redraw the diagram showing the angle that you require.

G is the centre of mass and θ is the angle FE makes with the vertical when the lamina is suspended from F.

$$\tan \theta = \frac{HG}{FH} = \frac{6 - \frac{25}{11}}{\frac{19}{11}} = \frac{41}{19}$$

$$\theta = 65.1° \text{ (3 s.f.)}$$

Exercise **2F**

1 a The diagram shows a uniform lamina.

The lamina is freely suspended from the point O and hangs in equilibrium.

Find the angle between OA and the downward vertical.

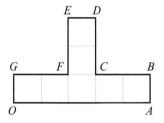

b The diagram shows a uniform lamina.

The lamina is freely suspended from the point O and hangs in equilibrium. Find the angle between OA and the downward vertical.

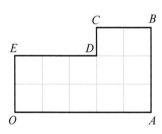

c The diagram shows a uniform lamina.

The lamina is freely suspended from the point O and hangs in equilibrium.

Find the angle between OA and the downward vertical.

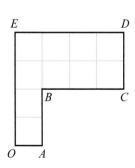

2 The diagram shows a uniform lamina.

The lamina is freely suspended from the point A and hangs in equilibrium.
Find the angle between AB and the downward vertical.

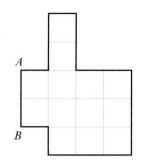

3 The diagram shows a uniform lamina.

The lamina is free to rotate about a fixed smooth horizontal axis perpendicular to the plane of the lamina, passing through the point A, and hangs in equilibrium.

Find the angle between AB and the horizontal.

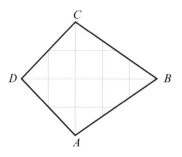

4 The framework in question 2, Exercise 2E is freely suspended from the point A and allowed to hang in equilibrium. Find the angle between OA and the downward vertical.

5 The shape in question 4, Exercise 2E is freely suspended from the point A and allowed to hang in equilibrium. Find the angle between OA and the horizontal.

(E) **6** $PQRS$ is a uniform lamina.

Find the distance of the centre of mass of the lamina from:

a PS **(4 marks)**

b PQ **(4 marks)**

The lamina is suspended from the point Q and allowed to hang in equilibrium.

c Find the angle that PQ makes with the vertical.   **(3 marks)**

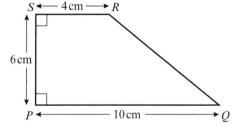

(E/P) **7** The L-shaped lamina shown has mass M kg.
Find the angle that BC makes with the vertical when a mass of $0.2M$ kg is attached to F and the lamina is freely suspended from C.   **(8 marks)**

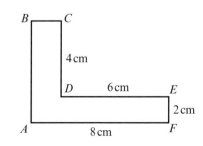

E/P **8** The lamina shown is a quarter circle with radius 4 cm and mass M kg. Find the angle that AB makes with the vertical when a mass of $0.5M$ kg is attached to C and the lamina is freely suspended from B. **(6 marks)**

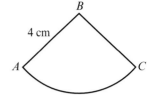
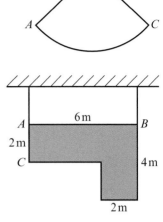

E/P **9** The diagram shows a uniform lamina suspended from a horizontal ceiling by two light inextensible vertical strings attached at points A and B, such that AB is horizontal.

a Find the distance of the centre of mass of the lamina from AC. **(2 marks)**

b Given that the mass of the lamina is 12 kg, find the tensions in the strings. **(4 marks)**

The string at A breaks, and the lamina hangs in equilibrium suspended from point B.

c Find, to the nearest degree, the angle that AB makes with the horizontal. **(5 marks)**

> **Problem-solving**
>
> For part **b**, you will need to consider the horizontal moments about A and B. Have a look at Example 19 for an example of this technique.

E/P **10** The diagram shows a uniform semicircular lamina of mass m kg. The lamina is suspended by two light inextensible vertical strings, such that PQ hangs vertically. The tensions in the strings are T_1 and T_2, as shown in the diagram. A mass of km kg is attached to the lamina at Q, where k is a constant, and the lamina hangs in equilibrium with PQ still vertical and $T_1 = 5T_2$.

Find k. **(6 marks)**

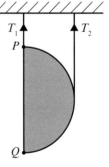

E/P **11** A sign is modelled as a uniform rectangular lamina $ABCDE$ which has a quarter circle removed from its bottom right corner as shown in the diagram.

a By taking A as the origin show that the centre of mass of the sign lies

at the point $\left(\dfrac{2a(38 - 3\pi)}{3(24 - \pi)}, \dfrac{104a}{3(24 - \pi)} \right)$. **(6 marks)**

The sign is suspended using vertical ropes attached to the sign at B and at C and hangs in equilibrium with BC horizontal. The weight of the sign is W and the ropes are modelled as light inextensible strings.

b Find the tension in each rope in terms of W and π. **(4 marks)**

The rope attached at B breaks and the sign hangs freely in equilibrium suspended from C.

c Find the angle CD makes with the vertical. **(4 marks)**

Challenge

The mobile of mass M shown in the diagram hangs in equilibrium from a ceiling on a 25 cm cable.

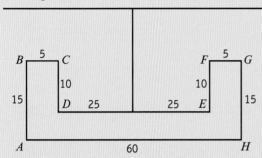

Work out, in terms of M, the smallest mass that can be fixed at A that will cause G to touch the ceiling.

2.7 Frameworks in equilibrium

Problems involving frameworks in equilibrium can be solved using the same methods that are used to solve problems involving laminas in equilibrium.

Example 19

The inverted T-shaped uniform framework of weight W shown in the diagram is freely suspended from D and G by two vertical strings, so that AH is horizontal. The strings and the framework lie in the same plane.

a Find, in terms of W, the tensions T_1 and T_2.

The string at D breaks and the framework hangs freely in equilibrium from G.

b Find the angle that DE makes with the vertical when the framework comes to rest in equilibrium.

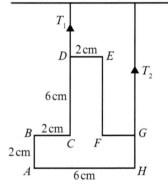

a Let AH lie on the x-axis and AB lie on the y-axis with the point A at the origin.

$$2\binom{0}{1} + 2\binom{1}{2} + 6\binom{2}{5} + 2\binom{3}{8} + 6\binom{4}{5}$$
$$+ 2\binom{5}{2} + 2\binom{6}{1} + 6\binom{3}{0} = 28\binom{\bar{x}}{\bar{y}}$$

$$28\binom{\bar{x}}{\bar{y}} = \binom{84}{88}$$

$$\binom{\bar{x}}{\bar{y}} = \binom{3}{\frac{22}{7}}$$

Res (↑) $T_1 + T_2 = W$

Taking moments about D gives

$W = 4T_2 \Rightarrow T_2 = \frac{1}{4}W$ and $T_1 = \frac{3}{4}W$

First find the centre of mass of the framework.

Problem-solving

The framework is in equilibrium, so the resultant moment must be 0. The centre of mass of the framework is a horizontal distance of 1 cm from D, so the clockwise moment is $1 \times W = W$. The anticlockwise moment is due to T_2 so is equal to $4T_2$.

b The centre of mass lies $\frac{22}{7}$ cm from AH.

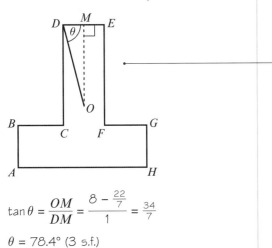

Redraw the diagram showing the angle that you require.

O is the centre of mass and θ is the angle DE makes with the vertical when suspended from D.

$$\tan \theta = \frac{OM}{DM} = \frac{8 - \frac{22}{7}}{1} = \frac{34}{7}$$

$$\theta = 78.4° \text{ (3 s.f.)}$$

Example 20

The inverted T-shaped uniform framework used in Example 19 has mass W kg.

a Find the location of the centre of mass when a mass of $\frac{1}{4}W$ kg is fixed at point H.

b Find the angle that DE makes with the vertical when the system is freely suspended from E.

a $$W\begin{pmatrix} 3 \\ \frac{22}{7} \end{pmatrix} + \frac{1}{4}W\begin{pmatrix} 6 \\ 0 \end{pmatrix} = \frac{5}{4}W\begin{pmatrix} \overline{x} \\ \overline{y} \end{pmatrix}$$

Recall from Example 19 that the centre of mass of the lamina is at $\left(3, \frac{22}{7}\right)$.

The centre of mass of the added mass is at $(6, 0)$.

$$\begin{pmatrix} 3 \\ \frac{22}{7} \end{pmatrix} + \begin{pmatrix} \frac{3}{2} \\ 0 \end{pmatrix} = \frac{5}{4}\begin{pmatrix} \overline{x} \\ \overline{y} \end{pmatrix}$$

$$\frac{5}{4}\begin{pmatrix} \overline{x} \\ \overline{y} \end{pmatrix} = \begin{pmatrix} \frac{9}{2} \\ \frac{22}{7} \end{pmatrix}$$

$$\begin{pmatrix} \overline{x} \\ \overline{y} \end{pmatrix} = \begin{pmatrix} \frac{18}{5} \\ \frac{88}{35} \end{pmatrix}$$

So the centre of mass is at the point $\left(\frac{18}{5}, \frac{88}{35}\right)$

Find the centre of mass of the framework and the added mass.

b

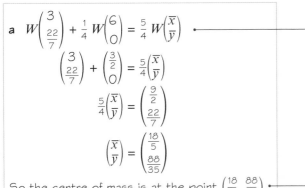

Redraw the diagram showing the angle that you require.

O is the centre of mass and θ is the angle DE makes with the vertical when the framework is suspended from E.

$$\tan \theta = \frac{PO}{PE} = \frac{8 - \frac{88}{35}}{6 - \frac{18}{5}} = \frac{16}{7}$$

$$\theta = 66.4° \text{ (3 s.f.)}$$

Exercise 2G

1 The uniform framework shown opposite is freely suspended from the point B and hangs in equilibrium. Find the angle between BC and the downward vertical.

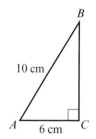

2 The uniform framework shown opposite is freely suspended from the point D and hangs in equilibrium. Find the angle between CD and the downward vertical.

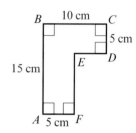

E/P 3 The uniform framework shown opposite is freely suspended from the point A and hangs in equilibrium.

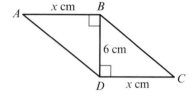

  a Given that the angle between AB and the downward vertical is $\arctan\frac{3}{8}$, work out the value of x. **(4 marks)**

  The framework has mass M kg.

  b A particle of mass kM kg is attached to A and the framework is then freely suspended from the point B so that it hangs in equilibrium. Given that the downward vertical now makes an angle of $\arctan\frac{8}{15}$ with BD, work out the value of k. **(4 marks)**

P 4 The uniform framework shown in question 1 has mass M kg. A particle of mass $0.75M$ kg is attached to A and the framework is then freely suspended from the point B so that it hangs in equilibrium. Find the angle between BC and the downward vertical.

P 5 The uniform framework shown in question 2 has mass M kg. A particle of mass $0.15M$ kg is attached to F and the framework is then freely suspended from the point B so that it hangs in equilibrium. Find the angle between BC and the downward vertical.

E/P 6 The uniform framework shown in the diagram is made of a square of side 4 cm and two quarter circles and has mass M kg. A particle of mass $0.1M$ kg is attached to C and the framework is then freely suspended from the point A so that it hangs in equilibrium. Find the angle between FE and the downward vertical. **(6 marks)**

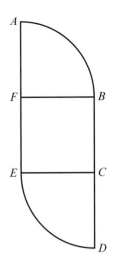

E/P 7 The uniform framework of weight W shown in the diagram is suspended by two vertical strings from A and D such that FE is horizontal. The strings and the framework lie in the same plane. The lengths shown are in metres and all the angles are right angles.

   a Work out, in terms of W, the tensions T_1 and T_2 **(6 marks)**

   The string at A snaps and the framework hangs in equilibrium suspended from D.

   b Find the angle that DE makes with the vertical. **(3 marks)**

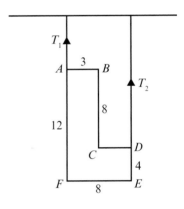

E/P 8 A uniform framework made from uniform rods is shown opposite. It has weight W and is suspended by one vertical string attached at A and one string angled at $60°$ to the vertical attached at D. The strings and the framework lie in the same plane. The framework hangs such that FE is horizontal. The lengths are $AB = 6$ cm, $AH = 4$ cm, $BC = 2$ cm, $CD = 3$ cm and $DE = 10$ cm. There is only one rod in the span BG.

   a Work out, in terms of W, the tensions T_1 and T_2 **(6 marks)**

   The string at D snaps.

   b Find the angle that AB makes with the vertical when the framework comes to rest in equilibrium. **(3 marks)**

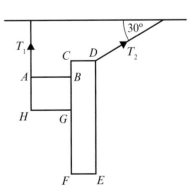

Challenge

A square uniform framework $ABCD$ is suspended by two light inextensible strings. the first string is perpendicular to AB and is attached at the point P, which lies on AB such that $AP:PB = 1:3$. The second string is attached at C and makes an angle of θ with the horizontal, as shown in the diagram.

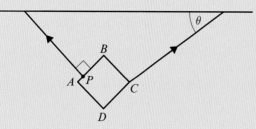

Given that the framework rests in equilibrium with AC horizontal, find θ, giving your answer in degrees, correct to one decimal place.

2.8 Non-uniform composite laminas and frameworks

You can use the methods developed in this chapter to solve problems involving non-uniform laminas and frameworks.

Example **21**

The lamina shown is made from a square piece of cardboard $ABCD$ that has had the corner C folded to the centre of the square. Find the position of the centre of mass of the lamina.

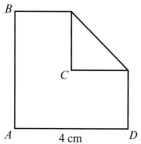

Problem-solving

You can model this situation as a composite lamina made up of three different uniform laminas. The folded section of card is modelled as a lamina with **twice** the density of the other sections.

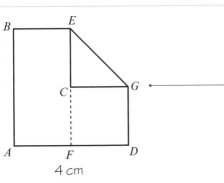

Redraw the diagram, splitting the lamina into its composite parts.

Let A be the origin so that AD lies on the x-axis and AB lies on the y-axis.

The coordinates of the centre of mass of the rectangle $ABEF$ are $(1, 2)$.

The coordinates of the centre of mass of the square $CGDF$ are $(3, 1)$.

The coordinates of the centre of mass of the triangle CEG are

$$\left(\frac{2 + 2 + 4}{3}, \frac{2 + 4 + 2}{3}\right) = \left(\frac{8}{3}, \frac{8}{3}\right)$$

Use COM of a triangle with vertices at (x_1, y_1), (x_2, y_2), (x_3, y_3) is found at:

$$\left(\frac{x_1 + x_2 + x_3}{3}, \frac{y_1 + y_2 + y_3}{3}\right)$$

Rectangle $ABEF$ has area $8\,\text{cm}^2$, square $CGDF$ has area $4\,\text{cm}^2$. Triangle CEG has area $2\,\text{cm}^2$, but this material is twice the density of the other material.

Remember that the card is folded over so the triangle CEG is made of two layers.

$$8\begin{pmatrix}1\\2\end{pmatrix} + 4\begin{pmatrix}3\\1\end{pmatrix} + 4\begin{pmatrix}\frac{8}{3}\\\frac{8}{3}\end{pmatrix} = 16\begin{pmatrix}\overline{x}\\\overline{y}\end{pmatrix}$$

Use 4 for the mass of the triangular section.

$$\begin{pmatrix}\overline{x}\\\overline{y}\end{pmatrix} = \begin{pmatrix}\frac{23}{12}\\\frac{23}{12}\end{pmatrix}$$

So the centre of mass of the lamina is at $\left(\frac{23}{12}, \frac{23}{12}\right)$.

Example 22

The triangular framework shown is made from three pieces of uniform wire, AB, AC and BC of mass $2M$, $3M$ and $5M$ respectively. Find the coordinates of the centre of mass of the framework.

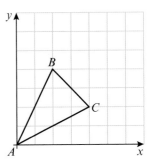

$$2M\begin{pmatrix}1\\2\end{pmatrix} + 3M\begin{pmatrix}2\\1\end{pmatrix} + 5M\begin{pmatrix}3\\3\end{pmatrix} = 10M\begin{pmatrix}\bar{x}\\\bar{y}\end{pmatrix}$$ ●————— Multiply the mass of each piece of wire by the position of its centre of mass.

$$\begin{pmatrix}\bar{x}\\\bar{y}\end{pmatrix} = \begin{pmatrix}2.3\\2.2\end{pmatrix}$$

So the coordinates of the centre of mass are at (2.3, 2.2).

Exercise 2H

(P) 1 The rectangular lamina shown is made from a rectangular piece of cardboard *ABCD* where *AB* is 6 cm and *BC* is 4 cm.

The rectangular lamina is folded along the line *EF* to make the square *AEFD*.

Find the position of the centre of mass of the square lamina.

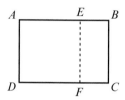

(P) 2 The rectangular lamina shown is made from a rectangular piece of cardboard *ABCD* where *AB* is 10 cm and *BC* is 6 cm.

The lamina is folded so that the side *AD* lies along the side *AB*. Find the position of the centre of mass of the resulting lamina.

(E/P) 3 The diagram shows a rectangular lamina *ABCD*, where *AB* = 10 cm and *BC* = 6 cm. The lamina is folded so that *D* lies exactly on top of *B*. It is then suspended freely from *A*. Find the angle that *AD* makes with the vertical. **(8 marks)**

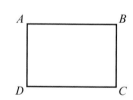

(P) 4 The framework shown is made from four pieces of uniform wire, *AB*, *BC*, *CD* and *DA* of masses 2*M*, 3*M*, *M* and 5*M* respectively. Find the coordinates of the centre of mass of the framework.

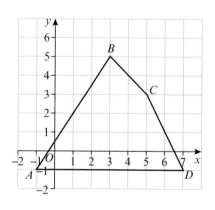

(P) 5 The framework shown is made from three pieces of uniform circular wire made from the same material. The wire used to make BC and AC is twice as thick as the wire used to make AB. Find the coordinates of the centre of mass of the framework.

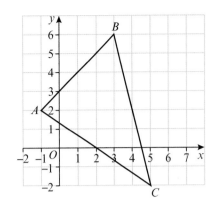

(E/P) 6 The framework shown is made from four pieces of uniform wire, AB, BC, CD and AD. AD and BC are made of the same material and AD has mass M. AB and CD each have mass $0.5M$. The framework is suspended freely from B. Find the angle that BC makes with the vertical. **(8 marks)**

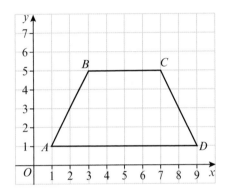

(E/P) 7 The diagram shows a sign for an ice-cream shop. The sign is made from a sheet of wood in the shape of an isosceles triangle, attached to a semicircular sheet of painted metal, as shown in the diagram. The triangle is modelled as a lamina of mass 4 kg, and the semicircle is modelled as a lamina of mass 16 kg.

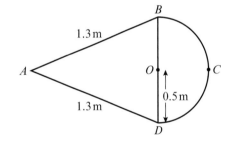

a Find the distance of the centre of mass of the composite lamina from BD. **(4 marks)**

The shop owner wants to suspend the sign by two inextensible vertical wires, so that the axis of symmetry AOC is horizontal. She attaches one wire to point B.

b State, with a reason, whether she should attach the other wire to point A or point C. **(1 mark)**

c Using your answer to part **b**, find the tension in each wire. **(3 marks)**

(E/P) 8 The diagram shows a mobile made from two different flat materials attached at one edge. The mobile is modelled as a square lamina $ABDE$ of density 20 g cm^{-2}, and an isosceles triangular lamina BCD of density 60 g cm^{-2}.

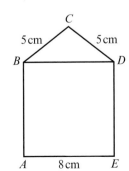

The mobile is suspended from point B and hangs in equilibrium.

a Find, correct to 1 d.p., the size of the acute angle that AB makes with the vertical. **(8 marks)**

A mass of 500 g is attached to point A.

b Find, correct to 1 d.p., the size of the new angle that AB makes with the vertical. **(5 marks)**

Challenge

The first two steps in constructing a paper aeroplane from a rectangular piece of paper are as follows:

1 Fold AB and BC to the centre line of the paper.

2 Fold DE and EF to the centre line of the paper.

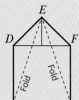

a Given that the resulting shape is an isosceles triangle, show that the sides of the original rectangle are in the ratio $2 : \sqrt{2} + 1$.

b Given that the width of the original rectangle $AC = x$ cm, find the position of the centre of mass of the folded isosceles triangle in terms of x.

Mixed exercise **2**

(E/P) **1** The diagram shows a uniform lamina consisting of a semicircle joined to a triangle ADC.

The sides AD and DC are equal.

a Find the distance of the centre of mass of the lamina from AC. **(4 marks)**

The lamina is freely suspended from A and hangs at rest.

b Find, to the nearest degree, the angle between AC and the vertical. **(2 marks)**

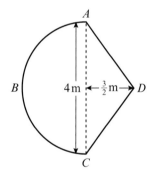

The mass of the lamina is M. A particle P of mass kM is attached to the lamina at D. When suspended from A, the lamina now hangs with its axis of symmetry, BD, horizontal.

c Find, to 3 significant figures, the value of k. **(6 marks)**

(E/P) **2** A uniform triangular lamina ABC is in equilibrium, suspended from a fixed point O by a light inextensible string attached to the point B of the lamina, as shown in the diagram.

Given that $AB = 9$ cm, $BC = 12$ cm and $A\hat{B}C = 90°$, find the angle between BC and the downward vertical. **(8 marks)**

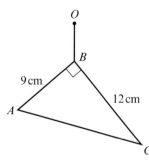

3 Four particles P, Q, R and S of masses 3 kg, 5 kg, 2 kg and 4 kg are placed at the points $(1, 6)$, $(-1, 5)$, $(2, -3)$ and $(-1, -4)$ respectively. Find the coordinates of the centre of mass of the particles. **(4 marks)**

E/P **4** A uniform rectangular piece of card $ABCD$ has $AB = 3a$ and $BC = a$. One corner of the rectangle is folded over to form a trapezium $ABED$ as shown in the diagram.

Find the distance of the centre of mass of the trapezium from:

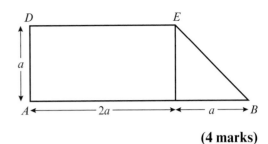

  a AD **(4 marks)**

  b AB **(4 marks)**

The lamina $ABED$ is freely suspended from E and hangs at rest.

  c Find the angle between DE and the horizontal. **(2 marks)**

The mass of the lamina is M. A particle of mass m is attached to the lamina at the point B. The lamina is freely suspended from E and it hangs at rest with AB horizontal.

  d Find m in terms of M. **(4 marks)**

E/P **5** A thin uniform wire of length $5a$ is bent to form the shape $ABCD$, where $AB = 2a$, $BC = 2a$, $CD = a$ and BC is perpendicular to both AB and CD, as shown in the diagram.

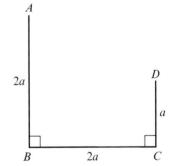

  a Find the distance of the centre of mass of the wire from:

    i AB **ii** BC **(4 marks)**

The wire is freely suspended from B and hangs at rest.

  b Find, to the nearest degree, the angle between AB and the vertical. **(2 marks)**

E **6** A uniform lamina consists of a rectangle $ABCD$, where $AB = 3a$ and $AD = 2a$, with a square hole $EFGA$, where $EF = a$, as shown in the diagram.

Find the distance of the centre of mass of the lamina from:

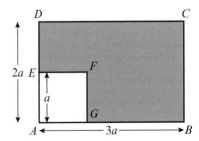

  a AD **(6 marks)**

  b AB **(2 marks)**

E/P **7** The rectangular lamina shown in the diagram is folded so that C lies exactly on top of A.

A mass of $0.25M$ is then attached at D and the lamina is freely suspended from B. Given that the lamina has mass M, find the angle that BC makes with the vertical. **(12 marks)**

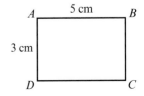

E/P 8 The lamina shown in the diagram is suspended from a point A.

    a Find the angle made by AH with the vertical.   **(8 marks)**

    A mass of $5M$ kg is now attached at point B. Given that the lamina has a mass of $10M$:

    b find the change in the angle made by AH with the vertical.   **(6 marks)**

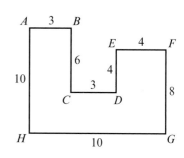

E/P 9 The framework shown is made from four pieces of uniform wire, AB, BC, CD and AD. AD and BC are made of the same material and AD has mass M. AB has mass $0.25M$ and CD has mass $0.5M$. The framework is suspended freely from B. Show that AB makes an angle of $\arctan\frac{125}{53}$ with the vertical.   **(12 marks)**

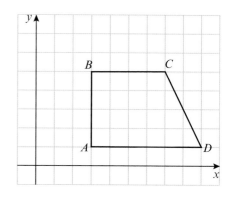

E/P 10 The rectangular lamina shown is made from two rectangular pieces of cardboard $ABCD$ and $BEFC$ where AB is $12\,\text{cm}$, BC is $8\,\text{cm}$ and BE is $4\,\text{cm}$. The two pieces of cardboard are attached to each other along the line BC.

The density of the cardboard used to make $BEFC$ is three times the density of the card used to make $ABCD$ and $ABCD$ has mass M.

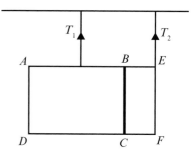

The lamina is supported by two vertical strings, one attached at the midpoint of AB and one attached at E. The lamina is positioned such that AD is vertical.

    a Work out the tensions T_1 and T_2 in terms of M and g, the acceleration due to gravity.   **(6 marks)**

The string at E snaps.

    b Work out the angle AB makes with the vertical when the lamina has come to rest in equilibrium.   **(4 marks)**

E/P 11 A piece of card is in the shape of an equilateral triangle ABC of mass $4M$ and side length $10\,\text{cm}$. The triangle is folded so that vertex C sits on the midpoint of AB, as shown in the diagram.

A mass of $2M$ is attached to the lamina at D. The lamina is suspended by two vertical strings attached at A and B causing AB to lie horizontally.

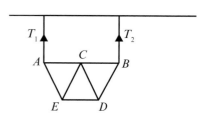

a Work out, in terms of M and g, the acceleration due to gravity, the values of T_1 and T_2

(6 marks)

The string at A snaps.

b Work out the angle AB makes with the vertical when the lamina has come to rest in equilibrium.

(4 marks)

Challenge

The shape shown is made from an isosceles triangular framework ABC of mass $8M$ and a square lamina $ACDE$ of mass $9M$.

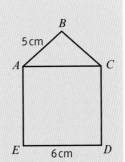

When the shape is suspended from A the vertical cuts the side CD at the point F.

The lamina is then folded along the line EF and allowed to hang freely from A again. Work out the angle AC makes with the vertical.

Summary of key points

1 The centre of mass of a large body is the point at which the whole mass of the body can be considered to be concentrated.

2 If a system of n particles with masses $m_1, m_2, \ldots, m_n$ are placed along the x-axis at the points $(x_1, 0), (x_2, 0), \ldots, (x_n, 0)$ respectively, then:

$$\sum_{i=1}^{n} m_i x_i = \bar{x} \sum_{i=1}^{n} m_i$$

where $(\bar{x}, 0)$ is the position of the centre of mass of the system.

3 If a system consists of n particles: mass m_1 with position vector $\mathbf{r}_1$, mass m_2 with position vector $\mathbf{r}_2, \ldots,$ mass m_n with position vector $\mathbf{r}_n$ then:

$$\sum m_i \mathbf{r}_i = \bar{\mathbf{r}} \sum m_i$$

where $\bar{\mathbf{r}}$ is the position vector of the centre of mass of the system.

4 If a question does not specify axes or coordinates you will need to choose your own axes and origin.

5 An object which has one dimension (its thickness) very small compared with the other two (its length and width) is modelled as a lamina. This means that it is regarded as being two-dimensional with area rather than volume.

6 If a uniform lamina has an axis of symmetry then its centre of mass must lie on the axis of symmetry. If the lamina has more than one axis of symmetry then it follows that the centre of mass must be at the point of intersection of the axes of symmetry.

7 The centre of mass of a uniform triangular lamina is at the intersection of the medians. This point is called the centroid of the triangle.

8 If the coordinates of the three vertices of a uniform triangular lamina are (x_1, y_1), (x_2, y_2) and (x_3, y_3) then the coordinates of the centre of mass are given by taking the average (mean) of the coordinates of the vertices:

$$G \text{ is the point } \left(\frac{x_1 + x_2 + x_3}{3}, \frac{y_1 + y_2 + y_3}{3} \right)$$

9 The centre of mass of a uniform plane lamina or framework will always lie on an axis of symmetry.

10 A framework consists of a number of rods joined together or a number of pieces of wire joined together.

11 When a lamina or framework is suspended freely from a fixed point or pivots freely about a horizontal axis it will rest in equilibrium in a vertical plane with its centre of mass vertically below the point of suspension or the pivot.

Further centres of mass

3

Objectives

After completing this chapter you should be able to:

● Use calculus to find the centre of mass of a lamina → pages 78–87

● Find centres of mass of uniform bodies → pages 87–98

● Use symmetry and calculus to find the centre of mass of a uniform
solid of revolution → pages 88–98

● Find centres of mass of non-uniform bodies → pages 98–103

● Solve rigid body problems in equilibrium → pages 103–109

● Determine whether a rigid body on an inclined plane will slide or
topple → pages 110–117

Prior knowledge check

1 A particle of mass m is
suspended by two inelastic
light strings as shown.

Find T_1 and T_2 in terms of
m and g.

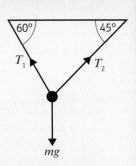

← **Statistics and Mechanics 2, Section 5.1**

2 The region bounded by the curve with equation
$y = \sqrt{x}\ln x$, the x-axis and the line $x = 4$ is rotated
through 360° about the x-axis. Show that volume
of the resulting solid is $\pi(A \ln 4 - B)$ where A and
B are positive rational constants to be found.

← **Core Pure 1, Section 5.1**

3 Given that $y = (1 + 2x)$, evaluate $\dfrac{\int_0^4 xy^2\,dx}{\int_0^4 y^2\,dx}$

← **Pure Year 1, Section 13.1**

When a gymnast balances,
they adjust their body
position so that their centre
of mass is in a favourable
position.

3.1 Using calculus to find centres of mass

A In the previous chapter you found the centres of mass of laminas by considering moments and symmetry. For a system of particles $m_1, m_2, \ldots$ positioned at $(x_1, y_1), (x_2, y_2), \ldots$ respectively in the plane:

- $\sum m_i x_i = \bar{x} \sum m_i$ and $\sum m_i y_i = \bar{y} \sum m_i$

You used this result to find the centre of mass of a composite lamina by considering the centres of mass of its component parts as particles. You can also use these results in conjunction with integration to find the position of the centre of mass of a uniform lamina.

Suppose you need to find the centre of mass of the uniform lamina bounded by the curve with equation $y = f(x)$, the x-axis and the lines $x = a$ and $x = b$ shown shaded in the diagram below.

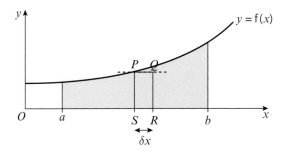

Consider the lamina as made up of small rectangular strips such as $PQRS$, where PQ is parallel to the x-axis. Let P have coordinates (x, y) and let the width of the strip be δx.

Notation The rectangle $PQRS$ of width δx is sometimes called an **elemental strip** of the lamina.

The height of the strip is y so its area is $y \, \delta x$.

The mass of the strip (m_i) is $\rho y \, \delta x$, where ρ is the mass per unit area of the lamina.

Watch out ρ is the Greek letter 'rho'. In this chapter ρ might represent mass **per unit area**, mass **per unit length** or **density** (mass per **unit volume**). You will be told which quantity it represents in the question.

As the lamina is uniform, as $\delta x \to 0$ the centre of mass of the strip $\to \left(x, \frac{1}{2}y\right)$.

Let the coordinates of the centre of mass of the whole lamina be the point $(\bar{x}, \bar{y})$.

$$\sum m_i x_i = \bar{x} \sum m_i$$

So $\sum ((\rho y \delta x) x) = \bar{x} \sum \rho y \delta x$ ———— The summation is taken across all the strips between $x = a$ and $x = b$.

$$\bar{x} = \frac{\sum ((\rho y \delta x) x)}{\sum \rho y \delta x}$$

A As $\delta x \to 0$ the summations become integrals, giving:

$$\bar{x} = \frac{\int_a^b \rho xy \, dx}{\int_a^b \rho y \, dx}$$

$$= \frac{\int_a^b x f(x) \, dx}{\int_a^b f(x) \, dx}$$ Because the lamina is uniform, ρ is a constant, so it cancels. Use the equation of the curve to write y as $f(x)$.

Similarly:

$$\sum m_i y_i = \bar{y} \sum m_i$$

So $\sum \left((\rho y \delta x) \frac{y}{2} \right) = \bar{y} \sum \rho y \delta x$ ——— The vertical coordinate of the centre of mass of each strip is **half** of its height.

$$\bar{y} = \frac{\sum \left((\rho y \delta x) \frac{y}{2} \right)}{\sum \rho y \delta x}$$

As $\delta x \to 0$ the summations become integrals, giving:

$$\bar{y} = \frac{\int_a^b \frac{1}{2} \rho y^2 \, dx}{\int_a^b \rho y \, dx}$$

$$= \frac{\int_a^b \frac{1}{2} (f(x))^2 \, dx}{\int_a^b f(x) \, dx}$$

- **The centre of mass of a uniform lamina may be found using the formulae:**

  - $$\bar{x} = \frac{\int_a^b xy \, dx}{\int_a^b y \, dx} \text{ and } \bar{y} = \frac{\int_a^b \frac{1}{2} y^2 \, dx}{\int_a^b y \, dx}$$

  - $$M\bar{x} = \int_a^b \rho xy \, dx \text{ and } M\bar{y} = \int_a^b \frac{1}{2} \rho y^2 \, dx, \text{ where } M = \int_a^b \rho y \, dx \text{ is the total mass of the lamina,}$$ **and ρ is the mass per unit area of the lamina.**

Example 1

A

Use calculus to find the position of the centre of mass of a
right-angled triangular lamina OPQ with base d and height h,
as shown in the diagram.

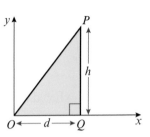

The equation of OP is $y = \frac{h}{d}x$

Find the equation of the line OP by calculating the gradient of OP and using $y = mx + c$, with $c = 0$.

The mass M of the triangular lamina

$$= \rho \times \text{area} = \rho \times \tfrac{1}{2}dh$$

Use area of triangle formula and let the mass per unit area be ρ.

Using the formula for $\bar{x}$, $M\bar{x} = \int_a^b \rho x y \, dx$

$$M\bar{x} = \int_0^d \rho x \frac{h}{d}x \, dx = \frac{h}{d}\rho \int_0^d x^2 \, dx$$

$$= \frac{h}{d}\rho \left[\tfrac{1}{3}x^3\right]_0^d = \tfrac{1}{3}\rho h d^2$$

So $\bar{x} = \dfrac{\tfrac{1}{3}\rho h d^2}{\tfrac{1}{2}\rho h d} = \tfrac{2}{3}d$

Use the formulae for the centre of mass of a lamina to find $\bar{x}$ and $\bar{y}$.

Also $M\bar{y} = \int_a^b \tfrac{1}{2}\rho y^2 \, dx = \int_0^d \tfrac{1}{2}\rho \left(\frac{h}{d}x\right)^2 dx$

$$= \tfrac{1}{2}\rho \left(\frac{h}{d}\right)^2 \left[\tfrac{1}{3}x^3\right]_0^d = \tfrac{1}{6}\rho h^2 d$$

So $\bar{y} = \dfrac{\tfrac{1}{6}\rho h^2 d}{\tfrac{1}{2}\rho h d} = \tfrac{1}{3}h$

So the centre of mass is at the point $\left(\tfrac{2}{3}d, \tfrac{1}{3}h\right)$.

Notice that the centre of mass is at
$$\left(\frac{x_1 + x_2 + x_3}{3}, \frac{y_1 + y_2 + y_3}{3}\right),$$
i.e. $\left(\dfrac{0 + d + d}{3}, \dfrac{0 + 0 + h}{3}\right).$ ← Section 2.3

Example 2

Find the coordinates of the centre of mass of the
uniform lamina bounded by the curve with
equation $y = 4 - x^2$, the x-axis and the y-axis,
as shown.

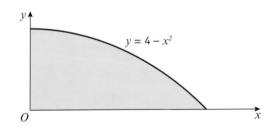

The curve meets the x-axis when $x = 2$.

Put $y = 0$ and solve $4 - x^2 = 0$ to obtain $x = 2$.

$$\bar{x} = \frac{\int_0^b xy\,dx}{\int_0^b y\,dx} = \frac{\int_0^2 x(4 - x^2)\,dx}{\int_0^2 (4 - x^2)\,dx}$$

Substitute $y = 4 - x^2$ into the formula for $\bar{x}$.

$$\int_0^2 x(4 - x^2)\,dx = \int_0^2 (4x - x^3)\,dx$$
$$= \left[2x^2 - \tfrac{1}{4}x^4\right]_0^2 = 8 - 4 = 4$$
$$\int_0^2 (4 - x^2)\,dx = \left[4x - \tfrac{1}{3}x^3\right]_0^2$$
$$= 8 - \tfrac{8}{3} = \tfrac{16}{3}$$

So $\bar{x} = \dfrac{4}{\frac{16}{3}} = \dfrac{3}{4}$

Integrate and evaluate $\bar{x}$.

$$\bar{y} = \frac{\int_a^b \tfrac{1}{2}y^2\,dx}{\int_a^b y\,dx} = \frac{\int_0^2 \tfrac{1}{2}(4 - x^2)^2\,dx}{\int_0^2 (4 - x^2)\,dx}$$

Substitute $y = 4 - x^2$ into the formula for $\bar{y}$.

$$\int_0^2 \tfrac{1}{2}(4 - x^2)^2\,dx = \tfrac{1}{2}\int_0^2 (16 - 8x^2 + x^4)\,dx$$
$$= \tfrac{1}{2}\left[16x - \tfrac{8}{3}x^3 + \tfrac{1}{5}x^5\right]_0^2$$
$$= \tfrac{1}{2}\left(32 - \tfrac{64}{3} + \tfrac{32}{5}\right) = \tfrac{128}{15}$$

Integrate and evaluate $\bar{y}$.

So $\bar{y} = \dfrac{\frac{128}{15}}{\frac{16}{3}} = \dfrac{8}{5}$

The coordinates of the centre of mass are $\left(\tfrac{3}{4}, \tfrac{8}{5}\right)$.

Example 3

A uniform semicircular lamina has radius r cm. Find the position of its centre of mass.

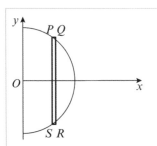

Take the diameter of the lamina as the y-axis, and the midpoint of the diameter as the origin.

Let $PQRS$ be an elemental strip with width δx, where P has coordinates (x, y).

A

The centre of mass of this strip is at the point $(x, 0)$. •──────────── The width of the strip may be ignored as it is small.

The centres of mass of all such strips are on the x-axis and so the centre of mass of the • lamina is also on the x-axis. Use a symmetry argument to explain why the centre of mass lies on the x-axis. This is the axis of symmetry of the lamina.

As point P lies on the circumference of the circle radius r, $x^2 + y^2 = r^2$, and so $y = \sqrt{r^2 - x^2}$

The area of the strip is $2y\,\delta x$ and so its mass is $2\rho y\,\delta x$, where ρ is the mass per unit area of the lamina. Note that the length of the strip is $2y$, due to the symmetry of the semicircle.

The mass M of the lamina is $\frac{1}{2}\pi r^2 \rho$ and $\bar{x}$ is obtained from: The area of the semicircle is $\frac{1}{2}\pi r^2$

$$M\bar{x} = \int_0^r 2\rho x \sqrt{(r^2 - x^2)}\,dx$$

$$= \rho \int_0^r 2x(r^2 - x^2)^{\frac{1}{2}}\,dx$$

$$= \rho\left[-\frac{2}{3}\left[r^2 - x^2\right)^{\frac{3}{2}}\right]_0^r$$ The integration may be done by substitution or by inspection using the chain rule in reverse.

$$= \frac{2}{3}\rho r^3$$

So $\bar{x} = \dfrac{\frac{2}{3}\rho r^3}{\frac{1}{2}\rho\pi r^2} = \dfrac{4r}{3\pi}$

The centre of mass is on the axis of symmetry at a distance of $\dfrac{4r}{3\pi}$ from the straight edge diameter.

Watch out You might be asked to prove this result using calculus in your exam. If you are not specifically asked to use calculus or integration, you may quote this result when solving problems.

Example **4**

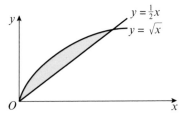

$y = \frac{1}{2}x$
$y = \sqrt{x}$

The diagram shows a uniform lamina occupying the shaded region bounded by the curve with equation $y = \sqrt{x}$, and the straight line with equation $y = \frac{1}{2}x$. Find the coordinates of the centre of mass of the lamina.

A

Consider an elemental strip such as $PQRS$, where P is the point (x, y_1), which lies on the curve $y = \sqrt{x}$, and Q is the point (x, y_2), which lies on the line $y = \frac{1}{2}x$.

The area of the strip is $(y_1 - y_2)\,\delta x$ and its mass is $\rho(y_1 - y_2)\,\delta x$ where ρ is the mass per unit area of the lamina.

> δx is the width of the strip.

The centre of mass of the strip lies at the point $\left(x, \frac{1}{2}(y_1 + y_2)\right)$

The line meets the curve when $\sqrt{x} = \frac{1}{2}x$, i.e. when $x = 0$ and $x = 4$.

> Square both sides and solve the resulting quadratic equation $x = \frac{1}{4}x^2$

The mass M of the lamina is given by:

$$M = \int_a^b \rho(y_1 - y_2)\,dx = \rho\int_0^4 (\sqrt{x} - \tfrac{1}{2}x)\,dx$$

> Sum the strips and let $\delta x \to 0$, so that the summations become integrals.

So $M = \rho\left[\frac{2}{3}x^{\frac{3}{2}} - \frac{1}{4}x^2\right]_0^4 = \rho\left(\frac{16}{3} - \frac{16}{4}\right) = \frac{4}{3}\rho$

Using $M\bar{x} = \int_a^b \rho x(y_1 - y_2)\,dx = \rho\int_0^4 (x^{\frac{3}{2}} - \tfrac{1}{2}x^2)\,dx$

> Use $M\bar{x} = \sum_{x=a}^{x=b} \rho x(y_1 - y_2)\,\delta x$ and let $\delta x \to 0$, so that the summation becomes an integral.

$$= \rho\left[\frac{2}{5}x^{\frac{5}{2}} - \frac{1}{6}x^3\right]_0^4 = \rho\left(\frac{64}{5} - \frac{64}{6}\right) = \frac{64}{30}\rho$$

So $\bar{x} = \frac{64}{30} \times \frac{3}{4} = \frac{8}{5}$ or 1.6

> Divide $\frac{64}{30}\rho$ by $\frac{4}{3}\rho$, as $m = \frac{4}{3}\rho$

Using $M\bar{y} = \int_a^b \frac{1}{2}\rho(y_1 + y_2)(y_1 - y_2)\,dx$

> Use $M\bar{y} = \sum_{x=a}^{x=b} \rho\frac{y_1 + y_2}{2}(y_1 - y_2)\,\delta x$ and let $\delta x \to 0$, so that the summation becomes an integral.

$$= \frac{1}{2}\rho\int_0^4 \left(\sqrt{x} + \tfrac{1}{2}x\right)\left(\sqrt{x} - \tfrac{1}{2}x\right)\,dx$$

$$= \frac{1}{2}\rho\int_0^4 (x - \tfrac{1}{4}x^2)\,dx$$

$$= \frac{1}{2}\rho\left[\frac{1}{2}x^2 - \frac{1}{12}x^3\right]_0^4 = \frac{1}{2}\rho\left(8 - \frac{64}{12}\right) = \frac{4}{3}\rho$$

So $\bar{y} = \frac{4}{3} \times \frac{3}{4} = 1$

> Divide $\frac{4}{3}\rho$ by $\frac{4}{3}\rho$, as $M = \frac{4}{3}\rho$

The centre of mass is at the point $(\frac{8}{5}, 1)$.

Example 5

Find the centre of mass of a uniform lamina in the form of a sector of a circle, radius r and centre O, which subtends an angle 2α at O.

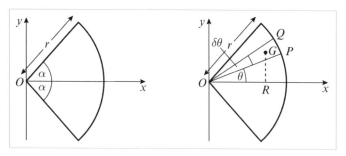

A

Divide the lamina into elements such as OPQ, which is a sector subtending an angle $\delta\theta$ at O.

The area of OPQ is $\frac{1}{2}r^2\,\delta\theta$ and its mass is $\frac{1}{2}r^2\rho\delta\theta$

The sector is approximately a triangle and so its centre of mass, G, is at a distance $\frac{2}{3}r$ from O.

The distance marked OR on the diagram is $\frac{2}{3}r\cos\theta$

The mass M of the whole sector is:

$\rho \times \frac{1}{2}r^2 2\alpha = \rho r^2\alpha$

Use $M\bar{x} = \sum\limits_{\theta=-\alpha}^{\theta=\alpha} \frac{1}{2}\rho r^2\,\delta\theta \times \frac{2}{3}r\cos\theta$

As $\delta\theta \to 0$, the summation becomes an integral

and $M\bar{x} = \int_{-\alpha}^{\alpha} \frac{1}{3}\rho r^3\cos\theta\,d\theta = \frac{1}{3}\rho r^3 \left[\sin\theta\right]_{-\alpha}^{\alpha}$

$\qquad = \frac{1}{3}\rho r^3 2\sin\alpha$

And so $\bar{x} = \dfrac{\frac{2}{3}\rho r^3\sin\alpha}{\rho r^2\alpha} = \dfrac{2r\sin\alpha}{3\alpha}$

The distance of the centre of mass from O is $\dfrac{2r\sin\alpha}{3\alpha}$ and it lies on the axis of symmetry.

This is the formula for the area of a sector of a circle.
← **Pure Year 2, Chapter 5**

The centre of mass of a triangle lies at the intersection of the medians. This is $\frac{2}{3}$ of the way along the line joining each vertex with the midpoint of the opposite side.

Point R is the foot of the perpendicular from G onto the x-axis.

As $\delta\theta$ is small, $\cos\left(\theta + \dfrac{\delta\theta}{2}\right) \approx \cos\theta$ and this is a reasonable approximation.

This formula is included in the formula booklet, but you should understand and learn how to derive it, as in this example.

Example 6

Find the centre of mass of a uniform wire in the form of an arc of a circle, radius r and centre O, which subtends an angle 2α at O.

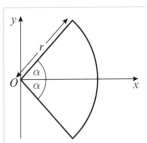
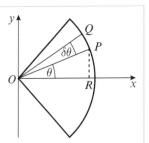

Divide the arc into elements and note that the arc PQ shown has length $r\delta\theta$ and mass $\rho r\delta\theta$

The length OR is $r\cos\theta$

The mass M of the whole wire is $\rho \times r2\alpha = 2\rho r\alpha$

Use $M\bar{x} = \sum\limits_{\theta=-\alpha}^{\theta=\alpha} \rho r\,\delta\theta \times r\cos\theta$

This is the formula for arc length.
← **Pure Year 2, Chapter 5**

R is the foot of the perpendicular from P to the x-axis.

A

As $\delta\theta \to 0$ the summation becomes an integral and

$$M\bar{x} = \int_{-\alpha}^{\alpha} \rho r^2 \cos\theta \, d\theta = \rho r^2 \, [\sin\theta]_{-\alpha}^{\alpha} = \rho r^2 \, 2\sin\alpha$$

And so $\bar{x} = \dfrac{2\rho r^2 \sin\alpha}{2\rho r\alpha} = \dfrac{r\sin\alpha}{\alpha}$

The centre of mass lies on the axis of symmetry, and is
at a distance $\dfrac{r\sin\alpha}{\alpha}$ from O.

This result is given in the formulae booklet, and you may quote it without proof.

Exercise 3A

1 Find, by integration, the centre of mass of the uniform triangular lamina enclosed by the lines $y = 6 - 3x$, $x = 0$ and $y = 0$.

2 Use integration to find the centre of mass of the uniform lamina occupying the finite region bounded by the curve with equation $y = 3x^2$, the x-axis and the line $x = 2$.

3 Use integration to find the centre of mass of the uniform lamina occupying the finite region bounded by the curve with equation $y = \sqrt{x}$, the x-axis and the line $x = 4$.

4 Use integration to find the centre of mass of the uniform lamina occupying the finite region bounded by the curve with equation $y = x^3 + 1$, the x-axis and the line $x = 1$.

5 Use integration to find the centre of mass of the uniform lamina occupying the finite region bounded by the curve with equation $y^2 = 4ax$, and the line $x = a$, where a is a positive constant.

6 Find the centre of mass of the uniform lamina occupying the finite region bounded by the curve with equation $y = \sin x$, $0 \leqslant x \leqslant \pi$ and the line $y = 0$.

7 Find the centre of mass of the uniform lamina occupying the finite region bounded by the curve with equation $y = \dfrac{1}{1 + x}$, $0 < x < 1$ and the lines $x = 0$, $x = 1$ and $y = 0$.

 8 Find, by integration, the centre of mass of a uniform lamina in the shape of a quadrant of a circle of radius r as shown.

(6 marks)

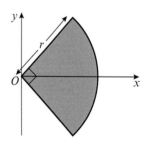

 9 The diagram shows a uniform lamina bounded by the curve $y = x^3$ and the line with equation $y = 4x$, where $x > 0$. Find the coordinates of the centre of mass of the lamina.

(10 marks)

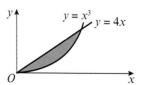

E/P **A** **10** The diagram shows a uniform lamina occupying the finite region bounded by the x-axis, the curve $y = \sqrt{24 - 4x}$ and the line with equation $y = 2x$. Find the coordinates of the centre of mass of the lamina.

(10 marks)

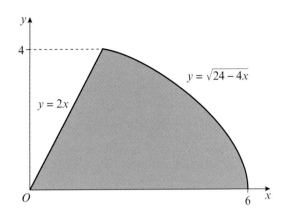

E/P **11** The diagram shows the uniform lamina bounded by the curves with equations $y = \dfrac{x + 5}{(x + 2)(2x + 1)}$ and $y = -\dfrac{x + 5}{(x + 2)(2x + 1)}$, and the lines $x = 0$ and $x = 4$.

Find the coordinates of the centre of mass of the lamina. **(7 marks)**

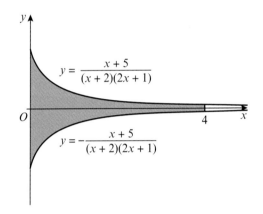

E/P **12** A uniform lamina is bounded by the curve $y = \dfrac{1}{\sqrt{x^2 + 4}}$ the x-axis, and the lines $x = 0$ and $x = 3$, as shown in the diagram.

Find the coordinates of the centre of mass of the lamina.

(10 marks)

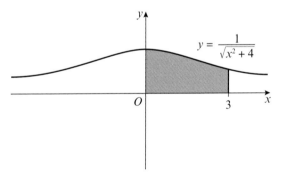

Challenge

A The diagram shows the curve C with parametric equations $x = t^2$, $y = t, t \in \mathbb{R}$

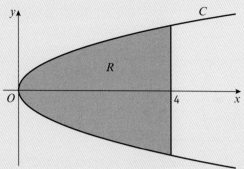

A pendant is modelled as a uniform lamina in the shape of the region R enclosed by the curve and the line $x = 4$.

The pendant is suspended in equilibrium from a string attached at a point P on its perimeter, such that no part of the pendant is higher than P.

Find the exact coordinates of the six possible positions of P.

3.2 Centre of mass of a uniform body

You can use symmetry to find the centre of mass of some uniform solids.

- ■ • **For a solid body the centre of mass is the point where the weight acts.**
  - • **For a uniform solid body the weight is evenly distributed through the body.**
  - • **The centre of mass will lie on any axis of symmetry.**
  - • **The centre of mass will lie on any plane of symmetry.**

Uniform solid sphere

- • **The centre of mass of a uniform solid sphere is at the centre of the sphere.**
  This point is the intersection of the infinite number of planes of symmetry and is the only point which lies on all of them.

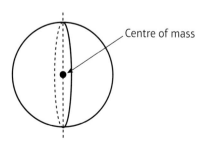

Centre of mass

A Uniform solid right circular cylinder

- **The centre of mass of a uniform solid right circular cylinder is at the centre of the cylinder.**
  This point is the intersection of the axis of symmetry and the plane of symmetry which bisects the axis and is parallel to the circular ends.

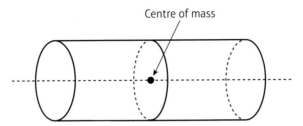

Centre of mass

There is another group of solids which are formed by rotating a region through 360° about the x-axis. It is possible to calculate the position of the centres of mass of these solids of revolution by using symmetry and calculus.

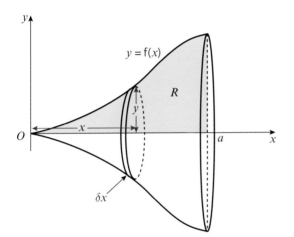

Links This is the solid of revolution formed by rotating the finite region enclosed by the curve $y = f(x)$, the coordinate axes and the line $x = a$ through 360° about the x-axis. It has volume

$$\pi \sum_{0}^{a} y^2 \, dx.$$

← **Core Pure 1, Chapter 5**

Suppose you need to find the centre of mass of the solid of revolution shown above.

Divide the volume up into a series of very thin circular discs of radius y and thickness δx.

Each disc has a volume of $\pi y^2 \delta x$ and each has a centre of mass that lies on the x-axis at a distance x from O.

So if the distance of the centre of mass of the whole solid of revolution from O is $\bar{x}$, then

$$\bar{x} \sum \rho \pi y^2 \, \delta x = \sum \rho \pi y^2 x \delta x$$

where ρ is the density, or mass per unit volume ← Use $\sum m_i x_i = \bar{x} \sum m_i$, applied to this volume.

A As $\delta x \to 0$, the number of discs becomes infinite and in the limit the sum is replaced by an integral:

$$\bar{x} = \frac{\int \rho \pi y^2 x \, dx}{\int \rho \pi y^2 \, dx}$$

If the rotation were about the y-axis you would use $\bar{y} = \dfrac{\int \rho \pi x^2 y \, dy}{\int \rho \pi x^2 \, dy}$

which may also be written

$$M\bar{x} = \int \rho \pi y^2 x \, dx$$

where M is the known mass of the solid.

The results from above are summarised below. Note that because the solid is uniform, ρ is constant so it cancels in the formulae for $\bar{x}$ and $\bar{y}$. π is also a constant so it cancels in these formulae.

- **For a uniform solid of revolution, where the revolution is about the x-axis, the centre of mass lies on the x-axis and its position on the axis is given by the formulae**

$$\bar{x} = \frac{\int y^2 x \, dx}{\int y^2 \, dx} \text{ or } M\bar{x} = \int \rho \pi y^2 x \, dx, \text{ where } M \text{ is the known mass of the solid and } \rho \text{ is its density.}$$

- **For a uniform solid of revolution, where the revolution is about the y-axis, the centre of mass lies on the y-axis and its position on the axis is given by the formulae**

$$\bar{y} = \frac{\int x^2 y \, dy}{\int x^2 \, dy} \text{ or } M\bar{y} = \int \rho \pi x^2 y \, dy, \text{ where } M \text{ is the known mass of the solid and } \rho \text{ is its density.}$$

   This is obtained from the previous result by interchanging x and y.

Example 7

Find the centre of mass of the uniform solid right circular cone with radius R and height h.

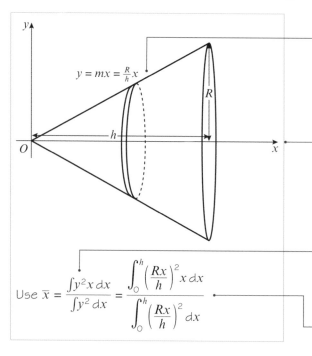

The gradient of the straight line through O is m, where $m = \dfrac{R}{h}$

$y = mx = \dfrac{R}{h}x$

The centre of mass lies on the axis of symmetry, which is the x-axis in the diagram.

Use $\bar{x} = \dfrac{\int y^2 x \, dx}{\int y^2 \, dx} = \dfrac{\displaystyle\int_0^h \left(\dfrac{Rx}{h}\right)^2 x \, dx}{\displaystyle\int_0^h \left(\dfrac{Rx}{h}\right)^2 \, dx}$

This is a formula you can use if you are finding the centre of mass of a volume of revolution.

The cone is generated by the straight line $y = \dfrac{R}{h}x$, which is rotated through 2π radians about the x-axis.

A

$$\text{so } \bar{x} = \frac{\dfrac{R^2}{h^2}\displaystyle\int_0^h x^3\,dx}{\dfrac{R^2}{h^2}\displaystyle\int_0^h x^2\,dx}$$

R and h are constant so take $\dfrac{R^2}{h^2}$ outside the integrals and cancel.

$$= \frac{\displaystyle\int_0^h x^3\,dx}{\displaystyle\int_0^h x^2\,dx}$$

$$= \frac{\left[\dfrac{x^4}{4}\right]_0^h}{\left[\dfrac{x^3}{3}\right]_0^h}$$

$$= \frac{\left(\dfrac{h^4}{4}\right)}{\left(\dfrac{h^3}{3}\right)}$$

$$= \tfrac{3}{4}h$$

Online Explore the centre of mass of a solid of revolution using GeoGebra.

- **The centre of mass of a uniform right circular cone lies on the axis of symmetry and is at a distance $\frac{3}{4}h$ from the vertex, or $\frac{1}{4}h$ from the circular base.**

Example 8

Find the centre of mass of the uniform solid hemisphere with radius R.

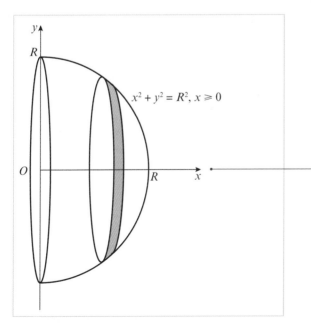

$x^2 + y^2 = R^2,\ x \geqslant 0$

The centre of mass lies on the axis of symmetry, which is the x-axis in the diagram.

A

Use $\bar{x} = \dfrac{\int y^2 x \, dx}{\int y^2 \, dx}$

$= \dfrac{\displaystyle\int_0^R (R^2 - x^2)x \, dx}{\displaystyle\int_0^R (R^2 - x^2) \, dx}$

$= \dfrac{\int (R^2 x - x^3) \, dx}{\int (R^2 - x^2) \, dx}$

$= \dfrac{\left[R^2 \dfrac{x^2}{2} - \dfrac{x^4}{4} \right]_0^R}{\left[R^2 x - \dfrac{x^3}{3} \right]_0^R}$

$= \dfrac{\left(\dfrac{R^4}{4} \right)}{\left(\dfrac{2R^3}{3} \right)}$

$= \dfrac{3}{8} R$

Divide the sphere up into a series of circular discs. Each disc has mass $\rho \pi y^2 \, \delta x$ and centre of mass at a distance x from O.

So if the distance of the centre of mass of the sphere from O is $\bar{x}$, then, as $\delta x \to 0$

$$\bar{x} = \dfrac{\int \rho \pi y^2 x \, dx}{\int \rho \pi y^2 \, dx} = \dfrac{\int y^2 x \, dx}{\int y^2 \, dx}$$

The sphere is generated by the circle $y^2 = R^2 - x^2$ which is rotated through 2π radians about the x-axis, so replace y^2 by $R^2 - x^2$

Watch out You might be asked to prove the results for the centre of mass of a cone and sphere using calculus. Otherwise, you can quote these results without proof when solving problems.

- **The centre of mass of a uniform solid hemisphere lies on the axis of symmetry and is at a distance $\frac{3}{8}R$ from the plane surface.**

Example **9**

Find the centre of mass of the uniform solid of revolution formed by rotating the finite region enclosed by the curve $y = x^2 + 1$, the x-axis and the lines $x = 0$ and $x = 3$ through 360° about the x-axis as shown in the diagram below.

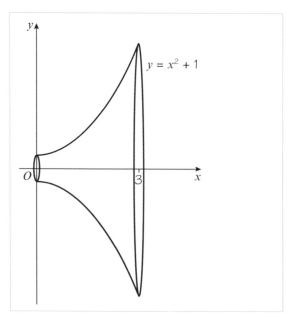

A

Use $\bar{x} = \dfrac{\int y^2 x \, dx}{\int y^2 \, dx}$

It is a good idea to write out the formula you are using before substituting.

$y = x^2 + 1$ so $y^2 = x^4 + 2x^2 + 1$

$y^2 = (x^2 + 1)^2$

$\bar{x} = \dfrac{\displaystyle\int_0^3 (x^5 + 2x^3 + x) \, dx}{\displaystyle\int_0^3 (x^4 + 2x^2 + 1) \, dx}$

$= \dfrac{\left[\frac{1}{6}x^6 + \frac{1}{2}x^4 + \frac{1}{2}x^2 \right]_0^3}{\left[\frac{1}{5}x^5 + \frac{2}{3}x^3 + x \right]_0^3}$

$= \dfrac{\left[\frac{1}{6}(3)^6 + \frac{1}{2}(3)^4 + \frac{1}{2}(3)^2 \right]}{\left[\frac{1}{5}(3)^5 + \frac{2}{3}(3)^3 + 3 \right]}$

$= \dfrac{\frac{333}{2}}{\frac{348}{5}} = \dfrac{555}{232}$

The centre of mass lies at the point $\left(\frac{555}{232}, 0 \right)$

Watch out In questions such as this you need to show full algebraic working. This means that you should show the integrated function together with the limits, and the step of substituting the limits. You should not use the numerical integration function on your calculator.

Example 10

Find the centre of mass of the uniform hemispherical shell with radius R.

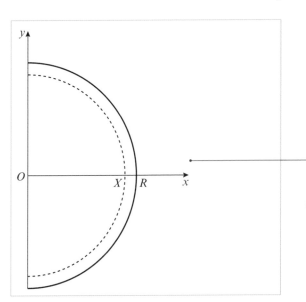

A hemispherical shell is a hollow hemisphere.

Problem-solving

You can obtain a hollow hemisphere by removing a solid concentric hemisphere of radius X from the solid hemisphere of radius R, and then considering what happens as $X \to R$.

A

Shape	Mass	Centre of mass
Solid hemisphere radius R	$\frac{2}{3}\rho\pi R^3$	$\left(\frac{3}{8}R, 0\right)$
Solid hemisphere radius X	$\frac{2}{3}\rho\pi X^3$	$\left(\frac{3}{8}X, 0\right)$
Hollow shell	$\frac{2}{3}\rho\pi(R^3 - X^3)$	$(\overline{x}, 0)$

From symmetry you can deduce that the centre of mass lies on the x-axis.

Taking moments about a horizontal axis through O in the plane face of the hemisphere:

$$\frac{2}{3}\rho\pi R^3 \times \frac{3}{8}R - \frac{2}{3}\rho\pi X^3 \times \frac{3}{8}X = \frac{2}{3}\rho\pi(R^3 - X^3) \times \overline{x}$$

So $\overline{x} = \frac{3}{8} \times \dfrac{R^4 - X^4}{R^3 - X^3} = \frac{3}{8} \times \dfrac{(R - X)(R + X)(R^2 + X^2)}{(R - X)(R^2 + RX + X^2)}$

$\qquad = \frac{3}{8}\dfrac{(R + X)(R^2 + X^2)}{(R^2 + RX + X^2)}$

As $X \to R$ you obtain the result for a hemispherical shell:

$$\overline{x} = \frac{3}{8} \times \frac{(2R)(2R^2)}{(3R^2)}$$

$$= \frac{1}{2}R$$

This result is given in the formulae booklet, and you may quote it without proof.

- **The centre of mass of a uniform hemispherical shell lies on the axis of symmetry and is at a distance $\frac{1}{2}R$ from the plane surface.**

Example 11

Show that the centre of mass of the uniform hollow right circular cone with radius R and height h is at a distance $\frac{1}{3}h$ from the base along the axis of symmetry.

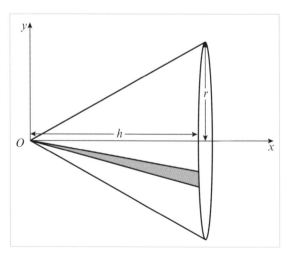

Divide the surface of the cone up into triangular strips with vertices at O and with bases on the circumference of the circular base of the cone. One is shown in the diagram.

Each of these triangles has centre of mass $\frac{2}{3}$ of the distance from O to the base of the cone.

So the centre of mass of the hollow cone is also $\frac{2}{3}$ of the distance from O to the base of the cone, but is on the axis of symmetry.

The distance of the centre of mass from the base is $h - \frac{2}{3}h = \frac{1}{3}h$

Use symmetry to obtain this result.

■ **The centre of mass of a conical shell lies on the axis of symmetry and is at a distance $\frac{1}{3}h$ from the base.**

Example 12

Find the position of the centre of mass of the frustum of a right circular uniform cone, of end radii 1 cm and 4 cm, and of height 7 cm.

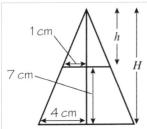

Let the large cone have height H and the small cone have height h.

From similar triangles $\frac{h}{H} = \frac{1}{4}$ or $H = 4h$

But $H = 7 + h$, so $h = \frac{7}{3}$, and $H = \frac{28}{3}$

Shape	Mass	Centre of mass
Large cone	$\frac{1}{3}\rho\pi 4^2 H$ $= \frac{64}{3}\rho\pi h$	$\left(\frac{1}{4}H, 0\right) = \left(\frac{7}{3}, 0\right)$
Small cone	$\frac{1}{3}\rho\pi 1^2 h = \frac{1}{3}\rho\pi h$	$\left(7 + \frac{1}{4}h, 0\right) = \left(\frac{91}{12}, 0\right)$
Frustum	$\frac{63}{3}\rho\pi h = 21\rho\pi h$	$(\bar{x}, 0)$

Taking moments about the base of the frustum:

$\frac{64}{3}\rho\pi h \times \frac{7}{3} - \frac{1}{3}\rho\pi h \times \frac{91}{12} = 21\rho\pi h \bar{x}$

So $\bar{x} = \left(\frac{448}{9} - \frac{91}{36}\right) \div 21 = \frac{9}{4}$

The centre of mass is $\frac{9}{4}$ cm above the base on the axis of symmetry.

A frustum is a portion of the cone lying between two parallel planes. It may be considered as a large cone with a small cone removed from the top.

Problem-solving

The masses are in the ratio $64:1:63$ and you can use these ratios in your moments equation to simplify the working.

So you would have $64 \times \frac{7}{3} - 1 \times \frac{91}{12} = 63\bar{x}$

Exercise **3B**

A 1 The finite region bounded by the curve
$y = x^2 - 4x$ and the x-axis is rotated
through 360° about the x-axis to form
a solid of revolution. Find the coordinates of its centre of mass.

> **Hint** In questions 1–4 use symmetry to find the
> coordinates of the centre of mass of the solid.

2 The finite region bounded by the curve $(x - 1)^2 + y^2 = 1$ is rotated through 180° about the
x-axis to form a solid of revolution. Find the coordinates of its centre of mass.

3 The finite region bounded by the curve $y = \cos x$, $\dfrac{\pi}{2} \leqslant x \leqslant \dfrac{3\pi}{2}$, and the x-axis, is rotated
through 360° about the x-axis to form a solid of revolution. Find the coordinates of its centre
of mass.

4 The finite region bounded by the curve $y^2 + 6y = x$ and the y-axis is rotated through 360° about
the y-axis to form a solid of revolution. Find the coordinates of its centre of mass.

5 Find, by integration, the coordinates of the
centre of mass of the solid formed when the
finite region bounded by the curve $y = 3x^2$,
the line $x = 1$ and the x-axis is rotated through 360° about the x-axis.

> **Hint** In questions 5–10 use integration to find
> the position of the centre of mass of the solid.

6 Find, by integration, the coordinates of the centre of mass of the solid formed when the finite
region bounded by the curve $y = \sqrt{x}$, the line $x = 4$ and the x-axis is rotated through 360° about
the x-axis.

7 Find, by integration, the coordinates of the centre of mass of the solid formed when the finite
region bounded by the curve $y = 3x^2 + 1$, the lines $x = 0$, $x = 1$ and the x-axis is rotated through
360° about the x-axis.

8 Find, by integration, the coordinates of the centre of mass of the solid formed when the
finite region bounded by the curve $y = \dfrac{3}{x}$, the lines $x = 1$, $x = 3$ and the x-axis is rotated
through 360° about the x-axis.

E 9 Find, by integration, the coordinates of the centre of mass of the solid formed when the finite
region bounded by the curve $y = 2e^x$, the lines $x = 0$, $x = 1$ and the x-axis is rotated through
360° about the x-axis. **(10 marks)**

E/P 10 Find, by integration, the coordinates of the centre of mass formed when the region bounded
by the curve $xe^y = 3$, the lines $y = 1$, $y = 2$ and the y-axis is rotated through 360° about the
y-axis. **(10 marks)**

A **E** **11** Find, by integration, the coordinates of the centre of mass formed when the region bounded by the curve $y = \dfrac{2}{1 + x}$, the lines $x = 0$, $x = 4$ and the x-axis is rotated through 360° about the x-axis. **(10 marks)**

E/P **12** Find the position of the centre of mass of the frustum of a right circular uniform solid cone, where the frustum has end radii 2 cm and 5 cm, and has height 4 cm. **(8 marks)**

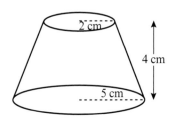

E/P **13** The diagram shows a frustum of a right circular uniform cone. The frustum has end diameters of 4 cm and 8 cm, and height 8 cm. A cylindical hole of diameter 2 cm is drilled through the frustum along its axis. Find the distance of the centre of mass of the resulting solid from the larger face of the frustum. **(12 marks)**

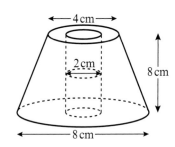

E/P **14** A thin uniform hemispherical shell has a circular base of the same material. Find the position of the centre of mass above the base in terms of its radius r. **(6 marks)**

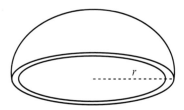

E/P **15** A thin uniform hollow cone has a circular base of the same material. Find the position of the centre of mass above the base, given that the radius of the cone is 3 cm and its height is 4 cm. **(6 marks)**

Hint The curved surface area of a cone of slant height l and base radius r has area $\pi r l$.

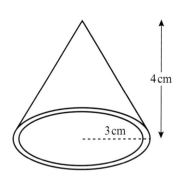

E/P **16** A cap of a sphere is formed by making a plane cut across the sphere. Show that the centre of mass of a cap of height h cut from a uniform solid sphere of radius a lies a perpendicular distance $\dfrac{h(4a - h)}{4(3a - h)}$ from the flat surface of the cap. **(12 marks)**

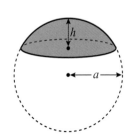

A **17** The finite region R, shown shaded on the diagram, is bounded
E/P by the curve with equation $y = 8 - x^3$ and the coordinate axes.

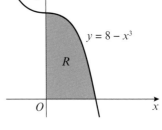

The region is rotated through 2π radians about the y-axis to
form a uniform solid of revolution, S.

a Find, using algebraic integration, the y coordinate of the
centre of mass of S. **(8 marks)**

A solid chocolate egg is formed by attaching a uniform solid
hemisphere to the base of the solid S, as shown in the diagram.
The units of measurement are cm.

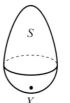

b Find the distance of the centre of mass of the chocolate
egg from its base, X, giving your answer correct to
3 significant figures. **(5 marks)**

E/P **18** A uniform solid is formed from two right circular cones. The cones each
have base radius r and heights x and kx respectively, where k is a constant.
The point O lies at the centre of the common plane face of both cones.

Given that the centre of mass of the solid is a distance $\frac{1}{10}x$ above O,
find the value of k. **(5 marks)**

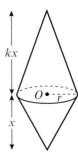

E/P **19** A uniform solid cylinder has height h and radius $2r$.
A hole is drilled in the cylinder in the shape of an
inverted cone of base radius r and height h.
The vertex of the cone lies on the base of the
cylinder, and the axes of the cone and the
cylinder are both vertical. The centre of the
top plane face of the cylinder, O, lies on the
circumference of the base of the cone.

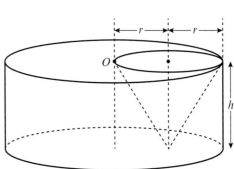

a Find the distance of the centre of mass of the
solid from its top plane face, giving your answer in the form kh where k is a rational constant
to be found. **(7 marks)**

b Fully describe the position of the centre of mass of the solid. **(5 marks)**

E/P **20** A game piece is modelled as a solid cone of base radius 5 cm
and height 12 cm, sitting on top of a solid cylinder of radius
5 cm and height 3 cm. The cone and cylinder are made of the
same uniform material.

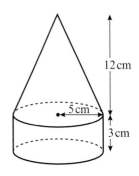

a Find the distance of the centre of mass of the game piece from the base of the cylinder.

(6 marks)

In order to save money, the manufacturer decides to make the game piece hollow. The game piece is now modelled as the curved surfaces of a cone and cylinder, and a single circular face on its base. All the surfaces are made from the same uniform material.

b Find the distance of the centre of mass of the hollow game piece from the base of the cylinder.

(6 marks)

Challenge

Using calculus, find the centre of mass of the uniform hemispherical shell with radius R.

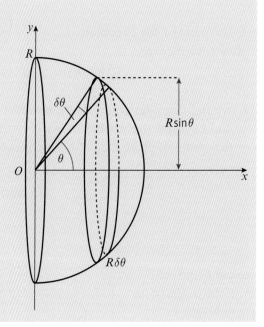

Hint Divide the shell into small elemental cylindrical rings, centred on the x-axis, with radius $R \sin \theta$, and height $R \delta \theta$, where θ is the angle between the radius R and the x-axis.

3.3 Non-uniform bodies

You can find the centres of mass of composite bodies made from materials of different densities.

Example 13

A uniform solid right circular cone of height $2R$ and base radius R is joined at its base to the base of a uniform solid hemisphere. The centres of their bases coincide at O and their axes are collinear. The radius of the hemisphere is $2R$.

Find the position of the centre of mass of the composite body if the cone has four times the density of the hemisphere.

A

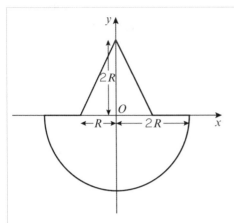

Let ρ be the mass per unit volume for the hemisphere and 4ρ be the mass per unit volume for the cone.

Shape	Mass	Ratio of masses	Distance from O to centre of mass
Cone	$\frac{1}{3}4\rho\pi R^2(2R)$	1	$\frac{1}{4}(2R) = \dfrac{R}{2}$
Hemisphere	$\frac{2}{3}\rho\pi(2R)^3$	2	$-\frac{3}{8}(2R) = -\dfrac{3R}{4}$
Composite body	$\frac{1}{3}4\rho\pi R^2(2R) + \frac{2}{3}\rho\pi(2R)^3$	3	$\bar{x}$

The centre of mass lies on the common axis of symmetry.
Take moments about an axis through O:

$1 \times \dfrac{R}{2} - 2 \times \dfrac{3R}{4} = 3 \times \bar{x}$

So $\bar{x} = -\frac{1}{3}R$

The centre of mass of the composite body is a distance $-\frac{1}{3}R$ below O, on the common axis of symmetry.

Problem-solving

Although each separate solid has uniform density, the composite body does not. Use ρ to represent the density (mass per unit volume) of the hemisphere, so that the cone has density 4ρ.

The masses are in the ratio 1:2:3 and you can use these ratios in your moments equation to simplify the working.

If you take moments about a point inside the body, be careful with your positive and negative signs. The centre of mass of the whole body is **below** O.

In some cases the density of a body may vary continuously along the length of the body. You can use integration to find centres of mass of **non-uniform rods**, where the mass per unit length of the rod is given as a function of distance.

This cylinder is modelled as a non-uniform rod. At a distance x along the rod, the mass per unit length of the rod is given by $\rho = f(x)$.

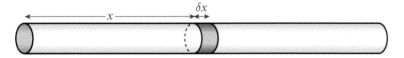

The mass of a section of length δx at a distance x from the end of the rod is $\rho\delta x = f(x)\delta x$. So the mass of the whole rod is $\sum f(x)\delta x$.

Problem-solving

Because the cylinder is modelled as a rod and density is given as mass per unit length, you do not need to consider the volume of the cylinder. You could use a similar model for any prism with a constant cross-section, where density varies as a function of height.

A If the centre of mass of the rod is a distance $\bar{x}$ from the end, then

$$\bar{x}\sum f(x)\delta x = \sum xf(x)\delta x \longleftarrow \text{The moment of one section of the cylinder is } x\rho\delta x = xf(x)\delta x$$

As $\delta x \to 0$ the summation can be replaced by an integral.

■ **For a non-uniform rod with a variable mass per unit length, $\rho = f(x)$, and length l, the distance of the centre of mass from the end of the rod is given by:**

$$\bar{x} = \frac{\int_0^l x\rho \, dx}{\int_0^l \rho \, dx} = \frac{\int_0^l xf(x) \, dx}{\int_0^l f(x) \, dx}$$

Example 14

A non-uniform telegraph pole is 10 m long. At any distance x metres from its base, the mass per unit length of the telegraph pole is given by $m = 20 - \frac{1}{5}x \, \text{kg m}^{-1}$. By modelling the telegraph pole as a rod, find:

a the mass of the telegraph pole

b the distance of the centre of mass of the telegraph pole from its top.

a Total mass $= \displaystyle\int_0^{10} \left(20 - \frac{1}{5}x\right) dx$

$= \left[20x - \frac{1}{10}x^2\right]_0^{10}$

$= \left[20(10) - \frac{1}{10}(10)^2\right]$

$= (200 - 10)$

$= 190 \, \text{kg}$ •————

b Taking moments about the base of the telegraph pole:

$190\bar{x} = \displaystyle\int_0^{10} x\left(20 - \frac{1}{5}x\right) dx$ •————

$= \left[10x^2 - \frac{1}{15}x^3\right]_0^{10}$

$= \left[10(10)^2 - \frac{1}{15}(10)^3\right]$

$= 1000 - \frac{200}{3}$

$= \frac{2800}{3}$ •————

$\bar{x} = \frac{280}{57} \, \text{m}$

So the centre of mass is $\frac{280}{57}$ m from the base of the telegraph pole and therefore $\frac{290}{57}$ m from the top of the telegraph pole.

Problem-solving

The mass per unit length of the telegraph pole varies along the length of the pole. To find the total mass you need to integrate the mass per unit length across the length of the whole pole.

Integrate then substitute in the limits.

Where $\bar{x}$ is the distance of the centre of mass from the base of the telegraph pole.

Integrate then substitute in the limits.

Exercise (3C)

1 A uniform solid right circular cone of height 10 cm and base radius 5 cm is joined at its base to the base of a uniform solid hemisphere. The centres of their bases coincide and their axes are collinear. The radius of the hemisphere is also 5 cm. The density of the hemisphere is twice the density of the cone. Find the position of the centre of mass of the composite body.

2 A solid is composed of a uniform solid right circular cylinder of height 10 cm and base radius 6 cm joined at its top plane face to the base of a uniform cone of the same radius and of height 5 cm. The centres of their adjoining circular faces coincide at point O and their axes are collinear. The mass per unit volume of the cylinder is three times that of the cone. Find the position of the centre of mass of the composite body.

3 **a** Prove that the centre of mass of a uniform right square-based pyramid of height h lies a distance $\dfrac{h}{4}$ from its base. **(6 marks)**

A solid is composed of a cube of side 8 cm joined at its top plane face to the base of a uniform square-based pyramid of side 8 cm and perpendicular height 6 cm. The centres of their adjoining square faces coincide at the point O and their axes are collinear.

   b Given that the cube has twice the density of the pyramid, find the position of the centre of mass of the composite body. **(4 marks)**

4 **a** Find the distance of the centre of mass of a solid tetrahedron from its base, giving your answer in terms of the height of the tetrahedron, h. **(6 marks)**

A solid is composed of two solid tetrahedrons of side 9 cm joined together. The centres of their adjoining faces coincide at the point O and their axes are collinear.

   b Given that the bottom tetrahedron has three times the density of the top tetrahedron, find the position of the centre of mass of the composite body. **(4 marks)**

5 A truncated square-based pyramid has a height of 5 cm and a base length 10 cm. The top face of the truncated square-based pyramid has an area of 25 cm².

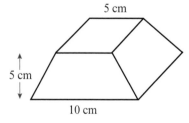

   a Find the height of the centre of mass of the truncated square-based pyramid above its base. **(6 marks)**

The truncated square-based pyramid and a cube of side length 5 cm are joined at the smaller square face of the truncated pyramid. The centres of their adjoining square faces coincide at the point O and their axes are collinear.

   b Given that the cube has twice the density of the truncated square-based pyramid, find the distance of the centre of mass of the composite body from O. **(4 marks)**

A **6** A non-uniform fence post is 1.2 m long. At any distance d metres from its base, the mass per

E/P unit length of the fence post is given by $m = 10\left(1 - \dfrac{d}{12}\right)$ kg m⁻¹. By modelling the fence post as a

rod, find:

a the mass of the fence post **(3 marks)**

b the distance of the centre of mass of the fence post from its base. **(4 marks)**

In reality, the post has a roughly circular cross-section. The model is refined so that the fence

post is modelled as a cylinder of radius 0.1 m with density $100\left(1 - \dfrac{d}{12}\right)$ kg m⁻³ at a distance d m

from its base.

c State, with reasons, whether your answers to parts **a** and **b** will change under the new model.

(2 marks)

E/P **7** A solid wooden bowl is modelled as a uniform solid
right circular cylinder with height $2r$ and radius $4r$.
The centre of one face is at O. A hemisphere of
diameter $2r$ is removed from the bowl. The centre
of the hemisphere lies on a diameter of the circular
face of the cylinder, and the point O lies on the
circumference of the circular face of the hemisphere.

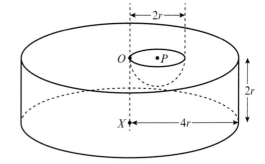

a Show that the centre of mass of the whole

bowl is at a vertical distance $\dfrac{381r}{376}$ from the

plane face containing O. **(6 marks)**

The hemisphere is filled with water. The density of the water is twice the density of the wood
from which the bowl is made.

b Find:

i the vertical distance of the centre of mass of the filled bowl from the plane face containing O

ii the horizontal distance from the axis of the cylinder, OX, to the centre of mass of the filled
bowl.

In each case, give your answer in the form kr where k is a rational number to be found. **(6 marks)**

P **8** A non-uniform rod of length l m is such that, at a distance x m from the end of the rod, the mass
per unit length of the rod is given by m kg m⁻¹. In each of the following cases, find the distance
of the centre of mass from the end of the rod, giving your answers to 3 significant figures where
appropriate.

a $l = 6$ and $m = (x + 1)^2$

b $l = 10$ and $m = 10 - \frac{1}{4}x$

c $l = 2$ and $m = \dfrac{1}{1 + x^2}$

d $l = 5$ and $m = e^{0.5x}$

9 The mast of a ship is 18 m high. The mast is modelled as a non-uniform rod, such that, at a
height h m above its base, the mass per unit length of the mast, m kg m^{-1}, is given by the
formula $m = 50\mathrm{e}^{-0.01h}$

Find, correct to 3 significant figures:

a the mass of the mast **(3 marks)**

b the height of the centre of mass of the mast above its base. **(6 marks)**

10 A non-uniform rod AB of length 2 m is such that, at a distance x m from A, the mass per unit
length of the rod is given by $(5 - px)$ kg m^{-1}.

Given that the mass of the rod is 7 kg, find:

a the value of p **(3 marks)**

b the distance of the centre of mass of the rod from A. **(4 marks)**

Challenge

A non-uniform post is 2 m long and has mass 10 kg. At any distance
x metres from its base, the mass per unit length of the post is given by
$m = a(1 - bx)$ kg m^{-1}. Given that the centre of mass of the post is 1.5 m
above its base, work out the possible values of a and b.

3.4 **Rigid bodies in equilibrium**

You can solve problems about rigid bodies
which are suspended in equilibrium.

Links You can combine resultant forces and moments
to solve problems involving ladders leaning against
walls. ← **Statistics and Mechanics Year 2, Chapter 7**

If a body is resting in equilibrium then there
is zero resultant force in any direction.
This means that the sum of the components of all the forces in any direction is zero, and the sum of
the moments of the forces about any point is zero.

- **When a lamina is suspended freely from a fixed point or pivots freely about a horizontal axis
  it will rest in equilibrium in a vertical plane with its centre of mass vertically below the point
  of suspension or pivot.**

This result is also true for a rigid body.

Let the body be suspended from a point A. The body rests in equilibrium and
the only forces acting on the body are its weight, W, and the force at point A, P.
This implies that the forces must be equal and opposite and act in the same
vertical line.

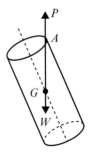

- **When a rigid body is suspended freely from a fixed point and rests in
  equilibrium then its centre of mass is vertically below the point
  of suspension.**

Example 15

A uniform solid hemisphere has radius r. It is suspended by a string attached to a point A on the rim of its base. Find the angle between the axis of the hemisphere and the downward vertical when the hemisphere is in equilibrium.

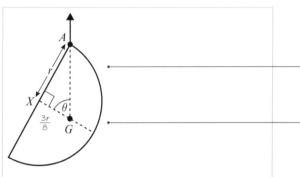

Draw a diagram showing the centre of mass, G, of the hemisphere below the point of suspension A.

Mark the angle between GA and the axis of the hemisphere and mark the radius and length XG.

The distance from the centre of mass to the base is $\dfrac{3r}{8}$ so $XG = \dfrac{3r}{8}$

Let $\angle XGA$ be θ

Use trigonometry to solve the problem.

Then $\tan\theta = \dfrac{r}{\frac{3}{8}r} = \dfrac{8}{3}$

So the required angle is $69.4°$ (3 s.f.)

Example 16

A non-uniform cylinder has height $20\,\text{cm}$ and radius $2\,\text{cm}$. At any height $h\,\text{cm}$ above its base, the density of the cylinder is given by $\rho = 3\left(2 - \dfrac{h}{50}\right)\text{g cm}^{-3}$. The cylinder is freely suspended from a point P that lies on the circumference of its top face. Find the angle between the downward vertical and the top face of the cylinder.

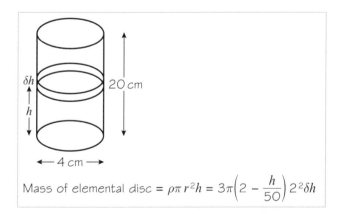

Mass of elemental disc $= \rho\pi r^2 h = 3\pi\left(2 - \dfrac{h}{50}\right)2^2\delta h$

Problem-solving

You need to work out the mass of a disc of height δx, then integrate to find the sum of these masses. Because the cylinder has a constant, uniform cross-section, and you know the centre of mass lies on its axis of symmetry, you could also tackle this problem by modelling the cylinder as a non-uniform rod with mass per unit length of $3\left(2 - \dfrac{h}{50}\right)\text{g cm}^{-1}$.

A

$$\text{Total mass} = 12\pi \int_0^{20} \left(2 - \frac{h}{50}\right) dh$$

$$= 12\pi \left[2h - \frac{h^2}{100}\right]_0^{20}$$

$$= 12\pi \left(2(20) - \frac{(20)^2}{100}\right)$$

$$= 12\pi(40 - 4)$$

$$= 432\pi$$

$$432\pi\bar{y} = 12\pi \int_0^{20} h\left(2 - \frac{h}{50}\right) dh$$

Use $M\bar{y} = \int \rho \pi r^2 h \, dh$ to find the position of $\bar{y}$.

where $\bar{y}$ is the distance of the centre of mass above the base of the cylinder.

Take moments from the base of the cylinder.

$$36\bar{y} = \int_0^{20} \left(2h - \frac{h^2}{50}\right) dh$$

$$= \left[h^2 - \frac{h^3}{150}\right]_0^{20}$$

$$= 20^2 - \frac{20^3}{150}$$

$$= 400 - \frac{160}{3}$$

$$= \frac{1040}{3}$$

$$\bar{y} = \frac{1040}{108} = \frac{260}{27}$$

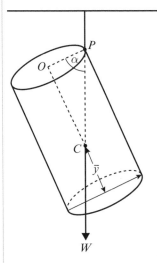

Draw a clear diagram to represent the situation. You need to find the angle between the top face and the vertical which is marked as α on this diagram.

$$\tan \alpha = \frac{OC}{OP}$$

where O is the centre of the top face of the cylinder and C is the centre of mass.

$$\tan \alpha = \frac{20 - \frac{260}{27}}{2} = \frac{140}{27}$$

$$\alpha = 79.1° \text{ (3 s.f.)}$$

Problem-solving

The density of the cylinder reduces as the height increases, so it is **bottom-heavy**. The centre of mass will be less than half-way up the cylinder, and the angle will be greater than $\arctan \frac{20}{4}$, the angle made between the top face and the diagonal.

Example 17

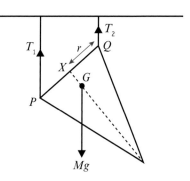

A The diagram shows a uniform solid right circular cone, of mass M with radius r and height $3r$. P and Q are points at opposite ends of a diameter on the circular plane face of the cone. The cone is suspended from a horizontal beam by two vertical inelastic strings fastened at P and Q. Given that the cone is in equilibrium with its axis at an angle of 45° to the horizontal, find the values of the tensions in the strings, giving your answers in terms of M and g.

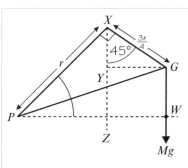

Let the tensions in the strings be T_1 and T_2.

Resolve vertically:

$$T_1 + T_2 = Mg \qquad (1)$$

Take moments about point P:

$$T_2 \times 2r\cos 45° = Mg\left(r\cos 45° + \frac{3r}{4}\sin 45°\right) \qquad (2)$$

Divide equation (2) through by $2r\cos 45°$, then $T_2 = \frac{7}{8}Mg$

Substitute into equation (1) to give $T_1 = \frac{1}{8}Mg$

So the tensions are $\frac{1}{8}Mg$ and $\frac{7}{8}Mg$

Moments could be taken about a number of points, but choosing point P eliminates T_1 and simplifies the algebra.

The distance PW is $PZ + ZW$.
Also $ZW = YG$.
So from the diagram:
$$PW = r\cos 45° + \frac{3r}{4}\sin 45°$$

Example 18

A Two smooth uniform spheres of radius 4 cm and mass 5 kg are suspended from the same point A by identical light inextensible strings of length 8 cm attached to their surfaces. The spheres hang in equilibrium, touching each other. What is the reaction between them?

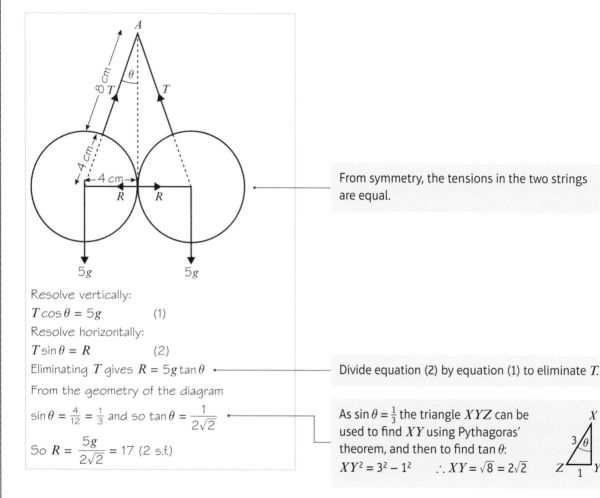

From symmetry, the tensions in the two strings are equal.

Resolve vertically:
$T \cos \theta = 5g$ (1)

Resolve horizontally:
$T \sin \theta = R$ (2)

Eliminating T gives $R = 5g \tan \theta$ Divide equation (2) by equation (1) to eliminate T.

From the geometry of the diagram

$\sin \theta = \frac{4}{12} = \frac{1}{3}$ and so $\tan \theta = \frac{1}{2\sqrt{2}}$

So $R = \dfrac{5g}{2\sqrt{2}} = 17$ (2 s.f.)

As $\sin \theta = \frac{1}{3}$ the triangle XYZ can be used to find XY using Pythagoras' theorem, and then to find $\tan \theta$:

$XY^2 = 3^2 - 1^2 \qquad \therefore XY = \sqrt{8} = 2\sqrt{2}$

Exercise 3D

1 A uniform solid right circular cone is suspended by a string attached to a point on the rim of its base. Given that the radius of the base is 5 cm and the height of the cone is 8 cm, find the angle between the vertical and the axis of the cone when it is in equilibrium.

2 A uniform solid cylinder is suspended by a string attached to a point on the rim of its base. Given that the radius of the base is 6 cm and the height of the cylinder is 10 cm, find the angle between the vertical and the circular base of the cylinder when it is in equilibrium.

3 A uniform hemispherical shell is suspended by a string attached to a point on the rim of its base. Find the angle between the vertical and the axis of the hemisphere when it is in equilibrium.

(8 marks)

A **E** **4 a** Find the position of the centre of mass of the frustum of a uniform solid right circular cone, of end radii 4 cm and 5 cm and of height 6 cm. (Give your answer to 3 s.f.) **(8 marks)**

This frustum is now suspended by a string attached to a point on the rim of its smaller circular face, and hangs in equilibrium.

b Find the angle between the vertical and the axis of the frustum. **(3 marks)**

E **5** The diagram shows a uniform solid cylinder of mass 2 kg, height 2 m and radius 0.5 m. The cylinder is suspended from two light, inelastic, vertical strings attached to the upper rim of the cylinder at points A and B. The line AB forms a diameter of the top face of the cylinder and is inclined at an angle of 10° to the horizontal.

Given that the cylinder is in equilibrium, work out the tension in each string. **(5 marks)**

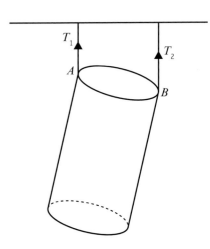

E **6** A non-uniform rod PQ of length 1 m is freely suspended from its ends by two light, inelastic strings which hang vertically. The tensions in the strings are T_1 and T_2 respectively, and PQ makes an angle of 45° with the vertical, as shown in the diagram.

At a distance h along the rod from Q, the mass per unit length of the rod is given by:

$$\rho = \frac{10}{\sqrt{1+h}}\,\text{kg}\,\text{m}^{-1}$$

a Find the distance of the centre of mass of the rod from Q. **(8 marks)**

b Find T_1 and T_2 **(4 marks)**

E/P **7** The diagram shows a non-uniform rod AB of length 10 m, which rests on two pivots P and Q that are positioned 1 m from each end of the rod.

At a distance x along the rod from A, the mass per unit length of the rod is $(1 + 3x)\,\text{kg}\,\text{m}^{-1}$.

a Find the reaction on the rod of each pivot. **(9 marks)**

A mass of m kg is attached to the rod at B so that the rod is on the point of turning about Q.

b Find the value of m. **(4 marks)**

A **8** A non-uniform solid cylinder has height 30 cm and base radius 10 cm. At a height h cm above
E/P the base of the cylinder, its density is given by $e^{0.1h}$ g cm^{-3}.

   a Show that the centre of mass of the cylinder lies a distance $\dfrac{10(2e^3 + 1)}{e^3 - 1}$ cm above the base of
   the cylinder. **(8 marks)**

   The cylinder is suspended from a point on its upper rim.

   b Find, to the nearest degree, the angle that the upper plane face of the cylinder makes with
   the vertical. **(4 marks)**

E **9** A solid is formed by removing a solid hemisphere of diameter
6 cm and centre O from a solid, uniform right circular
cylinder of height 10 cm and diameter 10 cm. The point O
lies at the centre of one circular face of the cylinder.

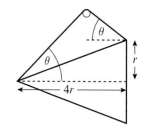

The cylinder is suspended by two light, inextensible
strings, which hang vertically. The strings are attached
to points A and B on the rims of the solid, such that
AB is parallel to the axis of the cylinder.

The solid has a density of $\dfrac{1}{\pi}$ kg m^3, and hangs in equilibrium
with AB horizontal.

   a Find the tensions in the strings. **(10 marks)**

   The string at B breaks, and the solid hangs in equilibrium
   from point A.

   b Find the angle between the circular plane face of the solid and the vertical. **(3 marks)**

E/P **10** A uniform solid right circular cone has base radius r, height $4r$
and mass m kg. One end of a light inextensible string is attached
to the vertex of the cone and the other end is attached to a
point on the rim of the base. The string passes over a smooth
peg and the cone rests in equilibrium with the axis horizontal,
and with the strings equally inclined to the horizontal at an
angle θ, as shown in the diagram.

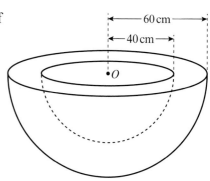

   a Show that angle θ satisfies the equation $\tan \theta = \frac{1}{2}$ **(4 marks)**

   b Find the tension in the string, giving your answer as an exact multiple of mg. **(4 marks)**

E/P **11** A plastic mould is formed by removing a solid hemisphere of
radius 40 cm and centre O from a solid hemisphere of metal
of radius 60 cm and centre O, as shown in the diagram.

The mould is filled to the brim with molten plastic,
which is allowed to solidify. The metal used to form
the mould has 10 times the density of the plastic.
After the mould is filled and set, it is suspended
from a point on the outer rim of its plane face,
and hangs in equilibrium.

Find the angle that the plane face makes with the vertical. **(10 marks)**

3.5 Toppling and sliding

A You can determine whether a body will remain in equilibrium or if equilibrium will be broken by sliding or toppling.

- **If a body rests in equilibrium on a rough inclined plane, then the line of action of the weight of the body must pass through the face of the body which is in contact with the plane.**

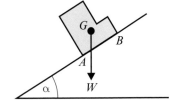

In the diagram, the weight of the body produces a clockwise moment about A which keeps it in contact with the plane.

If the angle of the plane is increased so that the line of action of the weight passes outside the face AB then the weight produces an anticlockwise moment about A. If the plane is sufficiently rough to preventy sliding, then the body will topple over.

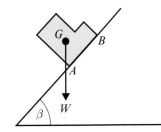

Online Explore toppling and sliding using GeoGebra.

The only forces acting on the body are its weight and the total reaction between the plane and the body. The total reaction between the plane and the body consists of a normal reaction force and a friction force. As the body is in equilibrium, these forces must be equal and opposite and act in the same vertical line.

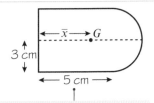

Example 19

A solid uniform cylinder of radius 3 cm and height 5 cm has a solid uniform hemisphere made from the same material, of radius 3 cm, joined to it so that the base of the hemisphere coincides with one circular end of the cylinder.

a Find the position of the centre of mass of the composite body.

The composite body is placed with the circular face of the cylinder on a rough inclined plane, which is inclined at an angle α to the horizontal. Given that the plane is sufficiently rough to prevent sliding,

b show that equilibrium is maintained provided that $\tan \alpha < \frac{28}{33}$

a

Shape	Mass	Ratio of masses	Height of centre of mass above base in cm
Cylinder	$\rho\pi 3^2 \times 5 = 45\rho\pi$	5	2.5
Hemisphere	$\frac{2}{3}\rho\pi 3^3 = 18\rho\pi$	2	$5 + \frac{3}{8} \times 3 = 6.125$
Composite body	$63\rho\pi$	7	$\bar{x}$

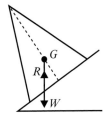

Draw a diagram and use a table to show masses and positions of centres of mass.

So $5 \times 2.5 + 2 \times 6.125 = 7\bar{x}$

and $\bar{x} = \dfrac{24.75}{7} = \dfrac{99}{28}$

Take moments to find the position of the centre of mass of the composite body.

b From the diagram, $\tan \alpha = \dfrac{3}{\frac{99}{28}} = 3 \times \dfrac{28}{99} = \dfrac{28}{33}$

If α is smaller than this value, G is above a point of contact and equilibrium is maintained. Equilibrium is maintained provided that $\tan \alpha < \dfrac{28}{33}$

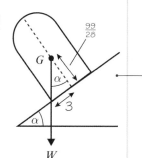

The limiting case is shown in the diagram, where the point vertically below the centre of mass, G, is on the edge of the area of contact.

If $\tan \alpha > \frac{28}{33}$ then the body will topple over because there will be a turning effect about the point A.

Example 20

A uniform solid cube of mass M and side $2a$ rests on a rough horizontal plane. The coefficient of friction is μ. A horizontal force of magnitude P is applied at the midpoint of an upper edge, perpendicular to that edge, as shown in the diagram. The reaction between the plane and the cube comprises a normal reaction force R and a friction force F and acts at a distance h from a lower edge of the cube as shown in the diagram.

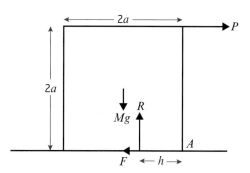

Find whether the cube remains in equilibrium, or whether the equilibrium is broken by sliding or toppling, in each of the following cases. Also determine the value of F and the value of h in each case, giving your answers in terms of M, g and a as appropriate:

a $P = 0$ **b** $P = \frac{1}{2}Mg$ and $\mu = \frac{3}{4}$ **c** $P = \frac{1}{4}Mg$ and $\mu = \frac{1}{5}$

For equilibrium:

Resolve horizontally $\rightarrow$

$\quad P - F = 0$, so $F = P$ (1)

Resolve vertically $\uparrow$

$\quad R - Mg = 0$ so $R = Mg$ (2)

Take moments about point A:

$\quad P \times 2a + R \times h = Mg \times a$ (3)

First use the conditions for equilibrium in the general case, i.e. resolve horizontally, resolve vertically and take moments.

A is the point on the bottom edge of the cube shown in the diagram.

a When $P = 0$

Substituting result from equation (2) into equation (3):

$\quad 0 + Mgh = Mga$

$\quad h = a$

Substituting $P = 0$ into equation (1) gives $F = 0$.

The cube remains in equilibrium. It does not slide and does not topple.

For part **a** substitute $P = 0$ in equations (1), (2) and (3).

In part **a** the normal reaction acts at the centre of the base.

A

b When $P = \frac{1}{2}Mg$

Substituting result from equation (2) into equation (3):

$$\frac{1}{2}Mg \times 2a + Mgh = Mga$$

$h = 0$, and the cube is about to topple.

Substituting $P = \frac{1}{2}Mg$ into equation (1) gives $F = \frac{1}{2}Mg$

But $\mu R = \frac{3}{4}Mg$ so $F < \mu R$ and the body does not slide.

> For part **b** substitute $P = \frac{1}{2}mg$ in equations (1), (2) and (3).

> In part **b** the normal reaction acts at A and so toppling is about to occur around the edge through A.

c When $P = \frac{1}{4}Mg$

Substituting result from equation (2) into equation (3):

$$\frac{1}{4}Mg \times 2a + Mgh = Mga$$

$h = \frac{1}{2}a$, and the cube does not topple.

Substituting $P = \frac{1}{4}Mg$ into equation (1) would give $F = \frac{1}{4}Mg$

But this is impossible as the maximum value that F can take is μR and $\mu R = \frac{1}{5}Mg$ so $F = \frac{1}{5}Mg$

The cube will slide if the force P is maintained.

> This implies that the body does not topple as the reaction is within the area of the base.

> Assuming equilibrium leads to a contradiction. This cube will slide as the force P exceeds the maximum friction force.

- **A body is on the point of sliding when $F = \mu R$**

- **A body is on the point of toppling when the reaction acts at the point about which turning can take place.**

Example 21

A uniform solid cube of mass M and side $2a$, is placed on a rough inclined plane which is at an angle α to the horizontal, where $\tan \alpha = \frac{1}{2}$. The coefficient of friction is μ.
Show that if $\mu < \frac{1}{2}$ the cube will slide down the slope.

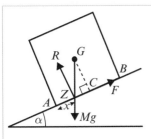

$R(\seararrow)$ $R - Mg\cos\alpha = 0$
$$R = Mg\cos\alpha$$
$R(\nearrow)$ $F - Mg\sin\alpha = 0$
$$F = Mg\sin\alpha$$

> Draw a diagram showing G the centre of mass of the cube, and points A and B on the edges of the cube in the same vertical plane as G.

> First use the conditions for equilibrium, i.e. resolve along and perpendicular to the plane and use $F \leqslant \mu R$, where F is the force of friction and R is the normal reaction.

For equilibrium $F < \mu R$, i.e. $\mu > \dfrac{F}{R}$

So $\mu > \dfrac{Mg\sin\alpha}{Mg\cos\alpha}$

i.e. $\mu > \tan\alpha$ and so $\mu > \frac{1}{2}$ for equilibrium.

> As $\tan\alpha = \frac{1}{2}$ the limiting case is when $\mu = \frac{1}{2}$

So if $\mu < \frac{1}{2}$ the cube will slide.

A

Let the point Z, vertically below the centre of mass G, be at a distance x from A up the plane.

Let C be the midpoint of AB.

From triangle GCZ, $\dfrac{a - x}{a} = \tan \alpha = \dfrac{1}{2}$

So $2a - 2x = a$

$x = \dfrac{1}{2}a$

So Z is between A and B and the cube does not topple.

Draw an enlarged triangle if it helps.

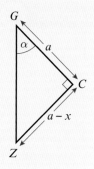

The normal reaction would act at point Z, which could also be established by taking moments and using the equilibrium conditions.

So the weight of the cube acts through a point within the area of contact.

Example 22

The cube in the previous example is now replaced by a cylinder of mass M with base radius a and height $6a$. The cylinder is placed on the rough inclined plane, which is inclined at an angle α to the horizontal, where $\tan \alpha = \dfrac{1}{2}$. The coefficient of friction is μ. Show that if $\mu > \dfrac{1}{2}$ the cylinder will topple about the lower edge.

As $\mu > \dfrac{1}{2}$ the cylinder will not slip.

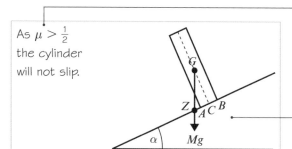

This condition was shown in Example 21, and the working here would be identical.
i.e. Resolve along and perpendicular to the plane and use $F \leqslant \mu R$.

The points A, B, C, G and Z are defined as in Example 21.

Let the point Z, vertically below the centre of mass G, be at a distance x from A down the plane.

Let C be the midpoint of AB.

From triangle GCZ, $\dfrac{a + x}{3a} = \tan \alpha = \dfrac{1}{2}$

So $2a + 2x = 3a$

$x = \dfrac{1}{2}a$

So Z is not between A and B and the cylinder will topple.

Draw an enlarged triangle GZC.

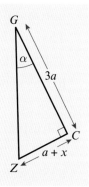

Another method you could use is to let the reaction act at point A and show that even in this position there will be a turning moment about A and therefore no equilibrium.

Exercise 3E

A

1 A uniform solid cylinder with base diameter 4 cm stands on a rough plane inclined at 40° to the horizontal. Given that the cylinder does not topple, find the maximum possible height of the cylinder.

2 A uniform solid right circular cylinder with base radius 3 cm and height 10 cm is placed with its circular plane base on a rough plane. The plane is gradually tilted until the cylinder topples over. Given that the cylinder does not slide down the plane before it topples,

    a find the angle that the plane makes with the horizontal at the point where the cylinder topples

    b find the minimum possible value of the coefficient of friction between the cylinder and the plane.

(P)

3 A uniform solid right circular cone with base radius 5 cm and height h cm is placed with its circular plane base on a rough plane. The coefficient of friction is $\frac{\sqrt{3}}{3}$. The plane is gradually tilted.

    When the plane makes an angle of θ with the horizontal, the cone is about to slide and topple at the same time. Find:

    a θ

    b h

(P)

4 A uniform solid right circular cone of mass M with base radius r and height $2r$ is placed with its circular plane base on a rough horizontal plane. A force P is applied to the vertex V of the cone at an angle of 60° above the horizontal as shown in the diagram.

    The cone begins to topple and to slide at the same time.

    a Find the magnitude of the force P in terms of M.

    b Calculate the value of the coefficient of friction.

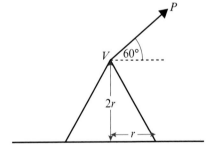

(P)

5 A frustum of a right circular solid cone has two plane circular end faces with radii r and $2r$ respectively. The distance between the end faces is $2r$.

    a Show that the centre of mass of the frustum is at a distance $\frac{11r}{14}$ from the larger circular face.

    b Find whether this solid can rest without toppling on a rough plane, inclined to the horizontal at an angle of 40°, if the face in contact with the inclined plane is:

      i the large circular end   **ii** the small circular end.

    c In order to answer part **b** you assumed that slipping did not occur. What does this imply about the coefficient of friction μ?

A
P
6 A uniform cube with edges of length $6a$ and weight W stands on a rough horizontal plane. The coefficient of friction is μ. A gradually increasing force P is applied at right angles to a vertical face of the cube at a point which is a distance a above the centre of that face.

   a Show that equilibrium will be broken by sliding or toppling depending on whether $\mu < \frac{3}{4}$ or $\mu > \frac{3}{4}$.

   b If $\mu = \frac{1}{4}$, and the cube is about to slip, find the distance from the point where the normal reaction acts, to the nearest vertical face of the cube.

E/P **7** Two uniform, rough, solid cuboids of dimensions $20\,\text{cm} \times 20\,\text{cm} \times 30\,\text{cm}$ are stacked as shown in the diagram shown opposite. Cuboid A has density ρ and cuboid B has density $k\rho$, for some constant k.

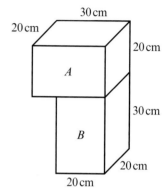

The cuboids are placed on a rough plane inclined at an angle of θ to the horizontal, as shown in the diagram opposite. The line of greatest slope of the plane is parallel to the plane faces shown in the diagram, and the angle of inclination of the plane is gradually increased.

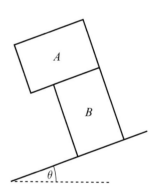

   a In the case when $k = 5$, find the angle of inclination at which the whole body begins to topple. **(8 marks)**

   b Given that cuboid A topples off cuboid B before the whole body topples, find the range of possible values of k. **(4 marks)**

E/P **8** A non-uniform cylinder is 1.5 m high and has radius 0.25 m. At any height, x m, above its base, the density of the cylinder is given by $\rho = \cosh x \text{ kg m}^{-3}$. The cylinder is placed on a rough slope. Assuming that the cylinder does not slide down the slope, what is the maximum possible angle of the slope to the horizontal, before the cylinder topples over? **(10 marks)**

E/P **9** A uniform solid cylinder of base radius r and height h has the same density as a uniform solid hemisphere of radius r. The plane face of the hemisphere is joined to a plane face of the cylinder to form the composite solid S shown. The point O is the centre of the plane base of S.

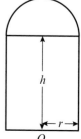

   a Show that the distance from O to the centre of mass of S is $\dfrac{6h^2 + 8hr + 3r^2}{4(3h + 2r)}$

                          (8 marks)

Chapter 3

A

The solid is placed on a rough plane which is inclined at an angle α to the horizontal. The plane base of S is in contact with the inclined plane.

b Given that $h = 3r$ and that S is on the point of toppling, find α to the nearest degree.

(4 marks)

c Given that the solid did not slip before it toppled, find the range of possible values for the coefficient of friction.

(4 marks)

(E/P) **10** A uniform solid paperweight is in the shape of a frustum of a cone. It is formed by removing a right circular cone of height h from a right circular cone of height $2h$ and base radius $2r$.

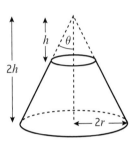

a Show that the centre of mass of the paperweight lies at a height of $\frac{11}{28}h$ from its base. **(6 marks)**

When placed with its curved surface on a horizontal plane, the paperweight is on the point of toppling.

b Find θ, the semi-vertical angle of the cone, to the nearest degree. **(4 marks)**

(E/P) **11** A uniform solid cone of mass M, height 8 cm and radius 3 cm, is placed with its circular base on a horizontal plane. The coefficient of friction is μ. A horizontal force of magnitude P is applied at the vertex of the cone.

a Find the value of P which will cause the cone to slide giving your answer in terms of μ, M and g. **(4 marks)**

b Find the value of P which will cause the cone to topple, giving your answer in terms of M and g. **(2 marks)**

c State whether the cone will slide or topple if:

   i $\mu = \frac{1}{4}$ **(1 mark)**

   ii $\mu = \frac{1}{2}$ **(1 mark)**

   iii $\mu = \frac{3}{8}$ **(1 mark)**

(E/P) **12** A uniform solid cylinder of mass 200 g, height 4 cm and radius 3 cm rests with its circular base on a plane. The plane makes an angle α with the horizontal where $\tan \alpha = \frac{3}{4}$. The coefficient of friction is $\frac{6}{17}$. A horizontal force P is applied to the highest point of the cylinder as shown in the diagram.

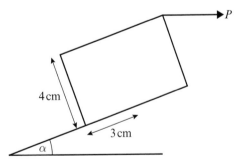

a Find the value of P which will just cause the cylinder to topple about the highest point of the base. **(4 marks)**

b Find the value of P which would cause the cylinder to slide up the plane. **(6 marks)**

c Show that the cylinder topples before it slides. **(1 mark)**

Challenge

A A child's toy is made from joining a right circular uniform solid cone, radius r and height h, to a uniform solid hemisphere of the same material and radius r. They are joined so that their plane faces coincide as shown in the diagram.

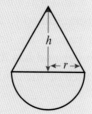

a Show that the distance of the centre of mass of the toy from the base of the cone is:

$$\left| \frac{h^2 - 3r^2}{4(2r + h)} \right|$$

The toy is placed with its hemisphere in contact with a horizontal plane and with its axis vertical. It is slightly displaced and released from rest.

b Given that the plane is sufficiently rough to prevent slipping, explain clearly, with reasons, what will happen in each of the following cases:

   i $h > r\sqrt{3}$ **ii** $h < r\sqrt{3}$ **iii** $h > r\sqrt{3}$

Mixed exercise ③

E/P **1** The curve shows a sketch of the region R bounded by the curve with equation $y^2 = 4x$ and the line with equation $x = 4$. The unit of length on both the axes is the centimetre. The region R is rotated through π radians about the x-axis to form a solid S.

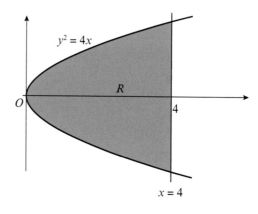

   a Show that the volume of the solid S is 32π cm³.

   (4 marks)

   b Given that the solid is uniform, find the distance of the centre of mass of S from O.

   (6 marks)

E/P **2** The region R is bounded by the curve with equation $y = \frac{1}{x}$, the lines $x = 1$, $x = 2$ and the x-axis, as shown in the diagram. The unit of length on both the axes is 1 m. A solid plinth is made by rotating R through 2π radians about the x-axis.

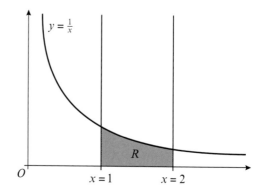

   a Show that the volume of the plinth is $\frac{\pi}{2}$ m³.

   (4 marks)

   b Find the distance of the centre of mass of the plinth from its larger plane face, giving your answer in cm to the nearest cm.

   (6 marks)

A **3** The diagram shows a cross-section of a uniform solid standing on horizontal
E/P ground. The solid consists of a uniform solid right circular cylinder, of
diameter 80 cm and height 40 cm, joined to a uniform solid hemisphere of
the same density. The circular base of the hemisphere coincides with the
upper circular end of the cylinder and has the same diameter as that
of the cylinder. Find the distance of the centre of mass of the solid
from the ground. **(6 marks)**

E/P **4** A simple wooden model of a rocket is made by taking a uniform cylinder,
of radius r and height $3r$, and carving away part of the top two-thirds to
form a uniform cone of height $2r$ as shown in cross-section in the diagram.
Find the distance of the centre of mass of the model from its plane
face. **(6 marks)**

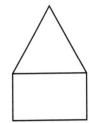

E/P **5** The diagram shows a cross-section containing the axis of symmetry of a
uniform body consisting of a solid right circular cylinder of base
radius r and height kr surmounted by a solid hemisphere of radius r.

a Given that the centre of mass of the body is at the centre C of the
common face of the cylinder and the hemisphere, find the value of k,
giving your answer to 2 significant figures. **(5 marks)**

b Explain briefly why the body remains at rest when it is placed with any part of its
hemispherical surface in contact with a horizontal plane. **(1 mark)**

E/P **6** A uniform lamina occupies the region R bounded by
the x-axis and the curve with equation
$y = \frac{1}{4}x(4 - x)$, $0 \leqslant x \leqslant 4$, as shown in Figure 1.

a Show by integration that the y-coordinate of the
centre of mass of the lamina is $\frac{2}{5}$. **(8 marks)**

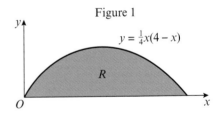
Figure 1

A uniform prism P has cross-section R. The prism is placed
with its rectangular face on a slope inclined at an angle θ
to the horizontal. The cross-section R lies in a vertical
plane as shown in Figure 2. The surfaces are sufficiently
rough to prevent P from sliding.

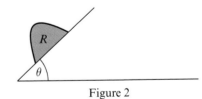
Figure 2

b Find the angle θ, for which P is about to topple. **(2 marks)**

E/P **7** A uniform semicircular lamina has radius $2a$ and the midpoint
of the bounding diameter AB is O.

a Using integration, show that the centre of mass of the

lamina is at a distance $\frac{8a}{3\pi}$ from O. **(8 marks)**

A The two semicircular laminas, each of radius a and with AO and OB as diameters, are cut away from the original lamina to leave the lamina $AOBC$ shown in the diagram, where OC is perpendicular to AB.

b Show that the centre of mass of the lamina $AOBC$ is at a distance $\frac{4a}{\pi}$ from O.   **(6 marks)**

The lamina $AOBC$ is of mass M and a particle of mass M is attached to the lamina at B to form a composite body.

c State the distance of the centre of mass of the body from OC and from OB.   **(1 mark)**

The body is smoothly hinged at A to a fixed point and rests in equilibrium in a vertical plane.

d Calculate, to the nearest degree, the acute angle between AB and the horizontal.   **(3 marks)**

(E/P) **8** A uniform wooden 'mushroom', used in a game, is made by joining a solid cylinder to a solid hemisphere. They are joined symmetrically, such that the centre O of the plane face of the hemisphere coincides with the centre of one of the ends of the cylinder. The diagram shows the cross-section through a plane of symmetry of the mushroom, as it stands on a horizontal table.

The radius of the cylinder is r, the radius of the hemisphere is $3r$, and the centre of mass of the mushroom is at the point O.

a Show that the height of the cylinder is $r\sqrt{\frac{81}{2}}$ **(6 marks)**

The table top, which is rough enough to prevent the mushroom from sliding, is slowly tilted until the mushroom is about to topple.

b Find, to the nearest degree, the angle with the horizontal through which the table top has been tilted.   **(3 marks)**

(E/P) **9** Figure 1 shows a finite region A which is bounded by the curve with equation $y^2 = 4ax$, the line $x = a$ and the x-axis. A uniform solid S_1 is formed by rotating A through 2π radians about the x-axis.

a Show that the volume of S_1 is $2\pi a^3$.   **(3 marks)**

b Show that the centre of mass of S_1 is a distance $\frac{2a}{3}$ from the origin O.   **(4 marks)**

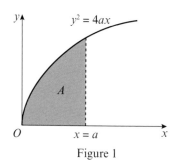

Figure 2 shows a cross-section of a uniform solid S which has been obtained by attaching the plane base of solid S_1 to the plane base of a uniform hemisphere S_2 of base radius $2a$.

c Given that the densities of solids S_1 and S_2 are ρ_1 and ρ_2 respectively, find the ratio $\rho_1 : \rho_2$ which ensures that the centre of mass of S lies in the common plane face of S_1 and S_2.   **(6 marks)**

d Given that $\rho_1 : \rho_2 = 6$, explain why the solid S may rest in equilibrium with any point of the curved surface of the hemisphere in contact with a horizontal plane.   **(2 marks)**

10 A mould for a right circular cone, base radius r and height h, is produced by making a conical hole in a uniform cylindrical block, base radius $2r$ and height $3r$. The axis of symmetry of the conical hole coincides with that of the cylinder, and AB is a diameter of the top of the cylinder, as shown in the diagram.

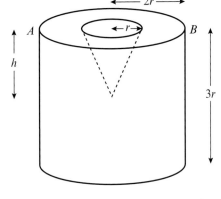

a Show that the distance from AB of the centre of

mass of the mould is $\dfrac{216r^2 - h^2}{4(36r - h)}$ **(8 marks)**

The mould is suspended from the point A, and hangs freely in equilibrium.

b In the case $h = 2r$, calculate, to the nearest degree, the angle between AB and the downward vertical. **(4 marks)**

11 A compound solid is made from an inverted frustum and a hemisphere as shown opposite.

The base of the frustum has a radius of $5\,\text{cm}$ and the hemisphere has a radius of $10\,\text{cm}$.

a Given that the density, ρ, of the hemisphere is three times the density of the frustum, find the height of the centre of mass of the compound solid from its base. **(8 marks)**

The compound solid is placed on a rough plane inclined at $45°$ to the horizontal.

b Given that the total mass of the compound solid is $m\,\text{kg}$, work out the minimum horizontal force, in terms of m, that must be applied at A to stop the solid from toppling over. The plane is sufficiently rough to prevent slipping. **(4 marks)**

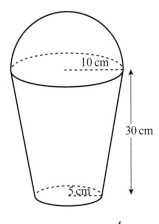

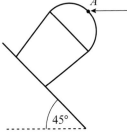

12 A solid is formed by rotating the region enclosed by the curve $y = \dfrac{2}{x + 3}$, the lines $x = 2$ and $x = 4$ and the x-axis, though $360°$ around the x-axis.

The solid is then placed on its smaller circular end on rough horizontal ground.

a Find the height of the centre of mass of the solid above the ground. **(10 marks)**

The solid is then placed on a rough inclined ramp angled at $\theta°$ to the horizontal.

b Assuming that the solid does not slip, work out the value of θ when the solid is on the point of tipping. **(4 marks)**

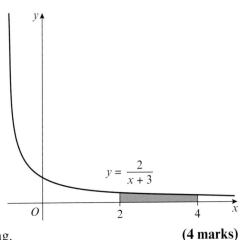

A 13 A concrete pile is used as a foundation for a bridge. The pile is modelled as a vertical non-
E/P uniform rod. The top of the pile is flush with the ground, and the base of the pile is 9 m deep.
The mass per unit length of the pile, m kg m^{-1}, at a depth of x m is given by the formula
$m = 1000 + 400\sqrt{x}$

    a Show that the pile has a mass of 16 200 kg. **(3 marks)**

    b Hence find the depth of the centre of mass of the pile. **(4 marks)**

E/P 14 A non-uniform solid right circular cylinder has base radius 4 cm and height 90 cm. The cylinder
is modelled as a non-uniform rod, with mass per unit length $(20 - \frac{1}{9}h)$ grams cm^{-1} at a height h
above its base.

    a Using this model, calculate the height of the centre of mass of the cylinder above its
    base. **(7 marks)**

The plane base of the cylinder is placed on a rough surface, which is inclined at an angle θ to
the horizontal. The cylinder rests in equilibrium on the point of toppling.

    b Comment on the suitability of the model for calculating:
      i the centre of mass of the cylinder
      ii the value of θ. **(2 marks)**

    c Find the value of θ in degrees to 3 significant figures. **(3 marks)**

E/P 15 A non-uniform rod AB of length 12 m hangs in horizontal equilibrium, supported by two light
inextensible strings, which hang vertically, attached at A and B. The mass per unit length,

m kg m^{-1}, of the rod at a distance x m from A is given by the formula $m = 2 + \dfrac{x^2}{4}$

Find the tension in each string. **(9 marks)**

E/P 16 The diagram shows a non-uniform rod AB resting on horizontally on two pegs. One peg is
positioned at A and the other peg is positioned three-quarters of the way along AB. At a
distance x m from A the mass per unit length of the rod is given by $(35 - \frac{1}{2}x)$ kg m^{-1}.

Given that the reactions of each peg on the rod are the same, find the length of the rod. **(10 marks)**

Challenge

A non-uniform solid right circular cone
has base radius h m and height h m.

At a distance x m above its base, the cone
has density $(x + 1)$ kg m^{-3}.

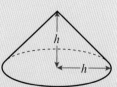

a Show that the mass of the cone is $\frac{1}{12}h^3(h + 4)$ kg.

b Given that the centre of the mass of the cone is a distance $\frac{1}{3}h$ above
its base, find the value of h.

c Show that as h varies the height of the centre of mass of the cone
above its base cannot exceed $\frac{2}{5}h$.

Summary of key points

A

1 **The centre of mass of a uniform lamina** may be found using the formulae:

- $$\bar{x} = \frac{\displaystyle\int_a^b xy\,dx}{\displaystyle\int_a^b y\,dx} \quad \text{and} \quad \bar{y} = \frac{\displaystyle\int_a^b \tfrac{1}{2}y^2\,dx}{\displaystyle\int_a^b y\,dx}$$

- $$M\bar{x} = \int_a^b \rho xy\,dx \quad \text{and} \quad M\bar{y} = \int_a^b \tfrac{1}{2}\rho y^2\,dx$$

  where $M = \displaystyle\int_a^b \rho y\,dx$, and is the total mass of the lamina, and ρ is the mass per unit area of the lamina.

2 **Standard results for uniform laminas and for arcs**

Lamina	Centre of mass along axis of symmetry
Semicircle, radius r	$\dfrac{4r}{3\pi}$ from the centre
Sector of circle, radius r, angle at centre 2α	$\dfrac{2r \sin \alpha}{3\alpha}$ from the centre
Circular arc, radius r, angle at centre 2α	$\dfrac{r \sin \alpha}{\alpha}$ from the centre

3 • For a solid body the centre of mass is the point where the weight acts.
   • For a uniform solid body the weight is evenly distributed through the body.
   • The centre of mass will lie on any axis of symmetry.
   • The centre of mass will lie on any plane of symmetry.

4 From symmetry, the centre of mass of some uniform bodies is at their geometric centre. These include the cube, the cuboid, the sphere, the right circular cylinder and the right circular prism.

5 For a **uniform solid of revolution**, where the revolution is about the x-axis, the centre of mass lies on the x-axis, by symmetry, and its position on the axis is given by the formulae

$$\bar{x} = \frac{\int \pi y^2 x\,dx}{\int \pi y^2\,dx} \quad \text{or} \quad M\bar{x} = \int \rho \pi y^2 x\,dx$$

where M is the known mass of the solid and ρ is its density.

6 For a **uniform solid of revolution**, where the revolution is about the y-axis, the centre of mass lies on the y-axis, by symmetry, and its position on the axis is given by the formulae

$$\bar{y} = \frac{\int \pi x^2 y\,dy}{\int \pi x^2\,dy} \quad \text{or} \quad M\bar{y} = \int \rho \pi x^2 y\,dy$$

where M is the known mass of the solid and ρ is its density.

7 Standard results for uniform bodies

Body	Centre of mass along axis of symmetry
Solid hemisphere, radius R	$\frac{3}{8}R$ from the centre
Hemispherical shell, radius R	$\frac{1}{2}R$ from the centre
Solid right circular cone, height h	$\frac{3}{4}h$ from the vertex, or $\frac{1}{4}h$ from the circular base
Conical shell, height h	$\frac{2}{3}h$ from the vertex, or $\frac{1}{3}h$ from the circular base

8 For a non-uniform rod with a variable mass per unit length, $\rho = f(x)$, and length l, the distance of the centre of mass from the end of the rod is given by:

$$\bar{x} = \frac{\int_0^l x\rho\,dx}{\int_0^l \rho\,dx} = \frac{\int_0^l xf(x)\,dx}{\int_0^l f(x)\,dx}$$

9 A rigid body is in equilibrium if:
- there is zero resultant force in any direction, i.e. the sum of the components of all the forces in any direction is zero, and
- the sum of the moments of the forces about any point is zero.

10 · **When a lamina is suspended freely** from a fixed point or pivots freely about a horizontal axis it will rest in equilibrium in a vertical plane with its centre of mass vertically below the point of suspension or pivot.
- **When a rigid body is suspended freely** from a fixed point and rests in equilibrium then its centre of mass is vertically below the point of suspension.

11 If a body rests in equilibrium on a rough inclined plane, then the line of action of the weight of the body must pass through the face of the body which is in contact with the plane.

12 You can establish whether equilibrium will be broken by sliding or by toppling by considering:
- a body is on the point of sliding when $F = \mu R$
- a body is on the point of toppling when the reaction acts at the point about which turning can take place.

Review exercise

E **1** A circular flywheel of diameter 7 cm is rotating about the axis through its centre and perpendicular to its plane with constant angular speed 1000 revolutions per minute.

Find, in m s⁻¹ to 3 significant figures, the speed of a point on the rim of the flywheel. **(2)**

← Section 1.1

E **2** A particle P of mass 0.5 kg is attached to one end of a light inextensible string of length 1.5 m. The other end of the string is attached to a fixed point A. The particle is moving, with the string taut, in a horizontal circle with centre O vertically below A. The particle is moving with constant angular speed 2.7 rad s⁻¹. Find:

a the tension in the string **(4)**

b the angle, to the nearest degree, that AP makes with the downward vertical. **(4)**

← Section 1.2

E/P **3**

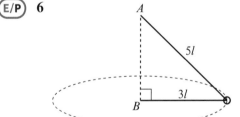

A particle P of mass m is attached to one end of a light string. The other end of the string is attached to a fixed point A. The particle moves in a horizontal circle with constant angular speed ω and with the string inclined at an angle of 60° to the vertical, as shown in the diagram above.

The length of the string is L.

a Show that the tension in the string is $2mg$. **(2)**

b Find ω in terms of g and L. **(4)**

← Section 1.3

E/P **4** A car moves round a bend which is banked at a constant angle of 10° to the horizontal. When the car is travelling at a constant speed of 18 m s⁻¹, there is no sideways frictional force on the car. The car is modelled as a particle moving in a horizontal circle of radius r metres. Calculate the value of r. **(6)**

← Section 1.3

E **5** A cyclist is travelling around a circular track which is banked at 25° to the horizontal. The coefficient of friction between the cycle's tyres and the track is 0.6. The cyclist moves with constant speed in a horizontal circle of radius 40 m, without the tyres slipping.

Find the maximum speed of the cyclist. **(9)**

← Section 1.3

E/P **6**

A light inextensible string of length $8l$ has its ends fixed to two points A and B, where A is vertically above B. A small smooth ring of mass m is threaded on the string. The ring is moving with constant speed in a horizontal circle with centre B and radius $3l$, as shown in the diagram. Find:

a the tension in the string **(3)**

b the speed of the ring. **(5)**

c State briefly in what way your solutions might no longer be valid if the ring were firmly attached to the string. **(1)**

← **Section 1.3**

(E) **7**

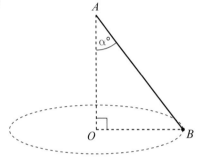

A metal ball B of mass m is attached to one end of a light inextensible string. The other end of the string is attached to a fixed point A. The ball B moves in a horizontal circle with centre O vertically below A, as shown in the diagram. The string makes a constant angle $\alpha°$ with the downward vertical and B moves with constant angular speed $\sqrt{2gk}$, where k is a constant. The tension in the string is $3mg$. By modelling B as a particle, find:

a the value of α **(4)**

b the length of the string. **(5)**

← **Section 1.3**

(E) **8** A particle P of mass m moves on the smooth inner surface of a spherical bowl of internal radius r. The particle moves with constant angular speed in a horizontal circle, which is at a depth $\frac{1}{2}r$ below the centre of the bowl. Find:

a the normal reaction of the bowl on P **(4)**

b the time for P to complete one revolution of its circular path. **(6)**

← **Section 1.3**

(E/P) **9**

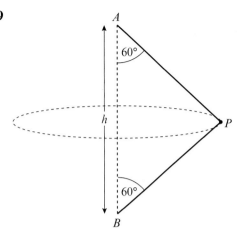

A particle P of mass m is attached to two light inextensible strings. The other ends of the string are attached to fixed points A and B. The point A is a distance h vertically above B. The system rotates about the line AB with constant angular speed ω. Both strings are taut and inclined at 60° to AB, as shown in the diagram. The particle moves in a circle of radius r.

a Show that $r = \dfrac{\sqrt{3}}{2}h$. **(3)**

b Find, in terms of m, g, h and ω, the tension in AP and the tension in BP. **(3)**

The time taken for P to complete one circle is T.

c Show that $T < \pi\sqrt{\dfrac{2h}{g}}$ **(2)**

← **Section 1.3**

(E) **10**

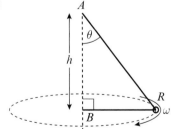

One end of a light inextensible string is attached to a fixed point A. The other end of the string is attached to a fixed point B, vertically below A, where $AB = h$. A small smooth ring R of mass m is threaded on the string. The ring R moves in a horizontal circle with centre B, as

125

shown in the diagram. The upper section of the string makes a constant angle θ with the downward vertical and R moves with constant angular speed ω. The ring is modelled as a particle.

a Show that $\omega^2 = \dfrac{g}{h}\left(\dfrac{1 + \sin\theta}{\sin\theta}\right)$. **(5)**

b Deduce that $\omega > \sqrt{\dfrac{2g}{h}}$. **(2)**

Given that $\omega = \sqrt{\dfrac{3g}{h}}$,

c find, in terms of m and g, the tension in the string. **(4)**

← **Section 1.3**

E/P **11**

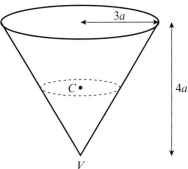

A hollow cone, of base radius $3a$ and height $4a$, is fixed with its axis vertical and vertex V downwards, as shown in the diagram. A particle moves in a horizontal circle with centre C, on the smooth inner surface of the cone with constant angular speed $\sqrt{\dfrac{8g}{9a}}$. Find the height of C above V. **(8)**

← **Section 1.3**

E **12**

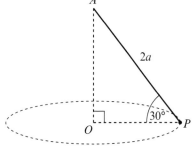

A particle P of mass m is attached to one end of a light inextensible string of length $2a$. The other end of the string is fixed

to a point A which is vertically above the point O on a smooth horizontal table. The particle P remains in contact with the surface of the table and moves in a circle with centre O and with angular speed $\sqrt{\dfrac{kg}{3a}}$, where k is a constant. Throughout the motion the string remains taut and $\angle APO = 30°$, as shown in the diagram.

a Show that the tension in the string is $\dfrac{2kmg}{3}$ **(3)**

b Find, in terms of m, g and k, the normal reaction between P and the table. **(3)**

c Deduce the range of possible values of k. **(2)**

The angular speed of P is changed to $\sqrt{\dfrac{2g}{a}}$. The particle P now moves in a horizontal circle above the table. The centre of this circle is X.

d Show that X is the midpoint of OA. **(7)**

← **Section 1.3**

E/P **13**

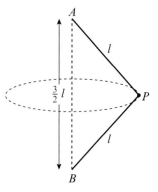

A particle P of mass m is attached to the ends of two light inextensible strings AP and BP, each of length l. The ends A and B are attached to fixed points, with A vertically above B and $AB = \frac{3}{2}l$, as shown in the diagram above. The particle P moves in a horizontal circle with constant angular speed ω. The centre of the circle is the midpoint of AB and both strings remain taut.

a Show that the tension in AP is $\frac{1}{6}m(3l\omega^2 + 4g)$. **(7)**

b Find, in terms of m, l, ω and g, an expression for the tension in BP. **(2)**

c Deduce that $\omega^2 \geq \dfrac{4g}{3l}$. **(2)**

← Section 1.3

A **a** Find:

  i the tension in the string

  ii the angle θ that the string makes with the vertical. **(5)**

b Show that $v = k\sqrt{gx}$, where k is a rational constant to be found. **(4)**

← Section 1.3

A **E/P** **14** A rough disc rotates in a horizontal plane with constant angular velocity ω about a fixed vertical axis. A particle P of mass m lies on the disc at a distance $\frac{4}{3}a$ from the axis. The coefficient of friction between P and the disc is $\frac{3}{5}$. Given that P remains at rest relative to the disc,

a prove that $\omega^2 \leq \dfrac{9g}{20a}$.

The particle is now connected to the axis by a horizontal light elastic string of natural length a and modulus of elasticity $2mg$. The disc again rotates with constant angular velocity ω about the axis and P remains at rest relative to the disc at a distance $\frac{4}{3}a$ from the axis.

b Find the greatest and least possible values of ω^2.

← Section 1.3

E/P **16** One end of a light inextensible string of length l is attached to a particle P of mass m. The other end is attached to a fixed point A. The particle is hanging freely at rest with the string vertical when it is projected horizontally with speed $\sqrt{\dfrac{5gl}{2}}$

a Find the speed of P when the string is horizontal. **(4)**

When the string is horizontal it comes into contact with a small smooth fixed peg which is at the point B, where AB is horizontal, and $AB < l$. Given that the particle then describes a complete semi-circle with centre B,

b find the least possible value of the length AB. **(9)**

← Section 1.4

E/P **15** The diagram shows a particle of mass 2 kg attached to a light elastic string of natural length x m and modulus of elasticity $10g$ N. The other end of the string is attached to a fixed point A, and the particle moves with constant speed v m s^{-1} in a horizontal circle with centre O, where O is vertically below A and $OA = x$ m.

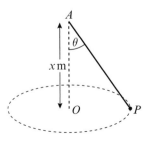

E/P **17** One end of a light inextensible string of length l is attached to a fixed point A. The other end is attached to a particle P of mass m, which is hanging freely at rest at point B. The particle P is projected horizontally from B with speed $\sqrt{3gl}$.

When AP makes an angle θ with the downward vertical and the string remains taut, the tension in the string is T.

a Show that $T = mg(1 + 3\cos\theta)$. **(6)**

b Find the speed of P at the instant when the string becomes slack. **(3)**

c Find the maximum height above the level of B reached by P. **(5)**

← Section 1.4

18

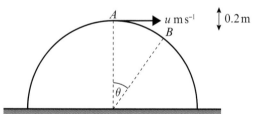

A particle of mass m is attached to one end of a light inextensible string of length l. The other end of the string is attached to a fixed point O. The particle is hanging at the point A, which is vertically below O. It is projected horizontally with speed u. When the particle is at the point P, $\angle AOP = \theta$, as shown in the diagram. The string oscillates through an angle α on either side of OA where $\cos \alpha = \frac{2}{3}$

a Find u in terms of g and l. **(3)**

When $\angle AOP = \theta$, the tension in the string is T.

b Show that $T = \dfrac{mg}{3}(9\cos\theta - 4)$. **(4)**

c Find the range of values of T during the oscillations. **(2)**

← **Section 1.4**

19

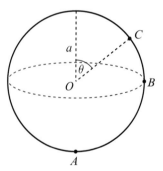

A smooth solid hemisphere, of radius $0.8\,\text{m}$ and centre O, is fixed with its plane face on a horizontal table. A particle of mass $0.5\,\text{kg}$ is projected horizontally with speed $u\,\text{m s}^{-1}$ from the highest point A of the hemisphere. The particle leaves the hemisphere at the point B, which is a vertical distance of $0.2\,\text{m}$ below the level of A. The speed of the particle at B is $v\,\text{m s}^{-1}$ and the angle between OA and OB is θ, as shown in the diagram.

a Find the value of $\cos\theta$. **(1)**

b Show that $v^2 = 5.88$. **(3)**

c Find the value of u. **(3)**

← **Section 1.5**

20 A smooth solid sphere, with centre O and radius a, is fixed to the upper surface of a horizontal table. A particle P is placed on the surface of the sphere at a point A, where OA makes an angle α with the upward vertical, and $0 < \alpha < \dfrac{\pi}{2}$. The particle is released from rest. When OP makes an angle θ with the upward vertical, and P is still on the surface of the sphere, the speed of P is v.

a Show that $v^2 = 2ga(\cos\alpha - \cos\theta)$. **(4)**

Given that $\cos\alpha = \frac{3}{4}$, find:

b the value of θ when P loses contact with the sphere **(5)**

c the speed of P as it hits the table. **(4)**

← **Section 1.5**

21

The diagram shows a fixed hollow sphere of internal radius a and centre O.

A particle P of mass m is projected horizontally from the lowest point A of the sphere with speed $\sqrt{\frac{7}{2}ag}$. It moves in a vertical circle, centre O, on the smooth inner surface of the sphere. The particle passes through the point B, which is in the same horizontal plane as O. It leaves the surface of the sphere at the point C, where OC makes an angle θ with the upward vertical.

a Find, in terms of m and g, the normal reaction between P and the surface of the sphere at B. **(4)**

← Section 1.5

A

b Show that $\theta = 60°$. **(7)**

After leaving the surface of the sphere, P meets it again at the point A.

c Find, in terms of a and g, the time P takes to travel from C to A. **(4)**

← Section 1.5

E/P **22**

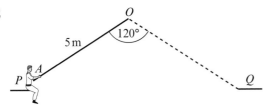

A trapeze artist of mass 60 kg is attached to the end A of a light inextensible rope OA of length 5 m. The artist must swing in an arc of a vertical circle, centre O, from a platform P to another platform Q, where PQ is horizontal. The other end of the rope is attached to the fixed point O which lies in the vertical plane containing PQ, with $\angle POQ = 120°$ and $OP = OQ = 5$ m, as shown in the diagram.

As part of her act, the artist projects herself from P with speed $\sqrt{15}$ m s^{-1} in a direction perpendicular to the rope OA and in the plane POQ. She moves in a circular arc towards Q. At the lowest point of her path she catches a ball of mass m kg which is travelling towards her with speed 3 m s^{-1} and parallel to QP. After catching the ball, she comes to rest at the point Q.

By modelling the artist and the ball as particles and ignoring her air resistance, find:

a the speed of the artist immediately before she catches the ball **(4)**

b the value of m **(6)**

c the tension in the rope immediately after she catches the ball. **(2)**

← Section 1.4

A **23**

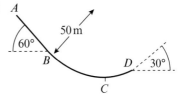

The diagram represents the path of a skier of mass 70 kg moving on a ski-slope $ABCD$. The path lies in a vertical plane. From A to B, the path is modelled as a straight line inclined at 60° to the horizontal. From B to D, the path is modelled as an arc of a vertical circle of radius 50 m. The lowest point of the arc BD is C.

At B, the skier is moving downwards with speed 20 m s^{-1}. At D, the path is inclined at 30° to the horizontal and the skier is moving upwards. By modelling the slope as smooth and the skier as a particle, find:

a the speed of the skier at C **(2)**

b the normal reaction of the slope on the skier at C **(2)**

c the speed of the skier at D **(2)**

d the change in the normal reaction of the slope on the skier as she passes B. **(4)**

The model is refined to allow for the influence of friction on the motion of the skier.

e State briefly, with a reason, how the answer to part **b** would be affected by using such a model. (No further calculations are expected.) **(1)**

← Section 1.4

E **24** A particle P of mass m is attached to one end of a light inextensible string of length a. The other end of the string is attached to a point O. The point A is vertically below O, and $OA = a$.

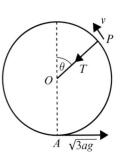

A The particle is projected horizontally from A with speed $\sqrt{(3ag)}$. When OP makes an angle θ with the upward vertical through O and the string is still taut, the tension in the string is T and the speed of P is v, as shown in the diagram.

a Find, in terms of a, g and θ, an expression for v^2. **(3)**

b Show that $T = (1 - 3\cos\theta)mg$. **(3)**

The string becomes slack when P is at the point B.

c Find, in terms of a, the vertical height of B above A. **(2)**

After the string becomes slack, the highest point reached by P is C.

d Find, in terms of a, the vertical height of C above B. **(5)**

← Section 1.5

E 25

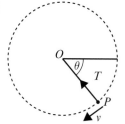

A particle P of mass m is attached to one end of a light inextensible string of length a. The other end of the string is fixed at a point O. The particle is held with the string taut and OP horizontal. It is then projected vertically downwards with speed u, where $u^2 = \frac{3}{2}ga$. When OP has turned through an angle θ and the string is still taut, the speed of P is v and the tension in the string is T, as shown in the diagram above.

a Find an expression for v^2 in terms of a, g and θ. **(2)**

b Find an expression for T in terms of m, g and θ. **(3)**

c Prove that the string becomes slack when $\theta = 210°$. **(2)**

d State, with a reason, whether P would complete a vertical circle if the string were replaced by a light rod. **(2)**

A After the string becomes slack, P moves freely under gravity and is at the same level as O when it is at the point A.

e Explain briefly why the speed of P at A is $\sqrt{\frac{3}{2}ga}$. **(1)**

The direction of motion of P at A makes an angle ϕ with the horizontal.

f Find ϕ. **(4)**

← Section 1.5

E 26

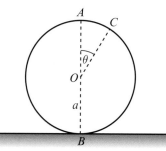

A particle is at the highest point A on the outer surface of a fixed smooth sphere of radius a and centre O. The lowest point B of the sphere is fixed to a horizontal plane. The particle is projected horizontally from A with speed u, where $u < \sqrt{(ag)}$. The particle leaves the sphere at the point C, where OC makes an angle θ with the upward vertical, as shown in the diagram above.

a Find an expression for $\cos\theta$ in terms of u, g and a. **(7)**

The particle strikes the plane with speed $\sqrt{\dfrac{9ag}{2}}$

b Find, to the nearest degree, the value of θ. **(7)**

← Section 1.5

E/P 27

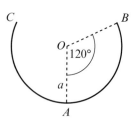

Part of a hollow spherical shell, centre O and radius a, is removed to form a bowl

A with a plane circular rim. The bowl is fixed with the circular rim uppermost and horizontal. The point A is the lowest point of the bowl. The point B is on the rim of the bowl and $\angle AOB = 120°$, as shown in the diagram above. A smooth small marble of mass m is placed inside the bowl at A and given an initial horizontal speed u. The direction of motion of the marble lies in the vertical plane AOB. The marble stays in contact with the bowl until it reaches B. When the marble reaches B, its speed is v.

a Find an expression for v^2. **(3)**

b For the case when $u^2 = 6ga$, find the normal reaction of the bowl on the marble as the marble reaches B. **(3)**

c Find the least possible value of u for the marble to reach B. **(3)**

The point C is the other point on the rim of the bowl lying in the vertical plane OAB.

d Find the value of u which will enable the marble to leave the bowl at B and meet it again at the point C. **(7)**

← Section 1.5

E **28** Three particles of masses $3m$, $5m$ and λm are placed at the points with coordinates $(4, 0)$, $(0, -3)$ and $(4, 2)$ respectively. The centre of mass of the three particles is at $(2, k)$.

a Show that $\lambda = 2$. **(4)**

b Calculate the value of k. **(2)**

← Section 2.2

E **29** Particles of masses $2M$, xM and yM are placed at points whose coordinates are $(2, 5)$, $(1, 3)$ and $(3, 1)$ respectively. Given that the centre of mass of the three particles is at the point $(2, 4)$, find the values of x and y. **(6)**

← Section 2.2

E **30** Three particles of masses $0.1\,\text{kg}$, $0.2\,\text{kg}$ and $0.3\,\text{kg}$ are placed at the points with position vectors $(2\mathbf{i} - \mathbf{j})\,\text{m}$, $(2\mathbf{i} + 5\mathbf{j})\,\text{m}$ and $(4\mathbf{i} + 2\mathbf{j})\,\text{m}$ respectively. Find the position vector of the centre of mass of the particles. **(5)**

← Section 2.2

E **31** Three particles of mass $2M$, M and kM, where k is a constant, are placed at points with position vectors $6\mathbf{i}\,\text{m}$, $4\mathbf{j}\,\text{m}$ and $(2\mathbf{i} - 2\mathbf{j})\,\text{m}$ respectively. The centre of mass of the three particles has position vector $(3\mathbf{i} + c\mathbf{j})\,\text{m}$, where c is a constant.

a Show that $k = 3$. **(4)**

b Hence find the value of c. **(3)**

← Section 2.2

E **32**

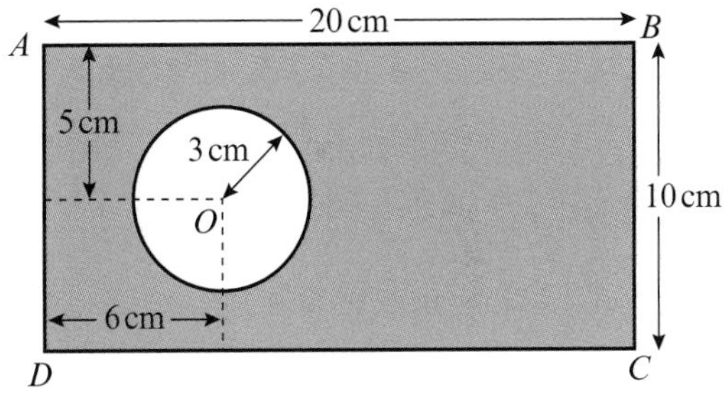

The figure shows a metal plate that is made by removing a circle of centre O and radius $3\,\text{cm}$ from a uniform rectangular lamina $ABCD$, where $AB = 20\,\text{cm}$ and $BC = 10\,\text{cm}$. The point O is $5\,\text{cm}$ from both AB and CD, and is $6\,\text{cm}$ from AD.

a Calculate, to 3 significant figures, the distance of the centre of mass of the plate from AD. **(6)**

The plate is freely suspended from A and hangs in equilibrium.

b Calculate, to the nearest degree, the angle between AB and the vertical. **(2)**

← Section 2.4, 2.6

131

E/P 33

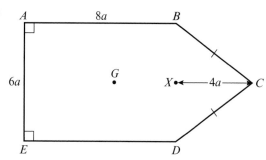

The figure shows a uniform lamina *ABCDE* such that *ABDE* is a rectangle, $BC = CD$, $AB = 8a$ and $AE = 6a$. The point *X* is the midpoint of *BD* and $XC = 4a$. The centre of mass of the lamina is at *G*.

a Show that $GX = \frac{44}{15}a$. **(6)**

The mass of the lamina is *M*. A particle of mass λM is attached to the lamina at *C*. The lamina is suspended from *B* and hangs freely under gravity with *AB* horizontal.

b Find the value of λ. **(3)**

← Sections 2.4, 2.6

E/P 34 A uniform square plate *ABCD* has mass $10M$ and the length of a side of the plate is $2l$. Particles of masses M, $2M$, $3M$ and $4M$ are attached at *A*, *B*, *C* and *D* respectively. Calculate, in terms of *l*, the distance of the centre of mass of the loaded plate from:

a *AB* **(7)**

b *BC* **(3)**

The loaded plate is freely suspended from the vertex *D* and hangs in equilibrium.

c Calculate, to the nearest degree, the angle made by *DA* with the downward vertical.

← Sections 2.4, 2.6

E/P 35

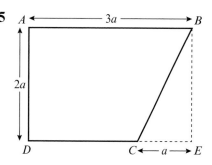

A uniform lamina *ABCD* is made by taking a uniform sheet of metal in the form of a rectangle *ABED*, with $AB = 3a$ and $AD = 2a$, and removing the triangle *BCE*, where *C* lies on *DE* and $CE = a$, as shown in the figure.

a Find the distance of the centre of mass of the lamina from *AD*. **(5)**

The lamina has mass *M*. A particle of mass *m* is attached to the lamina at *B*. When the loaded lamina is freely suspended from the midpoint of *AB*, it hangs in equilibrium with *AB* horizontal.

b Find *m* in terms of *M*. **(4)**

← Sections 2.4, 2.6

E/P 36

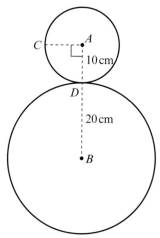

The figure shows a decoration which is made by cutting two circular discs from a sheet of uniform card. The discs are joined so that they touch at a point *D* on the circumference of both discs. The discs are coplanar and have centres *A* and *B* with radii 10 cm and 20 cm respectively.

a Find the distance of the centre of mass of the decoration from *B*. **(5)**

The point *C* lies on the circumference of the smaller disc and $\angle CAB$ is a right angle. The decoration is freely suspended from *C* and hangs in equilibrium.

b Find, in degrees to one decimal place, the angle between *AB* and the vertical. **(4)**

← Sections 2.4, 2.6

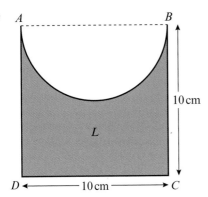

(E) 37

A uniform lamina L is formed by taking a uniform square sheet of material $ABCD$ of side 10 cm and removing a semicircle with diameter AB from the square, as shown in the figure.

a Find, in cm to 2 decimal places, the distance of the centre of mass of the lamina from the midpoint of AB. **(7)**

[The centre of mass of a uniform semicircular lamina, radius a, is at a distance $\dfrac{4a}{3\pi}$ from the centre of the bounding diameter.]

The lamina is freely suspended from D and hangs at rest.

b Find, in degrees to one decimal place, the angle between CD and the vertical. **(4)**

← Sections 2.4, 2.6

(E/P) 38

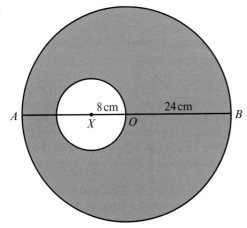

The figure shows a template T made by removing a circular disc, of centre X

and radius 8 cm, from a uniform circular lamina, of centre O and radius 24 cm. The point X lies on the diameter AOB of the lamina and $AX = 16$ cm. The centre of mass of T lies at the point G.

a Find AG. **(4)**

The template T is free to rotate about a smooth fixed horizontal axis, perpendicular to the plane of T, which passes through the midpoint of OB. A small stud of mass $\frac{1}{4}m$ is fixed at B, and T and the stud are in equilibrium with AB horizontal.

b Modelling the stud as a particle, find the mass of T in terms of m. **(3)**

← Sections 2.4, 2.6

(E/P) 39

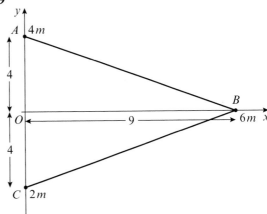

The figure shows a triangular lamina ABC. The coordinates of A, B and C are $(0, 4)$, $(9, 0)$ and $(0, -4)$ respectively. Particles of masses $4m$, $6m$ and $2m$ are attached at A, B and C respectively.

a Calculate the coordinates of the centre of mass of the three particles, without the lamina. **(4)**

The lamina ABC is uniform and of mass km. The centre of mass of the combined system consisting of the three particles and the lamina has coordinates $(4, \lambda)$.

b Show that $k = 6$. **(3)**

c Calculate the value of λ. **(2)**

The combined system is freely suspended from O and hangs at rest.

d Calculate, in degrees to one decimal place, the angle between AC and the vertical. **(3)**

← Sections 2.4, 2.6

40

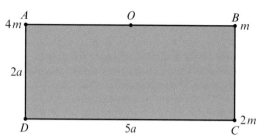

A loaded plate L is modelled as a uniform rectangular lamina $ABCD$ and three particles. The sides CD and AD of the lamina have lengths $5a$ and $2a$ respectively and the mass of the lamina is $3m$. The three particles have masses $4m$, m and $2m$ and are attached at the points A, B and C respectively, as shown in the figure.

a Show that the distance of the centre of mass of L from AD is $2.25a$. **(3)**

b Find the distance of the centre of mass of L from AB. **(2)**

The point O is the midpoint of AB. The loaded plate L is freely suspended from O and hangs at rest under gravity.

c Find, to the nearest degree, the size of the angle that AB makes with the horizontal. **(3)**

A horizontal force of magnitude P is applied at C in the direction CD. The loaded plate L remains suspended from O and rests in equilibrium with AB horizontal and C vertically below B.

d Show that $P = \frac{5}{4}mg$. **(4)**

e Find the magnitude of the force on L at O. **(4)**

← Sections 2.4, 2.6

41 A triangular frame ABC is made by bending a piece of wire of length 24 cm, so that AB, BC and AC are of lengths

6 cm, 8 cm and 10 cm respectively. Given that the wire is uniform, find the distance of the centre of mass of the frame from:

a AB **(4)**

b BC **(2)**

The frame is suspended from the corner A and hangs in equilibrium.

c Find, to the nearest degree, the acute angle made by AB with the downward vertical. **(3)**

← Sections 2.5, 2.7

42

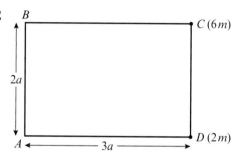

The figure shows four uniform rods joined to form a rectangular framework $ABCD$, where $AB = CD = 2a$ and $BC = AD = 3a$. Each rod has mass m. Particles of masses $6m$ and $2m$ are attached to the framework at points C and D respectively.

a Find the distance of the centre of mass of the loaded framework from:

i AB **ii** AD **(7)**

The loaded framework is freely suspended from B and hangs in equilibrium.

b Find the angle which BC makes with the vertical. **(3)**

← Sections 2.5, 2.7

43 Three uniform rods AB, BC and CA of masses $2m$, m and $3m$ respectively have lengths l, l and $l\sqrt{2}$ respectively. The rods are rigidly joined to form a right-angled triangular framework.

a Calculate, in terms of l, the distance of the centre of mass of the framework from:

i BC **ii** AB **(4)**

b Calculate the angle, to the nearest degree, that *BC* makes with the vertical when the framework is freely suspended from the point *B*. **(2)**

← Sections 2.5, 2.7

E 44

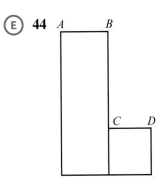

The figure shows the L-shaped lamina that is made from the rectangle *ABFG* and the square *CDEF*. *AB* has length *d* and *AG* has length 3*d*. *CD* and *DE* have length *d*. The density of the square is three times that of the rectangle.

Find, in terms of *d*, the distance of the centre of mass of the lamina from:

a *AG* **(4)**

b *GE* **(2)**

The lamina is suspended from the point *A* and hangs freely in equilibrium.

c Find, to the nearest degree, the angle that *AB* makes with the vertical. **(3)**

← Section 2.8

 45

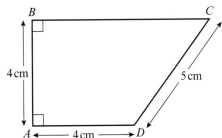

A framework is made from thin uniform wire of total length 20 cm. The framework is in the shape of a trapezium *ABCD*, where *AB* = *AD* = 4 cm, *CD* = 5 cm and *AB* is perpendicular to *BC* and *AD* as shown in the diagram.

AB, *BC* and *AD* are made from wire of mass 0.01 *M* kg per cm. *CD* is made from wire of mass 0.015 *M* kg per cm.

a Find the distance of the centre of mass of the framework from *AB*. **(6)**

The framework has mass *M*. A particle of mass *kM* is attached to the framework at *C*. When the framework is freely suspended from the midpoint of *BC*, the framework hangs in equilibrium with *BC* horizontal.

b Find the value of *k*. **(5)**

← Section 2.8

E/P 46 A shop sign of weight *W* N, that can be modelled as a lamina and is shown in the diagram below. The sign is suspended by two ropes that can be modelled as light inelastic strings.

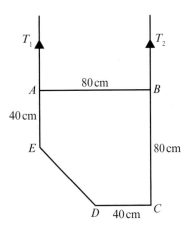

a Find the distance of the centre of mass of the lamina from *AE*. **(4)**

b Find the tension in each string. **(4)**

The rope attached at *A* will snap when the tension in it exceeds 10*W* N.

The rope attached at *B* will snap when the tension in it exceeds 8*W* N.

A particle of weight *kW* is attached to the sign at *C*. Given that neither rope breaks,

c find the largest possible value of *k*. **(4)**

← Section 2.8

A **47** A uniform solid is formed by rotating the region enclosed between the curve with equation $y = \sqrt{x}$, the x-axis and the line $x = 4$, through one complete revolution about the x-axis. Find the distance of the centre of mass of the solid from the origin O. **(5)**

E

← **Section 3.1**

E **48** **a** Use integration to show that the centre of mass of a uniform semicircular lamina, of radius a, is a distance $\frac{4a}{3\pi}$ from the midpoint of its straight edge, O. **(4)**

A semicircular lamina, of radius b with O as the midpoint of its straight edge, is removed from the first lamina.

b Show that the centre of mass of the resulting lamina is at a distance $\bar{x}$ from O, where
$$\bar{x} = \frac{4}{3\pi}\frac{(a^2 + ab + b^2)}{(a + b)}$$ **(6)**

c Hence find the position of the centre of mass of a uniform semicircular arc of radius a. **(2)**

← **Section 3.2**

E/P **49** A uniform triangular lamina ABC has $\angle ABC = 90°$ and $AB = c$.

a Using calculus, prove that the centre of mass of the lamina is at a distance $\frac{1}{3}c$ from BC. **(6)**

The diagram shows a uniform lamina in which $PQ = PS = 2a$, $SR = a$.

The centre of mass of the lamina is G.

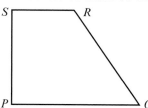

b Show that the distance of G from PS is $\frac{7}{9}a$. **(5)**

c Find the distance of G from PQ. **(5)**

← **Section 3.2**

A **50**

E/P

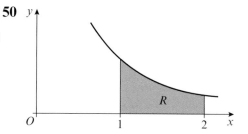

The shaded region R is bounded by the curve with equation $y = \frac{1}{2x^2}$, the x-axis and the lines $x = 1$ and $x = 2$, as shown above. The unit of length on each axis is 1 m. A uniform solid S has the shape made by rotating R through 360° about the x-axis.

a Show that the centre of mass of S is $\frac{2}{7}$ m from its larger plane face. **(8)**

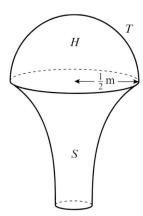

A sporting trophy T is a uniform solid hemisphere H joined to the solid S. The hemisphere has radius $\frac{1}{2}$ m and its plane face coincides with the larger plane face of S, as shown above. Both H and S are made of the same material.

b Find the distance of the centre of mass of T from its plane face. **(6)**

← **Sections 3.1, 3.2**

A 51 a Show, by integration, that the centre of
mass of a uniform solid hemisphere,
of radius R, is at a distance $\frac{3}{8}R$ from its
plane face.

The diagram shows a uniform solid top
made from a right circular cone of base
radius a and height ka and a hemisphere
of radius a. The circular plane faces of
the cone and hemisphere are coincident.
(6)

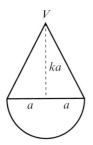

b Show that the distance of the centre of
mass of the top from the vertex V of
the cone is
$$\frac{(3k^2 + 8k + 3)a}{4(k + 2)}$$
(5)

The manufacturer requires the top to
have its centre of mass situated at the
centre of the coincident plane faces.

c Find the value of k for this
requirement. **(5)**

← Section 3.2

E/P 52 A bowl consists of a uniform solid metal
hemisphere, of radius a and centre
O, from which is removed the solid
hemisphere of radius $\frac{1}{2}a$ with the same
centre O.

a Show that the distance of the centre of
mass of the bowl from O is $\frac{45}{112}a$. **(6)**

The bowl is fixed with its plane face
uppermost and horizontal. It is now filled
with liquid. The mass of the bowl is M
and the mass of the liquid is kM, where k
is a constant. Given that the distance of
the centre of mass of the bowl and liquid
together from O is $\frac{17}{48}a$,

b find the value of k. **(6)**

← Section 3.3

A 53

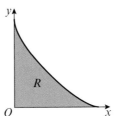

The shaded region R is bounded by part
of the curve with equation $y = \frac{1}{2}(x - 2)^2$,
the x-axis and the y-axis, as shown above.
The unit of length on both axes is 1 cm.
A uniform solid S is made by rotating
R through 360° about the x-axis. Using
integration,

a calculate the volume of the solid S,
leaving your answer in terms of π **(4)**

b show that the centre of mass of S
is $\frac{1}{3}$ cm from its plane face. **(7)**

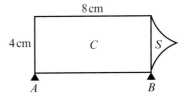

A tool is modelled as having two
components, a solid uniform cylinder
C and the solid S. The diameter of C is
4 cm and the length of C is 8 cm. One
end of C coincides with the plane face
of S. The components are made of
different materials. The weight of C is
$10W$ newtons and the weight of S is $2W$
newtons. The tool lies in equilibrium with
its axis of symmetry horizontal on two
smooth supports A and B, which are at
the ends of the cylinder, as shown above.

c Find the magnitude of the force of
the support A on the tool. **(5)**

← Sections 3.2, 3.4

A **54**

E

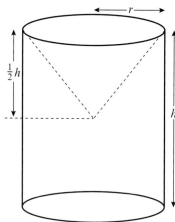

An ornament S is formed by removing a solid right circular cone, of radius r and height $\frac{1}{2}h$, from a solid uniform cylinder, of radius r and height h, as shown in the diagram.

a Show that the distance of the centre of mass of S from its plane face is $\frac{17}{40}h$. **(6)**

The ornament is suspended from a point on the circular rim of its open end. It hangs in equilibrium with its axis of symmetry inclined at an angle α to the horizontal. Given that $h = 4r$,

b find, in degrees to one decimal place, the value of α. **(4)**

← Sections 3.2, 3.4

E/P **55** **a** A uniform triangular lamina XYZ has $XY = XZ$ and the perpendicular distance of X from YZ is h. Prove, by integration, that the centre of mass of the lamina is at a distance $\frac{2h}{3}$ from X.

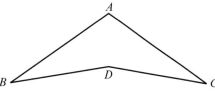

A uniform triangular lamina ABC has $AB = AC = 5a$, $BC = 8a$ and D is the centre of mass of the lamina. The triangle BCD is removed from the lamina, leaving the plate $ABDC$ shown in the diagram.

A

b Show that the distance of the centre of mass of the plate from A is $\frac{5a}{3}$. **(6)**

The plate, which is of mass M, has a particle of mass M attached at B. The loaded plate is suspended from C and hangs in equilibrium.

c Prove that in this position CB makes an angle of $\arctan\frac{1}{9}$ with the vertical. **(4)**

← Sections 3.2, 3.4

E/P **56** A closed container C consists of a thin uniform hollow hemispherical bowl of radius a, together with a lid. The lid is a thin uniform circular disc, also of radius a. The centre O of the disc coincides with the centre of the hemispherical bowl. The bowl and its lid are made of the same material.

a Show that the centre of mass of C is at a distance $\frac{1}{3}a$ from O. **(4)**

The container C has mass M. A particle of mass $\frac{1}{2}M$ is attached to the container at a point P on the circumference of the lid. The container is then placed with a point of its curved surface in contact with a horizontal plane. The container rests in equilibrium with P, O and the point of contact in the same vertical plane.

b Find, to the nearest degree, the angle made by the line PO with the horizontal. **(5)**

← Sections 3.2, 3.4

E/P **57**

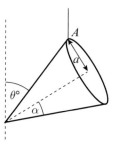

A uniform solid right circular cone has base radius a and semi-vertical angle α, where $\tan\alpha = \frac{1}{3}$. The cone is freely suspended by a string attached at a point A on the rim of its base, and hangs in

A equilibrium with its axis of symmetry making an angle of $\theta°$ with the upward vertical, as shown in the diagram. Find, to one decimal place, the value of θ. **(4)**

← Sections 3.2, 3.4

E/P **58** A uniform solid hemisphere H has base radius a and the centre of its plane circular face is C.

The plane face of a second hemisphere K, of radius $\frac{a}{2}$, and made of the same material as H, is stuck to the plane face of H, so that the centres of the two plane faces coincide at C, to form a uniform composite body S.

a Given that the mass of K is M, show that the mass of S is $9M$, and find, in terms of a, the distance of the centre of mass of the body S from C. **(5)**

A particle P, of mass M, is attached to a point on the edge of the circular face of H of the body S. The body S with P attached is placed with a point of the curved surface of the part H in contact with a horizontal plane and rests in equilibrium.

b Find the tangent of the acute angle made by the line PC with the horizontal. **(5)**

← Sections 3.2, 3.4

E **59 a** Prove, by integration, that the position of the centre of mass of a uniform solid right circular cone is one quarter of the way up the axis from the base. **(8)**

A solid is formed by removing a solid cone of height h and radius a from a solid cone of height H and radius a. The axes of the two cones coincide.

b Show that the centre of mass of the remaining solid S is a distance

$$\tfrac{1}{4}(3H - h)$$

from the vertex of the original cone. **(10)**

A The solid S is suspended by two vertical strings, one attached to the vertex and the other attached to a point on the bounding circular base.

c Given that S is in equilibrium, with its axis of symmetry horizontal, find, in terms of H and h, the ratio of the magnitude of the tension in the string attached to the vertex to that in the other string. **(4)**

← Sections 3.2, 3.4

E **60**

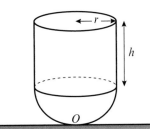

A child's toy consists of a uniform solid hemisphere attached to a uniform solid cylinder. The plane face of the hemisphere coincides with the plane face of the cylinder, as shown in the diagram above. The cylinder and the hemisphere each have radius r and the height of the cylinder is h. The material of the hemisphere is six times as dense as the material of the cylinder. The toy rests in equilibrium on a horizontal plane with the cylinder above the hemisphere and the axis of the cylinder vertical.

a Show that the distance d of the centre of mass of the toy from its lowest point O is given by

$$d = \frac{h^2 + 2hr + 5r^2}{2(h + 4r)} \qquad \textbf{(7)}$$

When the toy is placed with any point of the curved surface of the hemisphere resting on the plane it will remain in equilibrium.

b Find h in terms of r. **(3)**

← Sections 3.3, 3.4

A **61**
E

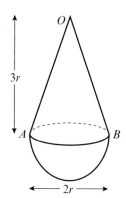

A child's toy consists of a uniform solid hemisphere, of mass M and base radius r, joined to a uniform solid right circular cone of mass m, where $2m < M$. The cone has vertex O, base radius r and height $3r$. Its plane face, with diameter AB, coincides with the plane face of the hemisphere, as shown in the diagram above.

a Show that the distance of the centre of mass of the toy from AB is

$$\frac{3(M-2m)}{8(M+m)}r \qquad \textbf{(5)}$$

The toy is placed with OA on a horizontal surface. The toy is released from rest and does not remain in equilibrium.

b Show that $M > 26m$ **(4)**

← Sections 3.3, 3.4

E/P **62**

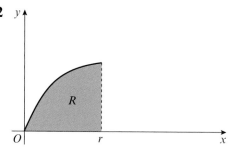

The diagram shows the region R bounded by the curve with equation $y^2 = rx$, where r is a positive constant, the x-axis and the line $x = r$. A uniform solid of revolution S is formed by rotating R through one complete revolution about the x-axis.

a Show that the distance of the centre of mass of S from O is $\frac{2}{3}r$. **(6)**

A The solid is placed with its plane face on a plane which is inclined at an angle α to the horizontal. The plane is sufficiently rough to prevent S from sliding. Given that S does not topple,

b find, to the nearest degree, the maximum value of α. **(4)**

← Sections 3.2, 3.5

E **63**

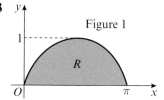

Figure 1

A uniform lamina occupies the region R bounded by the x-axis and the curve $y = \sin x$, $0 \leqslant x \leqslant \pi$, as shown in Figure 1.

a Show, by integration, that the y-coordinate of the centre of mass of the lamina is $\frac{\pi}{8}$ **(6)**

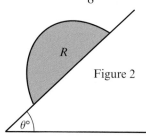

Figure 2

A uniform prism S has cross section R. The prism is placed with its rectangular face on a table which is inclined at an angle θ to the horizontal. The cross section R lies in a vertical plane as shown in Figure 2. The table is sufficiently rough to prevent S sliding. Given that S does not topple,

b find the largest possible value of θ. **(3)**

← Sections 3.2, 3.5

64

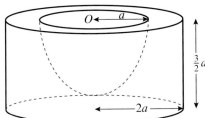

A uniform solid cylinder has radius $2a$ and height $\frac{3}{2}a$. A hemisphere of radius a is removed from the cylinder. The plane face of the hemisphere coincides with the upper plane face of the cylinder, and the centre O of the hemisphere is also the centre of this plane face, as shown in the diagram above. The remaining solid is S.

a Find the distance of the centre of mass of S from O. **(6)**

The lower plane face of S rests in equilibrium on a desk lid which is inclined at an angle θ to the horizontal. Assuming that the lid is sufficiently rough to prevent S from slipping, and that S is on the point of toppling when $\theta = \alpha$,

b find the value of α. **(3)**

Given instead that the coefficient of friction between S and the lid is 0.8, and that S is on the point of sliding down the lid when $\theta = \beta$,

c find the value of β. **(3)**

← Sections 3.2, 3.5

 65

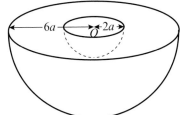

A uniform solid hemisphere, of radius $6a$ and centre O, has a solid hemisphere of radius $2a$, and centre O, removed to form a bowl B as shown above.

a Show that the centre of mass of B is $\frac{30}{13}a$ from O. **(6)**

A

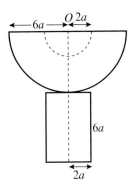

The bowl B is fixed to a plane face of a uniform solid cylinder made from the same material as B. The cylinder has radius $2a$ and height $6a$ and the combined solid S has an axis of symmetry which passes through O, as shown.

b Show that the centre of mass of S is $\frac{201}{61}a$ from O. **(6)**

The plane surface of the cylindrical base of S is placed on a rough plane inclined at $12°$ to the horizontal. The plane is sufficiently rough to prevent slipping.

c Determine whether or not S will topple. **(4)**

← Sections 3.2, 3.5

E/P **66**

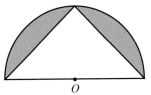

The diagram shows a cross section of a solid formed by the removal of a right circular cone, of base radius a and height a, from a uniform solid hemisphere of base radius a. The plane bases of the cone and the hemisphere are coincident, both having centre O.

a Show that G, the centre of mass of the solid, is at a distance $\frac{a}{2}$ from O. **(5)**

141

A

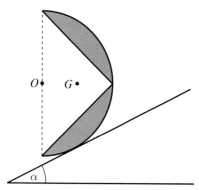

The second diagram shows a cross section of the solid resting in equilibrium with a point of its curved surface in contact with a rough inclined plane of inclination α. Given that O and G are in the same vertical plane through a line of greatest slope of the inclined plane, and that OG is horizontal,

b show that $\alpha = \dfrac{\pi}{6}$ **(4)**

Given that $\alpha = \dfrac{\pi}{6}$,

c find the smallest possible value of the coefficient of friction between the solid and the plane. **(4)**

← **Section 3.5**

E **67**

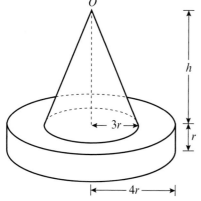

An experimental plastic traffic bollard B is made by joining a uniform solid cylinder to a uniform solid right circular cone of the same density. They are joined to form a symmetrical solid, in such a way that the centre of the plane face of the cone coincides with the centre of one of the plane faces of the cylinder, as shown in the diagram.

A

The cylinder has radius $4r$ and height r. The cone has vertex O, base radius $3r$ and height h.

a Show that the distance from O of the centre of mass of B is

$$\frac{32r^2 + 64rh + 9h^2}{4(16r + 3h)}$$ **(6)**

The bollard is placed on a rough plane which is inclined at an angle α to the horizontal. The circular base of B is in contact with the inclined plane. Given that $h = 4r$ and that B is on the point of toppling,

b find α, to the nearest degree. **(4)**

← **Sections 3.2, 3.5**

E **68**

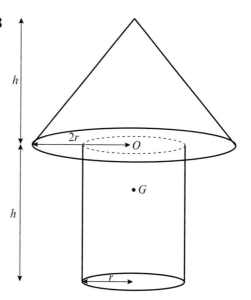

A model tree is made by joining a uniform solid cylinder to a uniform solid cone made of the same material. The centre O of the base of the cone is also the centre of one end of the cylinder, as shown in the diagram. The radius of the cylinder is r and the radius of the base of the cone is $2r$. The height of the cone and the height of the cylinder are each h. The centre of mass of the model is at the point G.

a Show that $OG = \frac{1}{14}h$. **(8)**

A

The model stands on a desk top with its plane face in contact with the desk top. The desk top is tilted until it makes an angle α with the horizontal, where $\tan \alpha = \frac{7}{26}$. The desk top is rough enough to prevent slipping and the model is about to topple.

b Find r in terms of h. **(4)**

← **Sections 3.2, 3.5**

E/P **69**

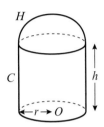

A body consists of a uniform solid circular cylinder C, together with a uniform solid hemisphere H which is attached to C. The plane face of H coincides with the upper plane face of C, as shown in the diagram. The cylinder C has base radius r, height h and mass $3M$. The mass of H is $2M$. The Sections O is the centre of the base of C.

a Show that the distance of the centre of mass of the body from O is

$$\frac{14h + 3r}{20} \quad \textbf{(5)}$$

The body is placed with its plane face on a rough plane which is inclined at an angle α to the horizontal, where $\tan \alpha = \frac{4}{3}$. The plane is sufficiently rough to prevent slipping. Given that the Sections is on the point of toppling,

b find h in terms of r. **(4)**

← **Sections 3.3, 3.5**

A **70**
E/P

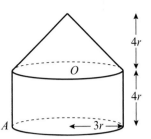

A toy is formed by joining a uniform solid right circular cone, of base radius $3r$ and height $4r$, to a uniform solid cylinder, also of radius $3r$ and height $4r$. The cone and cylinder are made from different materials, and the density of the cone is three times the density of the cylinder. The plane face of the cone coincides with a plane face of the cylinder, as shown in the diagram. The centre of this plane face is O.

a Find the distance of the centre of mass of the toy from O. **(6)**

The point A lies on the edge of the plane face of the cylinder which forms the base of the toy. The toy is suspended from A and hangs in equilibrium.

b Find, in degrees to one decimal place, the angle between the axis of symmetry of the toy and the vertical. **(4)**

The toy is placed with the curved surface of the cone on horizontal ground.

c Determine whether the toy will topple. **(4)**

← **Sections 3.3, 3.4, 3.5**

E **71** A non-uniform rod is 1.5 m long. At any distance h cm from its base, the mass per unit length of the rod is given by

$$m = \frac{3}{(1 + h)(2 + h)} \, \text{kg m}^{-1}. \text{ Find:}$$

a the mass of the rod **(4)**

b the distance of the centre of mass of the rod from its base. **(4)**

← **Sections 3.3, 3.4, 3.5**

A **72** A wooden oar is 2.4 m long.

E/P The oar is non-uniform rod, so that, at a distance x metres from its end, the mass per unit length of the rod, m kg m^{-1} is given by the formula $m = \dfrac{5}{1 + 4x^2}$

Find, to 2 decimal places:

a the mass of the oar **(4)**

b the distance of the centre of mass of the oar from its end. **(4)**

← **Section 3.3**

E/P **73** A non-uniform rod AB of length 20 m is such that, at a distance x m from the A, the mass per unit length of the rod is given by $(10 + kx)$ kg m^{-1}, where k is a positive constant.

a Without calculation, explain why the centre of mass of the rod will be closer to B than to A. **(1)**

Given that the mass of the rod is 750 kg, find:

b the value of k **(3)**

c the distance of the centre of mass of the rod from A. **(4)**

← **Section 3.3**

E/P **74** A non-uniform rod AB is suspended horizontally from two strings attached at A and B respectively. The two strings hang vertically. At a distance x m from A the mass per unit length of the rod is given by $(8 + x^2)$ kg m^{-1}

Given that the tension in the string at A is half the tension in the string at B, find the exact length of the rod. **(12)**

← **Sections 3.3, 3.4**

Challenge

1 The diagram shows two beads P_1 and P_2, of masses $5m$ and $7m$ respectively, threaded onto a smooth circular horizontal ring. The beads are projected in opposite directions at the same speed.

The beads collide with coefficient of restitution e at the point A, then collide again at the point B, where $\angle AOB = 90°$, as shown in the diagram.

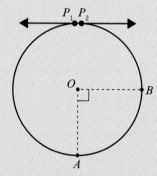

Find the value of e.

← **FM1, Chapter 4**

2 A roulette wheel is modelled as a circular ring of radius R resting on a smooth horizontal surface. A ball of mass m is held against the ring, and projected tangentially along the inside of the ring with initial speed u. The coefficient of friction between the ball and the ring is μ.

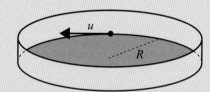

a Show that t seconds after the ball is projected its speed is $\dfrac{uR}{R + u\mu t}$.

Given that $R = 0.5$ m, $u = 40$ m s^{-1} and $\mu = 0.25$,

b find the time taken for the ball to complete its first complete revolution of the ring.

← **Section 1.2**

3 A piece of card is in the shape of an isosceles triangle ABC of mass $4M$ and side length $AB = 10$ cm, $BC = CA = 15$ cm. The triangle is folded so that vertex C sits on the midpoint of AB, as shown in the diagram.

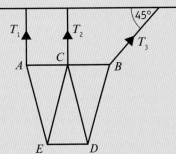

The lamina is suspended by two vertical strings attached at A and C and a string angled at $45°$ to the horizontal attached at B, causing AB to lie horizontally.

a Work out, in terms of M and g, the acceleration due to gravity, the values of T_1, T_2 and T_3.

A mass of $10M$ is attached to the lamina at B causing the strings attached at B and C to snap.

b Work out the angle AB makes with the vertical when the lamina has come to rest in equilibrium.

← **Section 2.8**

4 A non-uniform solid hemisphere has radius r cm. At a distance x cm from its plane face, the density of the hemisphere is $(5x + 2)$ g cm^{-3}.

The hemisphere is placed with is plane face on a rough slope inclined at an angle $\arctan 2$ to the horizontal.

Given that the hemisphere is on the point of tipping, find the exact value of r.

← **Sections 3.3, 3.4, 3.5**

4

Kinematics

Prior knowledge check

1   Integrate with respect to x

     a $\dfrac{8}{(2-3x)^3}$ **b** $4e^{3x}$ **c** $\sin 5\pi x$

     ← Pure Year 2, Sections 11.2, 11.4

2   $\dfrac{dy}{dx} = \dfrac{y}{(x+2)^2}$

     Find y in terms of x given $y = 3$ when
     $x = 1$ ← Pure Year 2, Section 11.10

3   Given that $\displaystyle\int_{1}^{2} \dfrac{1}{x^2+x}\,dx = \ln k$, find k,
     where k is a rational constant to be
     found.   ← Pure Year 2, Section 11.7

An object moving through a fluid, such as
water or air, experiences a frictional force
called **drag**, which increases as the object
moves faster. You can use differential
equations to solve problems where
acceleration is a function of velocity.

4.1 Acceleration varying with time

You can use calculus for a particle moving in a straight line with acceleration that varies with time.

- **To find the velocity from the displacement, you differentiate with respect to time.**
  To find the acceleration from the velocity, you differentiate with respect to time.

$$v = \frac{dx}{dt} \text{ and } a = \frac{dv}{dt} = \frac{d^2x}{dt^2}$$

- **To obtain the velocity from the acceleration, you integrate with respect to time.**
  To obtain the displacement from the velocity, you integrate with respect to time.

$$v = \int a \, dt \text{ and } x = \int v \, dt$$

These relationships are summarised in the following diagram.

```
        ┌────────────────────────┐
        │   Displacement (x)     │
        └────────────────────────┘
Differentiate ▼          ▲ Integrate
        ┌────────────────────────┐
        │     Velocity (v)       │
        └────────────────────────┘
Differentiate ▼          ▲ Integrate
        ┌────────────────────────┐
        │   Acceleration (a)     │
        └────────────────────────┘
```

Watch out When you integrate, it is important that you remember to include a constant of integration. Many questions include information which enables you to find the value of this constant.

Links If you are given $a = f(t)$ you can use direct integration to find expressions for v or x in terms of t. ← **Statistics and Mechanics 2, Chapter 8**

Example 1

A particle P starts from rest at a point O and moves along a straight line. At time t seconds the acceleration, $a \, \text{m s}^{-2}$, of P is given by

$$a = \frac{6}{(t + 2)^2}, \qquad t \geqslant 0.$$

a Find the velocity of P at time t seconds.

b Show that the displacement of P from O when $t = 6$ is $(18 - 12 \ln 2)\,\text{m}$.

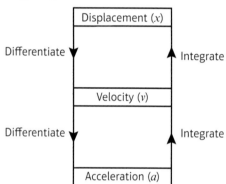

a $a = 6(t + 2)^{-2}$

$v = \int a \, dt = \int 6(t + 2)^{-2} \, dt = \frac{6(t + 2)^{-1}}{-1} + A$

$\quad = A - \frac{6}{t + 2}$

When $t = 0$, $v = 0$

$\quad 0 = A - \frac{6}{2} \Rightarrow A = 3$

$\quad v = 3 - \frac{6}{t + 2}$

To integrate, write $\frac{1}{(t + 2)^2}$ as $(t + 2)^{-2}$.

Find the velocity by integrating the acceleration with respect to time. It is important to include a constant of integration. The question includes information which enables you to find this constant.

P starts from rest. This means that $v = 0$ when $t = 0$. This initial condition enables you to find the constant of integration.

The velocity of P at time t seconds is

$\left(3 - \dfrac{6}{t+2}\right)$ m s^{-1}.

b Let the displacement of P from O at time t seconds be s metres.

$$s = \int v \, dt = \int \left(3 - \frac{6}{t+2}\right) dt$$

$$= 3t - 6\ln(t+2) + B$$

When $t = 0$, $s = 0$

$$0 = -6\ln 2 + B \Rightarrow B = 6\ln 2$$

$$s = 3t - 6\ln(t+2) + 6\ln 2$$

When $t = 6$

$$s = 18 - 6\ln 8 + 6\ln 2$$

$$= 18 - 6\ln\left(\frac{8}{2}\right) = 18 - 6\ln 4 = 18 - 12\ln 2$$

The displacement of P from O when $t = 6$ is $(18 - 12\ln 2)$ m, as required.

> Find the displacement by integrating the velocity with respect to time. Use a different letter for the constant of integration.

> As P starts at O, $s = 0$ when $t = 0$. This enables you to find the second constant of integration.

> Use the laws of logarithms to simplify your answer into the form asked for in the question. This can be done in more than one way. The working shown here uses
> $$\ln 8 - \ln 2 = \ln\left(\frac{8}{2}\right) = \ln 4$$
> and
> $$\ln 4 = \ln 2^2 = 2\ln 2.$$

← **Pure Year 1, Chapter 14**

Example 2

A particle P is moving along the x-axis. At time $t = 0$, the particle is at the origin O and is moving with speed 2 m s^{-1} in the direction Ox. At time t seconds, where $t \geqslant 0$, the acceleration of P is $4e^{-0.5t}$ m s^{-2} directed away from O.

a Find the velocity of P at time t seconds.

b Show that the speed of P cannot exceed 10 m s^{-1}.

c Sketch a velocity–time graph to illustrate the motion of P.

a $a = 4e^{-0.5t}$

Let the velocity of P at time t seconds be v m s^{-1}.

$$v = \int a \, dt = \int 4e^{-0.5t} \, dt$$

$$= -8e^{-0.5t} + C$$

When $t = 0$, $v = 2$

$$2 = -8 + C \Rightarrow C = 10$$

$$v = 10 - 8e^{-0.5t}$$

The velocity of P at time t seconds is $(10 - 8e^{-0.5t})$ m s^{-1}.

b For all x, $e^x > 0$ and so for all t, $8e^{-0.5t} > 0$.

It follows that $10 - 8e^{-0.5t} < 10$ for all t.

Hence, the speed of P cannot exceed 10 m s^{-1}.

> Use the rule $\int e^{kt} \, dt = \dfrac{1}{k}e^{kt} + C$
>
> Remember to include a constant of integration.

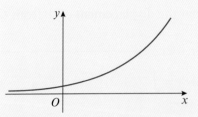

> This sketch of $y = e^x$ illustrates that $e^x > 0$, for all real values of x; both positive and negative.

> 10 minus a positive number must be less than 10.

c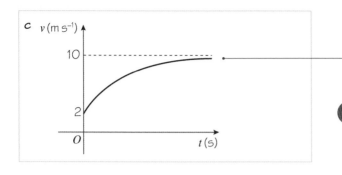

The curve approaches the line $v = 10$ but does not reach it. The line is an asymptote to the curve.

Notation

The velocity 10 m s^{-1} is the **terminal** or **limiting velocity** of P.

Example 3

A particle P is moving along the x-axis. At time t seconds, the velocity of P is $v \text{ m s}^{-1}$, where $v = 4 \sin(2\pi t)$. When $t = 0$, P is at O. Find:

a the magnitude of the acceleration of P when $t = \frac{2}{3}$

b the greatest distance from O attained by P during the motion.

a Let the acceleration of P at time t seconds be $a \text{ m s}^{-2}$.

$$a = \frac{dv}{dt} = 8\pi \cos(2\pi t)$$

When $t = \frac{2}{3}$

$$a = 8\pi \cos\left(\frac{4\pi}{3}\right) = 8\pi \times -\frac{1}{2} = -4\pi$$

The magnitude of the acceleration of P when $t = \frac{2}{3}$ is $4\pi \text{ m s}^{-2}$.

Find the acceleration by differentiating the velocity with respect to time.

$$\frac{d}{dt}(\sin kt) = k \cos kt$$

Here $k = 2\pi$.

When differentiating and integrating trigonometric functions, angles will always be measured in radians: $\cos\left(\frac{4\pi}{3}\right) = -\frac{1}{2}$

b Let the displacement of P at time t seconds be x metres.

$$x = \int v \, dt = -\frac{4}{2\pi} \cos(2\pi t) + C$$

$$= -\frac{2}{\pi} \cos(2\pi t) + C$$

When $t = 0$, $x = 0$

$$0 = -\frac{2}{\pi} + C \Rightarrow C = \frac{2}{\pi}$$

$$x = \frac{2}{\pi}(1 - \cos(2\pi t))$$

The greatest value of x occurs when $\cos(2\pi t) = -1$

The greatest value of x is $\frac{2}{\pi}(1 - (-1)) = \frac{4}{\pi}$

The greatest distance from O attained by P during the motion is $\frac{4}{\pi}$ m.

Find the displacement by integrating the velocity with respect to time. Use the initial condition in the question to find the constant of integration.

When $t = 0$, $\cos(2\pi t) = \cos 0 = 1$

The cosine of any function varies between $+1$ and -1 and so $1 - \cos(2\pi t)$ varies between 0 and 2. Its greatest value is therefore 2. You do not need to use calculus in this part of the question.

Example 4

A particle P is moving along the x-axis. Initially P is at the origin O. At time t seconds (where $t \geq 0$) the velocity, $v\,\text{m s}^{-1}$, of P is given by $v = te^{-\frac{t}{4}}$. Find the distance of P from O when the acceleration of P is zero.

$a = \dfrac{dv}{dt} = e^{-\frac{t}{4}} - \tfrac{1}{4}te^{-\frac{t}{4}} = e^{-\frac{t}{4}}\left(1 - \tfrac{1}{4}t\right)$

When $a = 0$, as $e^{-\frac{t}{4}} \neq 0$,

$1 - \tfrac{1}{4}t = 0 \Rightarrow t = 4$

$x = \displaystyle\int v\,dt = \int te^{-\frac{t}{4}}\,dt$

$\quad = -4te^{-\frac{t}{4}} + \displaystyle\int 4e^{-\frac{t}{4}}\,dt$

$\quad = -4te^{-\frac{t}{4}} - 16e^{-\frac{t}{4}} + A$

$\quad = A - e^{-\frac{t}{4}}(4t + 16)$

When $t = 0$, $x = 0$

$0 = A - 16 \Rightarrow A = 16$

Hence

$x = 16 - e^{-\frac{t}{4}}(4t + 16)$

When $t = 4$

$x = 16 - e^{-1}(4 \times 4 + 16) = 16(1 - 2e^{-1})$

When the acceleration of P is zero,

$OP = 16(1 - 2e^{-1})\,\text{m}$.

The first step is to find the value of t for which the acceleration is zero. Find the acceleration by differentiating the velocity using the product rule

$\dfrac{d}{dt}(uv) = v\dfrac{du}{dt} + u\dfrac{dv}{dt}$

with $u = t$ and $v = e^{-\frac{t}{4}}$.

Find the displacement by integrating the velocity with respect to time. You need to use integration by parts

$\displaystyle\int u\dfrac{dv}{dt}\,dt = uv - \int v\dfrac{du}{dt}\,dt$

with $u = t$ and $\dfrac{dv}{dt} = e^{-\frac{t}{4}}$.

Use the information that P is initially at the origin to find the value of the constant of integration A.

Substitute $t = 4$ into your expression for the displacement.

Example 5

A particle is moving along the x-axis. At time t seconds the velocity of P is $v\,\text{m s}^{-1}$ in the direction of x increasing, where

$$v = \begin{cases} 2t, & 0 \leq t \leq 2 \\ 2 + \dfrac{4}{t}, & t > 2 \end{cases}$$

When $t = 0$, P is at the origin O.

a Sketch a velocity–time graph to illustrate the motion of P in the interval $0 \leq t \leq 5$.

b Find the distance of P from O when $t = 5$.

a

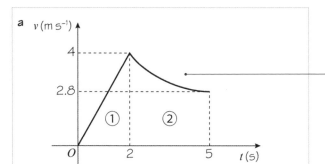

For $v = 2t$, when $t = 0$, $v = 0$ and when $t = 2$, $v = 4$. The graph is the line segment joining $(0, 0)$ to $(2, 4)$.

For $v = 2 + \dfrac{4}{t}$, when $t = 2$, $v = 4$

and when $t = 5$, $v = 2.8$. The graph is part of a reciprocal curve joining $(2, 4)$ to $(5, 2.8)$.

b The distance moved in the first two seconds is represented by the area labelled ①.

Let this area be A_1.

$A_1 = \frac{1}{2} \times 2 \times 4 = 4$

The distance travelled in the next three seconds is represented by the area labelled ②.

Let this area be A_2.

$A_2 = \displaystyle\int_2^5 \left(2 + \frac{4}{t}\right) dt$

$= [2t + 4 \ln t]_2^5$

$= (10 + 4 \ln 5) - (4 + 4 \ln 2)$

$= 6 + 4(\ln 5 - \ln 2) = 6 + 4 \ln \frac{5}{2}$

The distance of P from O when $t = 5$ is

$4 + 6 + 4 \ln \frac{5}{2} = \left(10 + 4 \ln \frac{5}{2}\right)$ m

The distance moved by P is represented by the area between the graph and the t-axis.

The area labelled ① can be found using the formula for the area of a triangle $\frac{1}{2} \times$ base $\times$ height.

The area labelled ② in the diagram can be found by definite integration.

Integrate the function $2 + \dfrac{4}{t}$ between the limits $t = 2$ and $t = 5$.

Example **6**

A particle P moves on the positive x-axis.

The velocity of P at time t seconds is $(2t^2 - 7t + 3)$ m s^{-1}, $t \geq 0$.

When $t = 0$, P is 10 m from the origin O.

Find:

a the values of t when P is instantaneously at rest

b the displacement of P when $t = 5$

c the total distance travelled by P in the interval $0 \leq t \leq 5$.

a $v = 2t^2 - 7t + 3$

$2t^2 - 7t + 3 = 0$

$(2t - 1)(t - 3) = 0$ so P is instantaneously at rest when $t = 0.5$ and $t = 3$

Set $v = 0$ to find the times when the particle is instantaneously at rest.

P is instantaneously at rest when $v = 0$.

b $s = \int (2t^2 - 7t + 3)\,dt = \frac{2}{3}t^3 - \frac{7}{2}t^2 + 3t + C$

When $t = 0$, $s = 10$ so $C = 10$

$s = \frac{2}{3}t^3 - \frac{7}{2}t^2 + 3t + 10$

When $t = 5$:

$s = \left(\frac{2}{3} \times 5^3\right) - \left(\frac{7}{2} \times 5^2\right) + (3 \times 5) + 10 = \frac{125}{6} = 20.83\,\text{m}$

c Velocity–time graph for the motion of the particle:

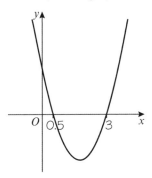

$s = \int (2t^2 - 7t + 3)\,dt$

Between $t = 0$ and $t = \frac{1}{2}$:

$\left[\frac{2}{3}t^3 - \frac{7}{2}t^2 + 3t\right]_0^{\frac{1}{2}}$

$= \left(\frac{2}{3} \times \left(\frac{1}{2}\right)^3 - \frac{7}{2} \times \left(\frac{1}{2}\right)^2 + 3 \times \frac{1}{2}\right) - 0 = \frac{17}{24}$

Between $t = \frac{1}{2}$ and $t = 3$:

$\left[\frac{2}{3}t^3 - \frac{7}{2}t^2 + 3t\right]_{\frac{1}{2}}^3$

$= \left(\left(\frac{2}{3} \times 3^3\right) - \left(\frac{7}{2} \times 3^2\right) + (3 \times 3)\right)$

$\quad - \left(\frac{2}{3} \times \left(\frac{1}{2}\right)^3 - \frac{7}{2} \times \left(\frac{1}{2}\right)^2 + 3 \times \frac{1}{2}\right)$

$= \left(-\frac{9}{2} - \frac{17}{24}\right) = -\frac{125}{24}$

Between $t = 3$ and $t = 5$:

$\left[\frac{2}{3}t^3 - \frac{7}{2}t^2 + 3t\right]_3^5$

$= \left(\left(\frac{2}{3} \times 5^3\right) - \left(\frac{7}{2} \times 5^2\right) + (3 \times 5)\right)$

$\quad - \left(\left(\frac{2}{3} \times 3^3\right) - \left(\frac{7}{2} \times 3^2\right) + (3 \times 3)\right)$

$= \frac{65}{6} - \left(-\frac{9}{2}\right) = \frac{46}{3}$

Total distance travelled by P in the interval $0 \leqslant t \leqslant 5$:

$\frac{17}{24} + \frac{125}{24} + \frac{46}{3} = 21.25\,\text{m}$

Integrate the expression for velocity to obtain the displacement.

Use the initial conditions given to find the value of C.

This is the expression for the **displacement** of P from the origin at time t.

Problem-solving

The particle **changes direction** twice in the interval $0 \leqslant t \leqslant 5$. If you were to find $\int_0^5 v\,dt$ you would be working out the **displacement** of the particle at time $t = 5$ from its position at time $t = 0$. To work out the **distance travelled** you need to find the total area enclosed by the velocity–time graph and the x-axis. Sketch the graph to show the critical points, and work out three separate integrals.

Distance travelled

$= \int_0^{0.5} v\,dt - \int_{0.5}^3 v\,dt + \int_3^5 v\,dt$

The negative term arises because the definite integral will be negative for an area below the x-axis.

The distance travelled between $t = \frac{1}{2}$ and $t = 3$ is $\frac{125}{24}$ m.

Exercise 4A

1 A particle P is moving in a straight line. Initially P is moving through a point O with speed $4\,\text{m s}^{-1}$. At time t seconds after passing through O the acceleration of P is $3\text{e}^{-0.25t}\,\text{m s}^{-2}$ in the direction OP. Find the velocity of the particle at time t seconds.

2 A particle P is moving along the x-axis in the direction of x increasing. At time t seconds, the velocity of P is $t\sin t\,\text{m s}^{-1}$. When $t = 0$, P is at the origin. Show that when $t = \dfrac{\pi}{2}$, P is 1 metre from O.

3 At time t seconds the velocity, $v\,\text{m s}^{-1}$, of a particle moving in a straight line is given by
$$v = \frac{4}{3 + 2t}, \qquad t \geqslant 0.$$
When $t = 0$, the particle is at a point A. When $t = 3$, the particle is at the point B. Find the distance between A and B.

4 A particle P is moving along the x-axis in the positive direction. At time t seconds the acceleration of P is $4\text{e}^{\frac{1}{2}t}\,\text{m s}^{-2}$ in the positive direction. When $t = 0$, P is at rest. Find the distance P moves in the interval $0 \leqslant t \leqslant 2$. Give your answer to 3 significant figures.

5 A particle P is moving along the x-axis. At time t seconds the displacement of P from O is $x\,\text{m}$ and the velocity of P is $4\cos 3t\,\text{m s}^{-1}$, both measured in the direction Ox. When $t = 0$ the particle P is at the origin O. Find:
 a the magnitude of the acceleration when $t = \dfrac{\pi}{12}$
 b x in terms of t
 c the smallest positive value of t for which P is at O.

6 A particle P is moving along a straight line. Initially P is at rest. At time t seconds P has velocity $v\,\text{m s}^{-1}$ and acceleration $a\,\text{m s}^{-2}$ where
$$a = \frac{6t}{(2 + t^2)^2}, \qquad t \geqslant 0.$$
Find v in terms of t.

(P) 7 A particle P is moving along the x-axis. At time t seconds the velocity of P is $v\,\text{m s}^{-1}$ in the direction of x increasing, where
$$v = \begin{cases} 4, & 0 \leqslant t \leqslant 3 \\ 5 - \dfrac{3}{t}, & 3 < t \leqslant 6 \end{cases}$$
When $t = 0$, P is at the origin O.
 a Sketch a velocity–time graph to illustrate the motion of P in the interval $0 \leqslant t \leqslant 6$.
 b Find the displacement of P from O when $t = 6$.

8 A particle P is moving in a straight line with acceleration $\sin \frac{1}{2} t \, \text{m s}^{-2}$ at time t seconds, $t \geqslant 0$. The particle is initially at rest at a point O. Find:

a the speed of P when $t = 2\pi$

b the displacement of P from O when $t = \dfrac{\pi}{2}$

(E/P) **9** A particle P is moving along the x-axis. At time t seconds P has velocity $v \, \text{m s}^{-1}$ in the direction x increasing and an acceleration of magnitude $4e^{0.2t} \, \text{m s}^{-2}$ in the direction x decreasing. When $t = 0$, P is moving through the origin with velocity $20 \, \text{m s}^{-1}$ in the direction x increasing. Find:

a v in terms of t **(3 marks)**

b the maximum value of x attained by P during its motion. **(3 marks)**

(P) **10** A car is travelling along a straight road. As it passes a sign S, the driver applies the brakes. The car is modelled as a particle. At time t seconds the car is x m from S and its velocity, $v \, \text{m s}^{-1}$, is modelled by the equation $v = \dfrac{3200}{c + kt}$, where c and k are constants.

Given that when $t = 0$, the speed of the car is $40 \, \text{m s}^{-1}$ and its deceleration is $0.5 \, \text{m s}^{-2}$, find:

a the value of c and the value of k

b x in terms of t.

(P) **11** A particle P is moving along a straight line. When $t = 0$, P is passing through a point A. At time t seconds after passing through A the velocity, $v \, \text{m s}^{-1}$, of P is given by

$$v = e^{2t} - 11e^t + 15t$$

Find:

a the values of t for which the acceleration is zero

b the displacement of P from A when $t = \ln 3$.

(P) **12** A particle P moves along a straight line. At time t seconds (where $t > 0$) the velocity of P is $(2t + \ln(t + 2)) \, \text{m s}^{-1}$. Find:

a the value of t for which the acceleration has magnitude $2.2 \, \text{m s}^{-2}$

b the distance moved by P in the interval $1 \leqslant t \leqslant 4$.

(E/P) **13** A particle P moves on the positive x-axis.

The velocity of P at time t seconds is $(3t^2 - 5t + 2) \, \text{m s}^{-1}$, $t \geqslant 0$

When $t = 0$, P is at the origin O.

Find:

a the values of t when P is instantaneously at rest **(2 marks)**

b the acceleration of P when $t = 5$ **(3 marks)**

c the total distance travelled by P in the interval $0 \leqslant t \leqslant 5$. **(5 marks)**

When $t = 0$, P is at the origin O.

d Show that P never returns to O, explaining your reasoning. **(3 marks)**

> **Problem-solving**
>
> In part **d**, form an expression for the displacement and show that $d \neq 0$ for any value of t except $t = 0$.

14 A particle moving in a straight line starts from rest at the point O at time $t = 0$. At time t seconds, the velocity v m s^{-1} of the particle is given by

$$v = 2t(t - 5), \quad 0 \leqslant t \leqslant 6$$
$$v = \frac{72}{t}, \quad 6 < t \leqslant 12$$

  a Sketch a velocity–time graph for the particle for $0 \leqslant t \leqslant 12$.   **(3 marks)**

  b Find the set of values of t for which the acceleration of the particle is positive.   **(2 marks)**

  c Find the total distance travelled by the particle in the interval $0 \leqslant t \leqslant 12$.   **(5 marks)**

Challenge

A truck travels along a straight road. The truck is modelled as a particle.

At time t seconds, $t \geqslant 2$, the acceleration is given by $\dfrac{60}{kt^2}$ m s^{-2} where k is a positive constant.

When $t = 2$ the truck is at rest and when $t = 5$ the speed of the truck is 9 m s^{-1}.

Show that the speed of the truck never reaches 15 m s^{-1}.

4.2   Acceleration varying with displacement

You can use calculus for a particle moving in a straight line with acceleration that varies with displacement.

When the acceleration of a particle is varying with time, the displacement (x), velocity (v) and acceleration (a) are connected by the relationships

$$a = \frac{dv}{dt} = \frac{d^2x}{dt^2}$$

Using the chain rule for differentiation

$$a = \frac{dv}{dt} = \frac{dv}{dx} \times \frac{dx}{dt}$$

As $v = \dfrac{dx}{dt}$

$$a = \frac{dv}{dx} \times v = v\frac{dv}{dx} \qquad (1)$$

Also, if you differentiate $\frac{1}{2}v^2$ implicitly with respect to x, you obtain

$$\frac{d}{dx}\left(\tfrac{1}{2}v^2\right) = \tfrac{1}{2} \times 2v \times \frac{dv}{dx} = v\frac{dv}{dx} \qquad (2)$$

Combining results (1) and (2), you obtain

■ $$a = v\frac{dv}{dx} = \frac{d}{dx}\left(\tfrac{1}{2}v^2\right)$$

You can use these two forms for acceleration to solve problems where the acceleration of a particle varies with displacement.

A For example, if you have an equation of the form

$$a = f(x)$$

you can write this as

$$\frac{d}{dx}\left(\tfrac{1}{2}v^2\right) = f(x)$$

Integrating both sides of the equation with respect to x,

$$\tfrac{1}{2}v^2 = \int f(x)\,dx$$

Example 7

A particle P is moving on the x-axis in the direction of x increasing. When the displacement of P from the origin O is x m and its speed is v m s^{-1}, the acceleration of P is $2x$ m s^{-2}. When P is at O, its speed is 6 m s^{-1}. Find v in terms of x.

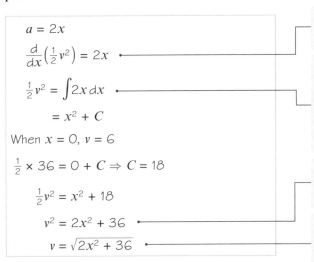

$$a = 2x$$

$$\frac{d}{dx}\left(\tfrac{1}{2}v^2\right) = 2x$$

$$\tfrac{1}{2}v^2 = \int 2x\,dx$$

$$= x^2 + C$$

When $x = 0$, $v = 6$

$$\tfrac{1}{2} \times 36 = 0 + C \Rightarrow C = 18$$

$$\tfrac{1}{2}v^2 = x^2 + 18$$

$$v^2 = 2x^2 + 36$$

$$v = \sqrt{2x^2 + 36}$$

Use $a = \dfrac{d}{dx}\left(\tfrac{1}{2}v^2\right)$

Integrate both sides of the equation with respect to x. As integration is the inverse process of differentiation,

$$\int \frac{d}{dx}\left(\tfrac{1}{2}v^2\right)dx = \tfrac{1}{2}v^2$$

Multiply by 2 throughout the equation and make v the subject of the formula.

As the question tells you that P is moving in the direction of x increasing, you do not need to consider the negative square root.

Example 8

A particle P is moving along a straight line. The acceleration of P, when it has displacement x m from a fixed point O on the line and velocity v m s^{-1}, is of magnitude $4x$ m s^{-2} and is directed towards O. At $x = 0$, $v = 20$. Find the values of x for which P is instantaneously at rest.

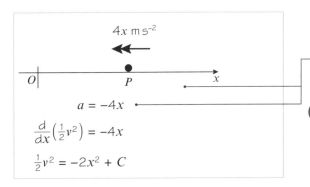

$$a = -4x$$

$$\frac{d}{dx}\left(\tfrac{1}{2}v^2\right) = -4x$$

$$\tfrac{1}{2}v^2 = -2x^2 + C$$

As its acceleration is towards O when P has positive displacement, the acceleration is in the direction of x decreasing, so the acceleration is negative.

Watch out When integrating, you must remember to include a constant of integration. In this question the information that at $x = 0$, $v = 20$ enables you to evaluate the constant.

A

At $x = 0$, $v = 20$

$\frac{1}{2} \times 20^2 = 0 + C \Rightarrow C = 200$

$\frac{1}{2}v^2 = -2x^2 + 200$

$v^2 = 400 - 4x^2$

When $v = 0$

$0 = 400 - 4x^2 \Rightarrow x^2 = 100$

$x = \pm 10$

The values of x for which P is instantaneously at rest are 10 and −10.

> The particle is instantaneously at rest when $v = 0$. Substitute $v = 0$ into this expression and solve the resulting equation for x.

> There are two points at which $v = 0$. The particle reverses direction at these points and will oscillate between them.

Example 9

A particle P is moving along the positive x-axis in the direction of x increasing. When $OP = x$ m, the velocity of P is v m s^{-1} and the acceleration of P is $\left(\dfrac{54}{x^3} - \dfrac{18}{x^5}\right)$ m s^{-2} where $x \geqslant 1$.
Given that $v = 6$ at $x = 1$, find v in terms of x.

$a = \dfrac{54}{x^3} - \dfrac{18}{x^5}$

$\dfrac{d}{dx}\left(\dfrac{1}{2}v^2\right) = \dfrac{54}{x^3} - \dfrac{18}{x^5}$

$= 54x^{-3} - 18x^{-5}$

$\dfrac{1}{2}v^2 = \dfrac{54x^{-2}}{-2} - \dfrac{18x^{-4}}{-4} + C$

$= C - \dfrac{27}{x^2} + \dfrac{9}{2x^4}$

At $x = 1$, $v = 6$

$18 = C - 27 + \dfrac{9}{2} \Rightarrow C = \dfrac{81}{2}$

$\dfrac{1}{2}v^2 = \dfrac{81}{2} - \dfrac{27}{x^2} + \dfrac{9}{2x^4}$

$v^2 = 81 - \dfrac{54}{x^2} + \dfrac{9}{x^4} = \left(9 - \dfrac{3}{x^2}\right)^2$

$v = 9 - \dfrac{3}{x^2}$

> Integrate both sides of this equation with respect to x using $\int \dfrac{d}{dx}\left(\dfrac{1}{2}v^2\right) dx = \dfrac{1}{2}v^2$ and $\int x^n \, dx = \dfrac{x^{n+1}}{n+1}$.
> Remember to include a constant of integration.

> Multiply throughout by 2 and then factorise the right hand side of the equation.

> Ttake the square root of both sides of this equation. As P is moving in the positive direction (the direction of x increasing), you can reject the other square root $v = -\left(9 - \dfrac{3}{x^2}\right)$. This expression is negative for $x \geqslant 1$.

Example 10

A particle P is moving along the x-axis. Initially P is at the origin O and is moving with velocity 1 m s^{-1} in the direction of x increasing. At time t seconds, P is x m from O, has velocity v m s^{-1} and acceleration of magnitude $\frac{1}{2}e^{-x}$ m s^{-2} directed towards O. Find:

a v in terms of x

b x in terms of t.

A

a $a = -\frac{1}{2}e^{-x}$

As the acceleration is directed toward O, it is in the direction of x decreasing and is negative.

$\frac{d}{dx}\left(\frac{1}{2}v^2\right) = -\frac{1}{2}e^{-x}$

$\frac{1}{2}v^2 = \frac{1}{2}e^{-x} + A$

In this example there are two different constants of integration. The initial conditions given in the question enable you to evaluate both constants.

At $x = 0$, $v = 1$

$\frac{1}{2} = \frac{1}{2} + A \Rightarrow A = 0$

$\frac{1}{2}v^2 = \frac{1}{2}e^{-x}$

The question requires you to make v the subject of the formula.

$v^2 = e^{-x}$

$v = e^{-\frac{x}{2}}$

b $\frac{dx}{dt} = e^{-\frac{x}{2}}$

This is a differential equation of the form $\frac{dx}{dt} = f(x)$.

You can solve it by separating the variables.

← **Pure Year 2, Section 11.10**

$e^{\frac{x}{2}}\frac{dx}{dt} = 1$

$\int e^{\frac{x}{2}}dx = \int 1\,dt$

$2e^{\frac{x}{2}} = t + B$

When $t = 0$, $x = 0$

$2 = 0 + B \Rightarrow B = 2$

$2e^{\frac{x}{2}} = t + 2$

$e^{\frac{x}{2}} = \frac{t}{2} + 1$

To make x the subject of this formula, take logarithms on both sides of the equation and use the property that $\ln e^{f(x)} = f(x)$.

$\frac{x}{2} = \ln\left(\frac{t}{2} + 1\right)$

$x = 2\ln\left(\frac{t}{2} + 1\right)$

Example 11

A particle P is moving along the positive x-axis. At $OP = x$ m, the velocity of P is v m s^{-1} and the acceleration of P is $\frac{k}{(2x + 3)^2}$ m s^{-2}, where k is a constant, directed away from O. At $x = 1$, $v = 10$ and at $x = 6$, $v = \sqrt{120}$.

a Find the value of k.

b Show that the speed of P cannot exceed $\sqrt{130}$ m s^{-1}.

a $a = \frac{k}{(2x + 3)^2}$

Integrate both sides of this equation with respect to x. For constants a and b and $n \neq -1$,

$\frac{d}{dx}\left(\frac{1}{2}v^2\right) = \frac{k}{(2x + 3)^2}$

$\int(ax + b)^n\,dx = \frac{1}{(n + 1)a}(ax + b)^{n+1} + C$

$\frac{1}{2}v^2 = A - \frac{k}{2(2x + 3)}$

$v^2 = B - \frac{k}{2x + 3}$

Multiply this equation throughout by 2. Twice one arbitrary constant is another arbitrary constant.

A

At $x = 1$, $v = 10$

$$100 = B - \frac{k}{5} \quad (1)$$

At $x = 6$, $v = \sqrt{120}$

$$120 = B - \frac{k}{15} \quad (2)$$

$(2) - (1)$

$$20 = -\frac{k}{15} - \left(-\frac{k}{5}\right) = \frac{2k}{15}$$

$$k = 20 \times \frac{15}{2} = 150$$

b Substituting $k = 150$ into (1)

$$100 = B - \frac{150}{5} \Rightarrow B = 130$$

$$v^2 = 130 - \frac{150}{2x + 3}$$

As x is moving along the positive x-axis $x > 0$, and so both $2x + 3$ and $\frac{150}{2x + 3}$ are positive.

Hence

$$v^2 = 130 - \frac{150}{2x + 3} < 130$$

The speed of P cannot exceed $\sqrt{130}$ m s⁻¹.

The conditions given in the question give you a pair of simultaneous equations in B and k. You find k to solve part **a**. You will also need to find B to solve part **b**.

As x increases, the velocity of P approaches $\sqrt{130}$ m s⁻¹ in the direction Ox asymptotically.

Exercise 4B

1 A particle P moves along the x-axis. At time $t = 0$, P passes through the origin O with velocity 5 m s⁻¹ in the direction of x increasing. At time t seconds, the velocity of P is v m s⁻¹ and $OP = x$ m. The acceleration of P is $\left(2 + \frac{1}{2}x\right)$ m s⁻², measured in the positive x direction. Find v^2 in terms of x.

2 A particle P moves along a straight line. When its displacement from a fixed point O on the line is x m and its velocity is v m s⁻¹, the deceleration of P is $4x$ m s⁻². At $x = 2$, $v = 8$. Find v in terms of x.

3 A particle P is moving along the x-axis in the direction of x increasing. At $OP = x$ m ($x > 0$), the velocity of P is v m s⁻¹ and its acceleration is of magnitude $\frac{4}{x^2}$ m s⁻² in the direction of x increasing. Given that at $x = 2$, $v = 6$, find the value of x for which P is instantaneously at rest.

P **4** A particle P moves along a straight line. When its displacement from a fixed point O on the line is x m and its velocity is v m s⁻¹, the acceleration of P is of magnitude $25x$ m s⁻² and is directed towards O. At $x = 0$, $v = 40$. In its motion P is instantaneously at rest at two points, A and B. Find the distance between A and B.

(A) **(P)** **5** A particle P is moving along the x-axis. At $OP = x$ m, the velocity of P is v m s^{-1} and its acceleration is of magnitude kx^2 m s^{-2}, where k is a positive constant in the direction of x decreasing. At $x = 0$, $v = 16$. The particle is instantaneously at rest at $x = 20$. Find:

    a the value of k

    b the velocity of P when $x = 10$.

6 A particle P is moving along the x-axis in the direction of x increasing. At $OP = x$ m, the velocity of P is v m s^{-1} and its acceleration is of magnitude $8x^3$ m s^{-2} in the direction PO. At $x = 2$, $v = 32$. Find the value of x for which $v = 8$.

7 A particle P is moving along the x-axis. When the displacement of P from the origin O is x m, the velocity of P is v m s^{-1} and its acceleration is $6 \sin \frac{x}{3}$ m s^{-2}. At $x = 0$, $v = 4$. Find:

    a v^2 in terms of x

    b the greatest possible speed of P.

(E) **8** A particle P is moving along the x-axis. At $x = 0$, the velocity of P is 2 m s^{-1} in the direction of x increasing. At $OP = x$ m, the velocity of P is v m s^{-1} and its acceleration is $(2 + 3e^{-x})$ m s^{-2}. Find the velocity of P at $x = 3$. Give your answer to 3 significant figures. **(6 marks)**

(E) **9** A particle P moves away from the origin O along the positive x-axis. The acceleration of P is of magnitude $\dfrac{4}{2x + 1}$ m s^{-2}, where $OP = x$ m, directed towards O. Given that the speed of P at O is 4 m s^{-1}, find:

    a the speed of P at $x = 10$ **(4 marks)**

    b the value of x at which P is instantaneously at rest. **(6 marks)**

    Give your answers to 3 significant figures.

(E) **10** A particle P is moving along the positive x-axis. At $OP = x$ m, the velocity of P is v m s^{-1} and its acceleration is $\left(x - \dfrac{4}{x^3}\right)$ m s^{-2}. The particle starts from the position where $x = 1$ with velocity 3 m s^{-1} in the direction of x increasing. Find:

    a v in terms of x **(4 marks)**

    b the least speed of P during its motion. **(6 marks)**

(E/P) **11** A particle P is moving along the x-axis. Initially P is at the origin O moving with velocity 15 m s^{-1} in the direction of x increasing. When the displacement of P from O is x m, its acceleration is of magnitude $\left(10 + \frac{1}{4}x\right)$ m s^{-2} directed towards O. Find the distance P moves before first coming to instantaneous rest. **(7 marks)**

(E/P) **12** A particle P is moving along the x-axis. At time t seconds, P is x m from O, has velocity v m s^{-1} and acceleration of magnitude $6x^{\frac{1}{3}}$ m s^{-2} in the direction of x increasing. When $t = 0$, $x = 8$ and $v = 12$. Find:

    a v in terms of x **(4 marks)**

    b x in terms of t. **(4 marks)**

Challenge

A particle P moves along the x-axis. At time $t = 0$, P passes through the origin moving in the positive x direction. At time t seconds, the velocity of P is v m s^{-1} and $OP = x$ metres. The acceleration of P is $\frac{1}{10}(25 - x)$. Given that the maximum speed of P is 12 m s^{-1}, find an expression for v^2 in terms of x.

4.3 Acceleration varying with velocity

When acceleration is given as a function of velocity, you can form and solve a differential equation to find an expression for velocity in terms of time.

■ **When the acceleration is a function of the velocity you can use**

$$a = \frac{dv}{dt}$$

for questions which involve working with time

Links This type of problem usually gives rise to a differential equation of the form $\frac{dv}{dt} = f(v)$, which can be solved by separating the variables.

← **Pure Year 2, Section 11.10**

Example 12

A van moves along a straight horizontal road. At time t seconds, $t \geqslant 0$, the acceleration of the van is $\frac{625 - v^2}{200}$ m s^{-2}, where v m s^{-1} is the velocity of the van. The van starts from rest.

a Find v in terms of t.

b Show that the speed of the van cannot exceed 25 m s^{-1}.

a $a = \dfrac{625 - v^2}{200}$

$\dfrac{dv}{dt} = \dfrac{625 - v^2}{200}$

Separate the variables by dividing both sides by $625 - v^2$:

$$\frac{1}{625 - v^2}\frac{dv}{dt} = \frac{1}{200}$$

$\displaystyle\int \frac{1}{625 - v^2}\,dv = \int \frac{1}{200}\,dt$

To integrate $\dfrac{1}{625 - v^2}$ you must factorise $625 - v^2 = (25 + v)(25 - v)$ and use partial fractions.

Let $\dfrac{1}{625 - v^2} = \dfrac{A}{(25 + v)} + \dfrac{B}{(25 - v)}$

Multiplying through by $(25 + v)(25 - v)$

$1 = A(25 - v) + B(25 + v)$

Let $v = 25$

$1 = 50B \Rightarrow B = \frac{1}{50}$

Let $v = -25$

$1 = 50A \Rightarrow A = \frac{1}{50}$

Find the values of A and B by setting $v = 25$ and $v = -5$.

Hence

$$\frac{1}{50}\int\frac{1}{25+v}\,dv + \frac{1}{50}\int\frac{1}{25-v}\,dv = \int\frac{1}{200}\,dt$$

$$\frac{1}{50}\ln(25+v) - \frac{1}{50}\ln(25-v) = \frac{t}{200} + C$$

$$\frac{1}{50}\ln\left(\frac{25+v}{25-v}\right) = \frac{t}{200} + C$$

$$\ln\left(\frac{25+v}{25-v}\right) = 50\left(\frac{t}{200} + C\right)$$

$$\left(\frac{25+v}{25-v}\right) = e^{\frac{t}{4}+50C}$$

So $\left(\dfrac{25+v}{25-v}\right) = De^{\frac{t}{4}}$

When $t = 0$, $v = 0$

$$\frac{25}{25} = De^0 \Rightarrow D = 1$$

Hence

$$\left(\frac{25+v}{25-v}\right) = e^{\frac{t}{4}}$$

$$25 + v = 25e^{\frac{t}{4}} - ve^{\frac{t}{4}}$$

$$v(e^{\frac{t}{4}} + 1) = 25(e^{\frac{t}{4}} - 1)$$

$$v = \frac{25(e^{\frac{t}{4}} - 1)}{e^{\frac{t}{4}} + 1}$$

b For all real t, $e^{\frac{t}{4}} - 1 < e^{\frac{t}{4}} + 1$

Hence $\left|\dfrac{e^{\frac{t}{4}} - 1}{e^{\frac{t}{4}} + 1}\right| < 1$

so $\left|\dfrac{25(e^{\frac{t}{4}} - 1)}{e^{\frac{t}{4}} + 1}\right| < 25$

So the speed of the van cannot exceed 25 m s⁻¹

Online Explore terminal or limiting velocity using GeoGebra.

Use $\ln a - \ln b = \ln\left(\dfrac{a}{b}\right)$ to simplify the left-hand side.

$e^{\frac{t}{4}+50C} = e^{\frac{t}{4}} \times e^{50C}$

C is a constant, so e^{50C} is a constant. You can simplify your working by writing $e^{50C} = D$.

Use the boundary conditions given to find the value of D.

Watch out Read the question carefully. You need to find v in terms of t so rearrange to make v the subject.

$x - 1 < x + 1$ for any real value of x

For $t > 0$ both the numerator and denominator of this fraction are positive and, as the numerator is less than the denominator, the value of the fraction must be less than 1.

In Example 12, if a graph of v against t is plotted, you obtain

The speed approaches 25 m s⁻¹ asymptotically but cannot exceed it. Such a speed is called a **terminal** or **limiting** speed.

In this context, as the motion is in one direction, the **terminal** or **limiting** speed is often called the **terminal** or **limiting** velocity.

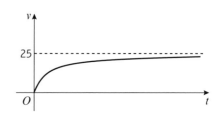

The terminal or limiting speed of a body occurs when the acceleration is zero.

In Example 12, the acceleration was given by

$$\frac{dv}{dt} = \frac{625 - v^2}{200}$$

Setting $\frac{dv}{dt} = 0$, and noting that v is positive, gives $v = 25$ and the terminal velocity is 25 m s^{-1}.

Watch out The terminal velocity is never actually attained by the van.

This result is confirmed by the answers to parts **a** and **b** above.

A ■ **If you have to find distances in problems where acceleration varies with velocity, you can use the relationship $a = v\dfrac{dv}{dx}$.**

Example 13

A particle P moves in a straight line. When the velocity of the particle is v m s^{-1}, the acceleration of the particle is given by $a = 4v$ m s^{-2}.

Find the distance moved by P as the velocity increases from 10 m s^{-1} to 15 m s^{-1}.

As the velocity is positive throughout the motion, you need to find the displacement of P, x m, from the point where it has a velocity of 10 m s^{-1}.

$a = 4v$

$v\dfrac{dv}{dx} = 4v$

$\dfrac{dv}{dx} = 4$

$v = \int 4dx$

$v = 4x + A$

When $x = 0$, $v = 10$

So $10 = 4 \times 0 + A \Rightarrow A = 10$

Hence $v = 4x + 10$

When $v = 15$

$15 = 4x + 10$

So $x = \dfrac{15 - 10}{4} = 1.25$

The distance moved by P as the velocity increases from 10 m s^{-1} to 15 m s^{-1} is 1.25 m.

Problem-solving

The question asks you to find a distance so use

$$a = v\frac{dv}{dx}$$

The distance is measured from the point where the velocity of P is 10 m s^{-1} so use $x = 0$ when $v = 10$ to find the constant of integration.

Watch out You could solve this problem by solving $\dfrac{dv}{dt} = 4v$ to obtain $v = 10e^{4t}$, and then integrating this expression. However, the correct integration would **not** be $\int_{10}^{15} 10e^{4t}\,dt$, as the limits would have to be values of t, not v. You would have to solve $15 = 10e^{4t}$ to find the appropriate upper limit.

Example (14)

A particle P moves along the positive x-axis. At time t seconds, the acceleration of the particle is $-(k^2 + v^2)$ where v m s^{-1} is the velocity of the particle and k is a positive constant. When $t = 0$, P is at O and $v = U$. The particle comes to rest at the point A. Find, in terms of k and U:

a the distance OA

b the time P takes to travel from O to A.

> As the acceleration is negative the particle is decelerating.

a $a = -(k^2 + v^2)$

$v\dfrac{dv}{dx} = -(k^2 + v^2)$

> You are asked to find the distance when P comes to rest so use $a = v\dfrac{dv}{dx}$.

Separating the variables and integrating

$\displaystyle\int \frac{v}{k^2 + v^2}\,dv = -\int dx$

> Use the reverse chain rule to integrate the left-hand side. As $\dfrac{d}{dv}(k^2 + v^2) = 2v$ then $\displaystyle\int \frac{v}{k^2+v^2}\,dv = \frac{1}{2}\ln(k^2+v^2)$

$\frac{1}{2}\ln(k^2 + v^2) = -x + B$

When $x = 0$, $v = U$

$\frac{1}{2}\ln(k^2 + U^2) = -0 + B \Rightarrow B = \frac{1}{2}\ln(k^2 + U^2)$

> Use the initial conditions to find the constant of integration.

Hence

$\frac{1}{2}\ln(k^2 + v^2) = -x + \frac{1}{2}\ln(k^2 + U^2)$

$x = \frac{1}{2}\ln(k^2 + U^2) - \frac{1}{2}\ln(k^2 + v^2)$

$x = \frac{1}{2}\ln\left(\dfrac{k^2 + U^2}{k^2 + v^2}\right)$

> **Watch out** In questions involving unknown constants the constant of integration might include these constants.

> From the laws of logarithms $\ln a - \ln b = \ln\left(\dfrac{a}{b}\right)$

At A, $v = 0$ and $x = OA$

$OA = \frac{1}{2}\ln\left(\dfrac{k^2 + U^2}{k^2}\right)$

> The particle comes to rest when $v = 0$.

b $a = -(k^2 + v^2)$

$\dfrac{dv}{dt} = -(k^2 + v^2)$

> You are asked to find the time when P comes to rest so use $a = \dfrac{dv}{dt}$

Separating the variables and integrating

$\displaystyle\int \frac{1}{k^2 + v^2}\,dv = -\int dt$

$\frac{1}{k}\arctan\left(\dfrac{v}{k}\right) = -t + C$

> You can use the standard result $\displaystyle\int \frac{1}{a^2+x^2}\,dx = \frac{1}{a}\arctan\frac{x}{a}$ from the formulae booklet. ← **Core Pure 2, Section 3.3**

When $t = 0$, $v = U$

$\frac{1}{k}\arctan\left(\dfrac{U}{k}\right) = -0 + C \Rightarrow C = \frac{1}{k}\arctan\left(\dfrac{U}{k}\right)$

Hence

$\frac{1}{k}\arctan\left(\dfrac{v}{k}\right) = -t + \frac{1}{k}\arctan\left(\dfrac{U}{k}\right)$

A

So

$$t = \frac{1}{k}\arctan\left(\frac{U}{k}\right) - \frac{1}{k}\arctan\left(\frac{v}{k}\right)$$

When $v = 0$

$$t = \frac{1}{k}\arctan\left(\frac{U}{k}\right)$$ ———————— Using arctan $0 = 0$.

The time P takes to travel from O to A is

$$\frac{1}{k}\arctan\left(\frac{U}{k}\right).$$

Exercise 4C

1 A particle moves in a straight horizontal line. When the velocity of the particle is v m s^{-1}, the acceleration of the particle is given by $a = e^{-v}$ m s^{-2}. The particle starts from rest. Find:

   a v in terms of t

   b the speed of the particle after 10 seconds.

2 A particle moves in a straight horizontal line When the velocity of the particle is v m s^{-1}, the acceleration of the particle is given by $a = -8v$ m s^{-2}. Find the time taken for the particle to slow down from 18 m s^{-1} to 6 m s^{-1}.

3 A particle P moves in a straight horizontal line with acceleration $a = -(3 + 0.6v)$ m s^{-2}, where v m s^{-1} is the speed of P at time t seconds. When $t = 0$, P is at a point A moving with speed 12 m s^{-1}. The particle P comes to rest at the point B. Find:

   a the time P takes to move from A to B

   b the distance AB.

(P) 4 A particle P falls vertically from rest. After t seconds it has speed v m s^{-1}. As it falls it has an acceleration of $g - 2v$.

   a Show that $2v = g(1 - e^{-2t})$

   b Find the distance that P falls in the first two seconds of its motion.

Problem-solving

In part **b**, integrate your expression for v with respect to t.

(E) 5 A lorry travels along a straight horizontal road. At time t seconds, the speed of the lorry is v m s^{-1} and the acceleration is $(3 - 0.25v)$ m s^{-2}. The lorry starts from rest. Find:

   a v in terms of t **(4 marks)**

   b the maximum speed of the lorry. **(4 marks)**

A 6 A particle moves along the x-axis in the direction of x increasing. When the speed of the particle is v m s^{-1}, the acceleration is $0.6v^2$ m s^{-2}. Initially the particle is at the origin and is moving with a speed of 14 m s^{-1}. Find the distance the particle moves before its speed is doubled.

A **7** A particle moves in a straight line with an initial velocity of u m s^{-1}. When the particle is moving
E/P with a velocity of v m s^{-1} the acceleration is $-(k + v^2)$ where k is a constant. Find the distance the
particle moves before coming to rest. **(5 marks)**

E/P **8** A particle P of mass m is moving along the x-axis Ox in the direction of x increasing. At time
t seconds, the velocity of P is v m s^{-1} and the acceleration is $-(a^2 + v^2)$ m s^{-2}, where a is a constant.
At time $t = 0$, P is at O and its speed is U m s^{-1}. At time $t = T$, $v = \frac{1}{2}U$.

a Show that $T = \dfrac{1}{k}\left(\arctan\dfrac{U}{k} - \arctan\dfrac{U}{2k}\right)$ **(4 marks)**

b Find the distance travelled by P as its speed is reduced from U to $\frac{1}{2}U$. **(4 marks)**

E/P **9** A car is moving along a straight horizontal road. At time t seconds, the speed of the car is v m s^{-1}
$(0 < v < 30)$ and the acceleration of the car is: $\dfrac{1600 - v^2}{64v}$ m s^{-2}.

The time taken for the car to accelerate from 10 m s^{-1} to 20 m s^{-1} is T seconds.

Show that $T = 32\ln\frac{5}{4}$. **(4 marks)**

Mixed exercise 4

E **1** A particle P is moving along the x-axis. At time t seconds, the displacement of P from the origin O
is x m and the velocity of P is $4e^{0.5t}$ m s^{-1} in the direction Ox. When $t = 0$, P is at O. Find:
a x in terms of t **(6 marks)**
b the acceleration of P when $t = \ln 9$. **(3 marks)**

E **2** A particle P moves along the x-axis in the direction of x increasing. At time t seconds, the velocity
of P is v m s^{-1} and its acceleration is $20te^{-t^2}$ m s^{-2}. When $t = 0$ the speed of P is 8 m s^{-1}. Find:
a v in terms of t **(3 marks)**
b the limiting velocity of P. **(2 marks)**

E **3** A particle P moves along a straight line. Initially P is at rest at a point O on the line. At time
t seconds, where $t \geq 0$, the acceleration of P is $\dfrac{18}{(2t + 3)^3}$ m s^{-2} directed away from O.

Find the value of t for which the speed of P is 0.48 m s^{-1}. **(4 marks)**

E **4** A car moves along a horizontal straight road. At time t seconds the acceleration of the car is
$\dfrac{100}{(2t + 5)^2}$ m s^{-2} in the direction of motion of the car. When $t = 0$, the car is at rest. Find:
a an expression for v in terms of t **(3 marks)**
b the distance moved by the car in the first 10 seconds of its motion. **(3 marks)**

(E) 5 A particle P is moving in a straight line with acceleration $\cos^2 t\,\text{m s}^{-2}$ at time t seconds. The particle is initially at rest at a point O.

    a Find the speed of P when $t = \pi$. **(4 marks)**

    b Show that the distance of P from O when $t = \dfrac{\pi}{4}$ is $\dfrac{1}{64}(\pi^2 + 8)\,\text{m}$. **(4 marks)**

(E/P) 6 A particle P is moving along the x-axis. At time t seconds, the velocity of P is $v\,\text{m s}^{-1}$ in the direction of x increasing, where

$$v = \begin{cases} \frac{1}{2}t^2, & 0 \leqslant t \leqslant 4 \\ 8e^{4-t}, & t > 4 \end{cases}$$

When $t = 0$, P is at the origin O. Find:

    a the acceleration of P when $t = 2.5$ **(2 marks)**

    b the acceleration of P when $t = 5$ **(2 marks)**

    c the distance of P from O when $t = 6$. **(3 marks)**

(E) 7 A particle P is moving along the x-axis. At time t seconds, P has velocity $v\,\text{m s}^{-1}$ in the direction of x increasing and an acceleration of magnitude $\dfrac{2t + 3}{t + 1}\,\text{m s}^{-2}$ in the direction of x increasing. When $t = 0$, P is at rest at the origin O. Find:

    a v in terms of t **(5 marks)**

    b the distance of P from O when $t = 2$. **(3 marks)**

(E/P) 8 A particle moving in a straight line starts from rest at the point O at time $t = 0$. At time t seconds, the velocity $v\,\text{m s}^{-1}$ of the particle is given by

$$v = \begin{cases} 3t^2 - 14t + 8, & 0 \leqslant t \leqslant 5 \\ 18 - \dfrac{t^2}{5}, & 5 < t \leqslant T \end{cases}$$

where T is the first time the particle comes to momentary rest when travelling with velocity $18 - \dfrac{t^2}{5}\,\text{m s}^{-1}$.

    a Find the value of T. **(2 marks)**

    b Sketch a velocity–time graph for the particle for $0 \leqslant t \leqslant T$. **(3 marks)**

    c Find the set of values of t for which the acceleration of the particle is positive. **(2 marks)**

    d Find the total distance travelled by the particle in the interval $0 \leqslant t \leqslant T$. **(5 marks)**

(E/P) 9 A particle P moves on the x-axis. At time t seconds the velocity of P is $v\,\text{m s}^{-1}$ in the direction of x increasing, where $v = (t - 4)(3t - 8)$, $t \geqslant 0$.

When $t = 0$, P is at the origin O.

    a Find the acceleration of P at time t seconds. **(2 marks)**

    b Find the total distance travelled by P in the first 3 seconds of its motion. **(3 marks)**

    c Show that P never returns to O, explaining your reasoning. **(3 marks)**

(E/P) **10** A particle moves on a horizontal plane with acceleration $-kv^2$ where $v\,\text{m s}^{-1}$ is the speed of the particle and k is a positive constant. At time $t = 0$ the particle has speed $U\,\text{m s}^{-1}$.

Find, in terms of k and U, the time at which the particle's speed is $\frac{2}{3}U$. **(5 marks)**

(E/P) **11** A particle P moves on the positive x-axis with an acceleration at time t seconds of $(3t - 4)\,\text{m s}^{-2}$. The particle starts from O with a velocity of $2\,\text{m s}^{-1}$.

Find:

a the values of t when P is instantaneously at rest **(4 marks)**

b the total distance travelled by P in the interval $0 \leqslant t \leqslant 4$. **(4 marks)**

(A)
(E/P) **12** A particle P moves along a straight line. When the displacement of P from a fixed point on the line is x m, its velocity is $v\,\text{m s}^{-1}$ and its acceleration is of magnitude $\dfrac{6}{x^2}\,\text{m s}^{-2}$ in the direction of x increasing. At $x = 3$, $v = 4$.

Find v in terms of x. **(4 marks)**

(E) **13** A particle is moving along the x-axis. At time $t = 0$, P is passing through the origin O with velocity $8\,\text{m s}^{-1}$ in the direction of x increasing. When P is x m from O, its acceleration is $\left(3 + \frac{1}{4}x\right)\text{m s}^{-2}$ in the direction of x decreasing.

Find the positive value of x for which P is instantaneously at rest. **(5 marks)**

(E) **14** A particle P is moving on the x-axis. When P is a distance x metres from the origin O, its acceleration is of magnitude $\dfrac{15}{4x^2}\,\text{m s}^{-2}$ in the direction OP. Initially P is at the point where $x = 5$ and is moving toward O with speed $6\,\text{m s}^{-1}$.

Find the value of x where P first comes to rest. **(6 marks)**

(E) **15** A particle P is moving along the x-axis. At time t seconds, the velocity of P is $v\,\text{m s}^{-1}$ and the acceleration of P is $(3 - x)\,\text{m s}^{-2}$ in the direction of x increasing. Initially P is at the origin O and is moving with speed $4\,\text{m s}^{-1}$ in the direction of x increasing. Find:

a v^2 in terms of x **(3 marks)**

b the maximum value of v. **(3 marks)**

(E) **16** A particle P is moving along the x-axis. At time $t = 0$, P passes through the origin O. After t seconds the speed of P is $v\,\text{m s}^{-1}$, $OP = x$ metres and the acceleration of P is $\dfrac{x^2(5 - x)}{2}\,\text{m s}^{-2}$ in the direction of x increasing. At $x = 10$, P is instantaneously at rest. Find:

a an expression for v^2 in terms of x **(4 marks)**

b the speed of P when $t = 0$. **(2 marks)**

A **E** 17 A particle P moves away from the origin along the positive x-axis. At time t seconds,

the acceleration of P is $\dfrac{20}{5x + 2}$ m s^{-2}, where $OP = x$ m, directed away from O. Given that the

speed of P is 3 m s^{-1} at $x = 0$, find, giving your answers to 3 significant figures,

a the speed of P at $x = 12$ **(4 marks)**

b the value of x when the speed of P is 5 m s^{-1}. **(3 marks)**

E/P 18 A particle P is moving along the x-axis. When $t = 0$, P is passing through O with velocity 3 m s^{-1} in the direction of x increasing. When $0 \leqslant x \leqslant 4$ the acceleration is of magnitude $\left(4 + \frac{1}{2}x\right)$ m s^{-2} in the direction of x increasing. At $x = 4$, the acceleration of P changes.

For $x > 4$, the magnitude of the acceleration remains $\left(4 + \frac{1}{2}x\right)$ m s^{-2} but it is now in the direction of x decreasing.

a Find the speed of P at $x = 4$. **(4 marks)**

b Find the positive value of x for which P is instantaneously at rest.
Give your answer to 2 significant figures. **(3 marks)**

E 19 A particle P is moving along the x-axis. At time t seconds P is x m from O, has velocity v m s^{-1} and acceleration of magnitude $(4x + 6)$ m s^{-2} in the direction of x increasing. When $t = 0$, P is passing through O with velocity 3 m s^{-1} in the direction of x increasing. Find:

a v in terms of x **(3 marks)**

b x in terms of t. **(4 marks)**

E/P 20 At time t seconds a particle P is moving in a straight line with speed v m s^{-1} and acceleration $-k(U^2 + v^2)$ where k is a positive constant and U is the speed of P when $t = 0$.

Show that, as v decreases from U to $\frac{1}{2}U$, P travels a distance $\dfrac{1}{2k} \ln \frac{8}{5}$. **(5 marks)**

E 21 A cyclist travels along a straight horizontal road with acceleration $\dfrac{6000 - v^3}{8000v}$ m s^{-2} where v m s^{-1}

is the speed of the cyclist. Find the distance the cyclist travels as her speed increases from 4 m s^{-1} to 8 m s^{-1}. **(5 marks)**

Challenge

A rocket is launched straight upwards from the Earth's surface with an initial velocity of 32 500 km h^{-1}. The flight of the rocket can be modelled as a particle with an acceleration of $-\dfrac{c}{x^2}$ km s^{-2} where $c = 4 \times 10^5$ and x km is the distance from the centre of the Earth. The radius of the Earth is 6370 km. Work out the maximum height above the surface of the Earth that the rocket will reach.

Summary of key points

1 To find the velocity from the displacement, you differentiate with respect to time.
   To find the acceleration from the velocity, you differentiate with respect to time.

$$v = \frac{dx}{dt} \text{ and } a = \frac{dv}{dt} = \frac{d^2x}{dt^2}$$

2 To obtain the velocity from the acceleration, you integrate with respect to time.
   To obtain the displacement from the velocity, you integrate with respect to time.

$$v = \int a \, dt \text{ and } x = \int v \, dt$$

3 When the acceleration is a function of the displacement you can use

$$a = v\frac{dv}{dx} = \frac{d}{dx}\left(\tfrac{1}{2}v^2\right)$$

4 When the acceleration is a function of the velocity you can use $a = \dfrac{dv}{dt}$ for questions which involve working with time.

5 If you have to find distances in problems where acceleration varies with velocity, you can use the relationship $a = v\dfrac{dv}{dx}$.

Dynamics

5

Prior knowledge check

1 A car of mass 1500 kg moves in a straight line. The total resistance to motion of the car is modelled as a constant force of magnitude 80 N. The car brakes with a constant force, bringing it to rest from a speed of 30 m s⁻¹ in a distance of 60 m.

  a Find the magnitude of the braking force.

  b Find the total work done in bringing the car to rest. ← **FM1, Section 2.1**

2 A particle travels along the positive x-axis with acceleration $a = \dfrac{e^{2t} - 3t^2}{5}$ at time t s. The particle starts from rest at the origin, O. Find:

  a the velocity of the particle when $t = 3$ s

  b the displacement from O when $t = 2$ s.
     ← **Statistics and Mechanics Year 2, Section 8.3**

3 An elastic string of natural length 2.5 m is fixed at one end and is stretched to a length of 3.4 m by a force of 5 N. Find the modulus of elasticity of the string. ← **FM1, Section 3.1**

These toys can be modelled as particles hanging vertically by an elastic string. The toys will experience **simple harmonic motion**, and their displacement–time graphs will be in the shape of a sine curve. → **Section 5.5**

5.1 Motion in a straight line with variable force

A You can use calculus to apply Newton's second law $F = ma$ to a particle moving in a straight line when the applied force is variable.

Links When the acceleration is a function of x, t or v, it can be written as $\dfrac{dv}{dt}$, $v\dfrac{dv}{dx}$ or $\dfrac{d^2x}{dt^2}$ ← **Chapter 4**

The applied force F can be a function of the displacement x, time t or velocity v.

Suppose F is a function of time, then using $a = \dfrac{dv}{dt}$

$$m\frac{dv}{dt} = F$$

$$\int m\, dv = \int F\, dt$$ ——————————————— Separate the variables.

$$mv = \int F\, dt$$ ——————————————— Integrate both sides. Mass is assumed to be constant.

When you work out $\int F\, dt$ you must remember to add a **constant of integration**. You will often be given boundary conditions that allow you to work out the value of this constant.

Example 1

A particle P of mass $0.5\,\text{kg}$ is moving along the x-axis. At time t seconds the force acting on P has magnitude $(5t^2 + e^{0.2t})\,\text{N}$ and acts in the direction OP. When $t = 0$, P is at rest at O. Calculate:

a the speed of P when $t = 2$

b the distance OP when $t = 3$.

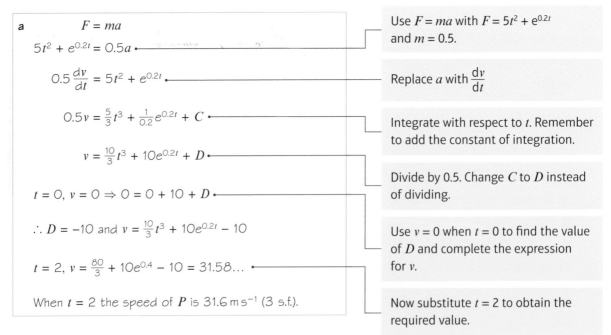

a $\qquad\qquad F = ma$

$\qquad 5t^2 + e^{0.2t} = 0.5a$ ——— Use $F = ma$ with $F = 5t^2 + e^{0.2t}$ and $m = 0.5$.

$\qquad 0.5\dfrac{dv}{dt} = 5t^2 + e^{0.2t}$ ——— Replace a with $\dfrac{dv}{dt}$

$\qquad 0.5v = \dfrac{5}{3}t^3 + \dfrac{1}{0.2}e^{0.2t} + C$ ——— Integrate with respect to t. Remember to add the constant of integration.

$\qquad v = \dfrac{10}{3}t^3 + 10e^{0.2t} + D$ ——— Divide by 0.5. Change C to D instead of dividing.

$\qquad t = 0, v = 0 \Rightarrow 0 = 0 + 10 + D$

$\qquad \therefore D = -10$ and $v = \dfrac{10}{3}t^3 + 10e^{0.2t} - 10$ ——— Use $v = 0$ when $t = 0$ to find the value of D and complete the expression for v.

$\qquad t = 2, v = \dfrac{80}{3} + 10e^{0.4} - 10 = 31.58\ldots$ ——— Now substitute $t = 2$ to obtain the required value.

When $t = 2$ the speed of P is $31.6\,\text{m s}^{-1}$ (3 s.f.).

b $\quad v = \frac{10}{3}t^3 + 10e^{0.2t} - 10$

$\frac{dx}{dt} = \frac{10}{3}t^3 + 10e^{0.2t} - 10$ ———————————————— Replace v with $\frac{dx}{dt}$

$x = \frac{10}{12}t^4 + \frac{10}{0.2}e^{0.2t} - 10t + K$ ———————————— Integrate with respect to t. Use a different letter for the constant of integration.

$t = 0, x = 0 \Rightarrow 0 = 0 + 50 - 0 + K$

$\therefore K = -50$ and $x = \frac{10}{12}t^4 + \frac{10}{0.2}e^{0.2t} - 10t - 50$ ——— Use the initial conditions to find K.

$t = 3 \Rightarrow x = \frac{10}{12} \times 3^4 + \frac{10}{0.2}e^{0.6} - 10 \times 3 - 50 = 78.60...$ ——— Now substitute $t = 3$ to obtain the required value.

When $t = 3$, OP is 78.6 m (3 s.f.).

Example 2

A pebble of mass 0.2 kg is moving on a smooth horizontal sheet of ice. At time t seconds (where $t \geqslant 0$) a horizontal force of magnitude $2t^2$ N and constant direction acts on the pebble. When $t = 0$ the pebble is moving in the same direction as the force and has speed 6 m s^{-1}. When $t = T$ the pebble has speed 36 m s^{-1}. Calculate the value of T.

$F = ma$

$0.2\frac{dv}{dt} = 2t^2$ ——————————————— Use $F = ma$ with $F = 2t^2$ and $m = 0.2$

As F is a function of t replace a with $\frac{dv}{dt}$

$0.2v = \frac{2}{3}t^3 + C$

$t = 0, v = 6 \Rightarrow 0.2 \times 6 = 0 + C$ ——————— Integrate with respect to t and use the initial conditions to find the value of C.

$\therefore C = 1.2$ and $0.2v = \frac{2}{3}t^3 + 1.2$

$t = T, v = 36 \Rightarrow 0.2 \times 36 + \frac{2}{3}T^3 + 1.2$ ——————— Substitute $t = T$ and solve for T.

$T^3 = \frac{3}{2} \times 6$

$T = 2.080...$

$T = 2.08$ (3 s.f.)

Suppose F is a function of displacement, x, then using $a = \frac{d}{dx}\left(\frac{1}{2}v^2\right) = v\frac{dv}{dx}$,

$mv\frac{dv}{dx} = F$

$\int mv\,dv = \int F\,dx$ ————————————————— Separate the variables.

$\frac{1}{2}mv^2 = \int F\,dx + C$

where C is the constant of integration ——————— m is constant so $\int mv\,dv = \frac{1}{2}mv^2 + C$

Example 3

A A particle P of mass 1.5 kg is moving in a straight line. The force acting on P has magnitude $(8 - 2\cos x)$ N, where x metres is the distance OP, and acts in the direction OP. When P passes through O its speed is $4\,\mathrm{m\,s^{-1}}$. Calculate the speed of P when $x = 2$.

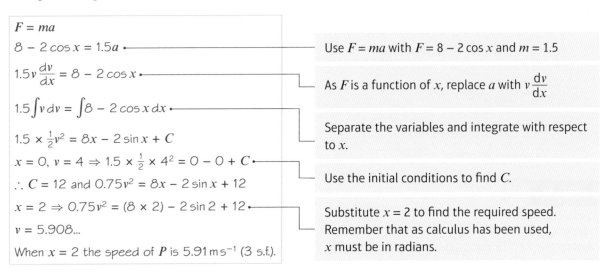

$F = ma$

$8 - 2\cos x = 1.5a$ — Use $F = ma$ with $F = 8 - 2\cos x$ and $m = 1.5$

$1.5v\dfrac{dv}{dx} = 8 - 2\cos x$ — As F is a function of x, replace a with $v\dfrac{dv}{dx}$

$1.5\displaystyle\int v\,dv = \int 8 - 2\cos x\,dx$ — Separate the variables and integrate with respect to x.

$1.5 \times \frac{1}{2}v^2 = 8x - 2\sin x + C$

$x = 0,\ v = 4 \Rightarrow 1.5 \times \frac{1}{2} \times 4^2 = 0 - 0 + C$ — Use the initial conditions to find C.

$\therefore C = 12$ and $0.75v^2 = 8x - 2\sin x + 12$

$x = 2 \Rightarrow 0.75v^2 = (8 \times 2) - 2\sin 2 + 12$ — Substitute $x = 2$ to find the required speed. Remember that as calculus has been used, x must be in radians.

$v = 5.908...$

When $x = 2$ the speed of P is $5.91\,\mathrm{m\,s^{-1}}$ (3 s.f.).

Example 4

A stone S of mass 0.5 kg is moving in a straight line on a smooth horizontal floor. When S is a distance x metres from a fixed point on the line, A, a force of magnitude $(5 + 7\cos x)$ N acts on S in the direction AS. Given that S passes through A with speed $2\,\mathrm{m\,s^{-1}}$, calculate:

a the speed of S as it passes through the point B, where $x = 3$

b the work done by the force in moving S from A to B.

a $F = ma$

$0.5v\dfrac{dv}{dx} = 5 + 7\cos x$ — Use $F = ma$ with $F = 5 + 7\cos x$ and $m = 0.5$. As F is a function of x, replace a with $v\dfrac{dv}{dx}$

$\displaystyle\int 0.5v\,dv = \int (5 + 7\cos x)\,dx$

$0.5 \times \frac{1}{2}v^2 = 5x + 7\sin x + C$ — Separate the variables and integrate. Use the initial conditions to find the value of C.

$x = 0,\ v = 2 \Rightarrow 0.5 \times \frac{1}{2} \times 2^2 = C$

$\therefore C = 1$ and $\frac{1}{4}v^2 = 5x + 7\sin x + 1$

$x = 3 \Rightarrow v^2 = 4(15 + 7\sin 3 + 1)$

$\qquad v^2 = 67.95...$

$\qquad v = 8.243...$ — Substitute $x = 3$ to find the value of v.

S passes through B with speed $8.24\,\mathrm{m\,s^{-1}}$ (3 s.f.)

b Work done = increase in K.E.

$= \frac{1}{2} \times 0.5 \times 67.95 - \frac{1}{2} \times 0.5 \times 2^2$ — Use the work–energy principle. The work done is equal to the increase in kinetic energy. $\leftarrow$ **FM1, Chapter 2**

$= 15.98...$

The work done is $16.0\,\mathrm{J}$ (3 s.f.).

A You can use calculus when a particle moves in a straight line against a resistance which varies with the speed of the particle.

- **In forming an equation of motion, forces that tend to decrease the displacement are negative and forces that tend to increase the displacement are positive.**

Example 5

A particle P of mass $0.5\,\text{kg}$ moves in a straight horizontal line. When the speed of P is $v\,\text{m s}^{-1}$, the resultant force acting on P is a resistance of magnitude $3v\,\text{N}$.

Find the distance moved by P as it slows down from $12\,\text{m s}^{-1}$ to $6\,\text{m s}^{-1}$.

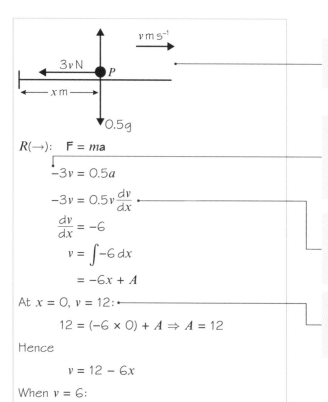

You measure the displacement of P, $x\,\text{m}$, from the point where it has a speed of $12\,\text{m s}^{-1}$.

$R(\rightarrow)$:　$F = ma$

$$-3v = 0.5a$$

$$-3v = 0.5v\frac{dv}{dx}$$

$$\frac{dv}{dx} = -6$$

$$v = \int -6\,dx$$

$$= -6x + A$$

As the resistance to motion is acting in the direction which decreases the displacement $x\,\text{m}$, the term $3v$, representing the resistance in the equation of motion, has a negative sign.

When the question asks you to relate distance to speed, choose the expression $a = v\dfrac{dv}{dx}$ for the acceleration.

At $x = 0$, $v = 12$:

$$12 = (-6 \times 0) + A \Rightarrow A = 12$$

The displacement is measured from the point where the speed of P is $12\,\text{m s}^{-1}$. Evaluate the constant of integration using $x = 0$ when $v = 12$.

Hence

$$v = 12 - 6x$$

When $v = 6$:

$$6 = 12 - 6x$$

$$x = \frac{12 - 6}{6} = 1$$

The distance moved by P as it slows down from $12\,\text{m s}^{-1}$ to $6\,\text{m s}^{-1}$ is $1\,\text{m}$.

Example 6

A A particle P of mass m is moving along Ox in the direction of x increasing. At time t, the only force acting on P is a resistance of magnitude $mk(c^2 + v^2)$, where v is the speed of P and c and k are positive constants. When $t = 0$, P is at O and $v = U$. The particle P comes to rest at the point A. Find:

a the distance OA

b the time P takes to travel from O to A.

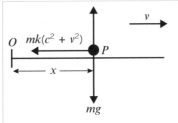

As the resistance to motion is acting in the direction which decreases the displacement x, the term $mk(c^2 + v^2)$, representing the resistance in the equation of motion, has a negative sign.

a $R(\rightarrow):$ $\qquad F = ma$

$$-mk(c^2 + v^2) = ma$$

$$-mk(c^2 + v^2) = mv\frac{dv}{dx}$$

P comes to rest when its speed is 0. Time is not involved, so you use $a = v\dfrac{dv}{dx}$

Separating the variables and integrating

$$\int \frac{v}{c^2 + v^2}\,dv = -\int k\,dx$$

$$\frac{1}{2}\ln(c^2 + v^2) = -kx + B$$

$$kx = B - \frac{1}{2}\ln(c^2 + v^2)$$

$\displaystyle\int \frac{f'(x)}{f(x)}\,dx = \ln f(x) + \text{a constant}$

As $\dfrac{d}{dv}(c^2 + v^2) = 2v$, then

$$\int \frac{v}{c^2 + v^2}\,dv = \frac{1}{2}\ln(c^2 + v^2)$$

At $x = 0$, $v = U$:

$$0 = B - \frac{1}{2}\ln(c^2 + U^2) \Rightarrow B = \frac{1}{2}\ln(c^2 + U^2)$$

You need only put the arbitrary constant on one side of the equation.

← **Pure Year 2, Chapter 11**

Hence

$$kx = \frac{1}{2}\ln(c^2 + U^2) - \frac{1}{2}\ln(c^2 + v^2)$$

$$x = \frac{1}{2k}\ln\left(\frac{c^2 + U^2}{c^2 + v^2}\right)$$

Use the initial conditions to find the constant of integration.

At A, $v = 0$ and $x = OA$:

$$OA = \frac{1}{2k}\ln\left(\frac{c^2 + U^2}{c^2}\right)$$

Using the law of logarithms

$$\ln a - \ln b = \ln\left(\frac{a}{b}\right)$$

b $R(\rightarrow):$ $\qquad F = ma$

$$-mk(c^2 + v^2) = ma$$

$$-mk(c^2 + v^2) = m\frac{dv}{dt}$$

The particle comes to rest where $v = 0$.

P comes to rest when its speed is 0. Distance is not involved, so you use $a = \dfrac{dv}{dt}$

Separating the variables and integrating

$$\int \frac{1}{c^2 + v^2}\, dv = -\int k\, dt$$

$$\frac{1}{c} \arctan\left(\frac{v}{c}\right) = -kt + D$$

Use the following standard result from the formulae booklet:

$$\int \frac{1}{a^2 + x^2}\, dx = \frac{1}{a} \arctan\left(\frac{x}{a}\right)$$

You need only put the arbitrary constant on one side of the equation.

Hence

$$kt = D - \frac{1}{c} \arctan\frac{v}{c}$$

When $t = 0$, $v = U$:

$$0 = D - \frac{1}{c} \arctan\left(\frac{U}{c}\right) \Rightarrow D = \frac{1}{c} \arctan\left(\frac{U}{c}\right)$$

Hence

$$kt = \frac{1}{c} \arctan\left(\frac{U}{c}\right) - \frac{1}{c} \arctan\left(\frac{v}{c}\right)$$

When $v = 0$:

$$kt = \frac{1}{c} \arctan\left(\frac{U}{c}\right)$$

Using arctan 0 = 0

$$t = \frac{1}{ck} \arctan\left(\frac{U}{c}\right)$$

The time P takes to travel from O to A is

$$\frac{1}{ck} \arctan\left(\frac{U}{c}\right)$$

Example 7

A car of mass 800 kg travels along a straight horizontal road. The engine of the car produces a constant driving force of magnitude 2000 N. At time t seconds, the speed of the car is $v\,\mathrm{m\,s^{-1}}$. As the car moves, the total resistance to the motion of the car is of magnitude $(400 + 4v^2)$ N. The car starts from rest.

a Find v in terms of t.

b Show that the speed of the car cannot exceed $20\,\mathrm{m\,s^{-1}}$.

a

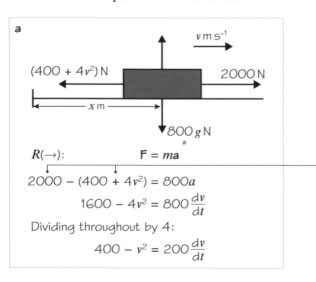

$R(\rightarrow)$: $\qquad$ $F = ma$

$$2000 - (400 + 4v^2) = 800a$$

$$1600 - 4v^2 = 800\frac{dv}{dt}$$

Dividing throughout by 4:

$$400 - v^2 = 200\frac{dv}{dt}$$

The driving force is in the direction of x increasing and so the term representing the driving force, 2000, is positive in the equation of motion.

The resistance is in the direction of x decreasing and so the term representing the resistance, $400 + 4v^2$, is negative in the equation of motion.

A

Separating the variables:

$$\int \frac{1}{400 - v^2}\, dv = \frac{1}{200} \int 1\, dt$$

Let

$$\frac{1}{400 - v^2} = \frac{1}{(20 + v)(20 - v)} = \frac{A}{20 + v} + \frac{B}{20 - v}.$$

> To integrate $\frac{1}{400 - v^2}$ you must factorise $400 - v^2 = (20 + v)(20 - v)$ and use partial fractions.

Multiplying throughout by $(20 + v)(20 - v)$:

$$1 = A(20 - v) + B(20 + v)$$

Let $v = -20$

$$1 = 40A \Rightarrow A = \frac{1}{40}$$

Let $v = 20$

$$1 = 40B \Rightarrow B = \frac{1}{40}$$

Hence

$$\frac{1}{40} \int \left(\frac{1}{20 + v} + \frac{1}{20 - v} \right) dv = \frac{1}{200} \int 1\, dt$$

$$\frac{1}{40} (\ln (20 + v) - \ln (20 - v)) = \frac{1}{200} t + C$$

$$\ln \left(\frac{20 + v}{20 - v} \right) = \frac{1}{5} t + D, \text{ where } D = 40C$$

> $\ln a - \ln b = \ln \left(\frac{a}{b} \right)$ so
> $\ln (20 + v) - \ln (20 - v) = \ln \left(\frac{20 + v}{20 - v} \right)$

$$\frac{20 + v}{20 - v} = e^{\frac{t}{5} + D} = e^D e^{\frac{t}{5}} = F e^{\frac{t}{5}}$$

> Replace the arbitrary constant e^D with a different arbitrary constant, F.

When $t = 0$, $v = 0$:

$$\frac{20 + 0}{20 - 0} = Fe^0 \Rightarrow F = 1$$

Hence

$$\frac{20 + v}{20 - v} = e^{\frac{t}{5}}$$

> To complete part **a**, you must make v the subject of this formula.

$$20 + v = 20 e^{\frac{t}{5}} - v e^{\frac{t}{5}}$$

$$v\left(e^{\frac{t}{5}} + 1\right) = 20\left(e^{\frac{t}{5}} - 1\right)$$

$$v = \frac{20\left(e^{\frac{t}{5}} - 1\right)}{\left(e^{\frac{t}{5}} + 1\right)}$$

b For all real t, $e^{\frac{t}{5}} - 1 < e^{\frac{t}{5}} + 1$

> $x - 1 < x + 1$ for all real values of x.

Hence

$$\left| \frac{e^{\frac{t}{5}} - 1}{e^{\frac{t}{5}} + 1} \right| < 1$$

and

$$\left| \frac{20\left(e^{\frac{t}{5}} - 1\right)}{\left(e^{\frac{t}{5}} + 1\right)} \right| < 20$$

> For $t > 0$, both the numerator and denominator of this fraction are positive and, as the numerator is less than the denominator, the value of the fraction must be between 0 and 1.

So the speed of the car cannot exceed $20\,\text{m}\,\text{s}^{-1}$, as required.

> $20\,\text{m}\,\text{s}^{-1}$ is called the terminal or limiting speed. ← **Chapter 4**

A 1 A particle P of mass $0.2\,\text{kg}$ is moving on the x-axis. At time t seconds P is x metres from the origin O. The force acting on P has magnitude $2\cos t\,\text{N}$ and acts in the direction OP. When $t = 0$, P is at rest at O. Calculate:

   a the speed of P when $t = 2$

   b the speed of P when $t = 3$

   c the time when P first comes to instantaneous rest

   d the distance OP when $t = 2$

   e the distance OP when P first comes to instantaneous rest.

2 A van of mass $1200\,\text{kg}$ moves along a horizontal straight road. At time t seconds, the resultant force acting on the car has magnitude $\dfrac{60\,000}{(t + 5)^2}\,\text{N}$ and acts in the direction of motion of the van. When $t = 0$, the van is at rest. The speed of the van approaches a limiting value $V\,\text{m\,s}^{-1}$. Find:

   a the value of V

   b the distance moved by the van in the first 4 seconds of its motion.

3 A particle P of mass $0.8\,\text{kg}$ is moving along the x-axis. At time $t = 0$, P passes through the origin O, moving in the positive x direction. At time t seconds, $OP = x$ metres and the velocity of P is $v\,\text{m\,s}^{-1}$. The resultant force acting on P has magnitude $\frac{1}{6}(15 - x)\,\text{N}$, and acts in the positive x direction. The maximum speed of P is $12\,\text{m\,s}^{-1}$.

   a Explain why the maximum speed of P occurs when $x = 15$.

   b Find the speed of P when $t = 0$.

4 A particle P of mass $0.75\,\text{kg}$ is moving in a straight line. At time t seconds after it passes through a fixed point on the line, O, the distance OP is x metres and the force acting on P has magnitude $(2e^{-x} + 2)\,\text{N}$ and acts in the direction OP. Given that P passes through O with speed $5\,\text{m\,s}^{-1}$, calculate the speed of P when:

   a $x = 3$

   b $x = 7$

   c Calculate the work done by the force in moving the particle from the point where $x = 3$ to the point where $x = 7$.

5 A particle P of mass $0.5\,\text{kg}$ moves away from the origin O along the positive x-axis. When $OP = x$ metres the force acting on P has magnitude $\dfrac{3}{x + 2}\,\text{N}$ and is directed away from O. When $x = 0$ the speed of P is $1.5\,\text{m\,s}^{-1}$. Find the value of x when the speed of P is $2\,\text{m\,s}^{-1}$.

E/P 6 A particle P of mass $250\,\text{g}$ moves along the x-axis in the positive direction. At time $t = 0$, P passes through the origin with speed $10\,\text{m\,s}^{-1}$. At time t seconds, the distance OP is x metres and the speed of P is $v\,\text{m\,s}^{-1}$. The resultant force acting on P is directed towards the origin and has magnitude $\dfrac{8}{(t + 1)^2}\,\text{N}$.

   a Show that $v = 2\left(\dfrac{16}{t + 1} - 11\right)$.   **(5 marks)**

   b Find the value of x when $t = 5$.   **(5 marks)**

7 A particle P of mass $0.6\,\text{kg}$ moves along the x-axis in the positive direction. A single force acting on P is directed towards the origin, O, and has magnitude $\dfrac{k}{(x+2)^2}\,\text{N}$ where

Problem-solving

The boundary conditions give the velocity for two different displacements. You need to set up and solve two simultaneous equations.

$OP = x$ metres and k is a constant.
At time $t = 0$, P passes through the origin. When $x = 3$ the speed of P is $5\,\text{m s}^{-1}$, when $x = 8$ the speed of P is $\sqrt{3}\,\text{m s}^{-1}$. Find the value of k. **(6 marks)**

8 A particle P of mass $2.5\,\text{kg}$ moves in a straight horizontal line. When the speed of P is $v\,\text{m s}^{-1}$, the resultant force acting on P is a resistance of magnitude $10v\,\text{N}$. Find the time P takes to slow down from $24\,\text{m s}^{-1}$ to $6\,\text{m s}^{-1}$.

9 A particle P of mass $0.8\,\text{kg}$ is moving along the axis Ox in the direction of x increasing. When the speed of P is $v\,\text{m s}^{-1}$, the resultant force acting on P is a resistance of magnitude $0.4v^2\,\text{N}$. Initially P is at O and is moving with speed $12\,\text{m s}^{-1}$. Find the distance P moves before its speed is halved.

10 A particle P of mass $0.5\,\text{kg}$ moves in a straight horizontal line against a resistance of magnitude $(4 + 0.5v)\,\text{N}$, where $v\,\text{m s}^{-1}$ is the speed of P at time t seconds. When $t = 0$, P is at a point A moving with speed $12\,\text{m s}^{-1}$. The particle P comes to rest at the point B. Find:
  a the time P takes to move from A to B　　　　**b** the distance AB.

11 A particle of mass m is projected along a rough horizontal plane with velocity $u\,\text{m s}^{-1}$. The coefficient of friction between the particle and the plane is μ. When the particle is moving with speed $v\,\text{m s}^{-1}$, it is also subject to an air resistance of magnitude $kmgv^2$, where k is a constant. Find the distance the particle moves before coming to rest.

12 A particle P of mass m is moving along the axis Ox in the direction of x increasing. At time t seconds, the velocity of P is v. The only force acting on P is a resistance of magnitude $k(a^2 + v^2)$. At time $t = 0$, P is at O and its speed is U. At time $t = T$, $v = \frac{1}{2}U$.
  a Show that $T = \dfrac{m}{ak}\left(\arctan\left(\dfrac{U}{a}\right) - \arctan\left(\dfrac{U}{2a}\right)\right)$. **(6 marks)**
  b Find the distance travelled by P as its speed is reduced from U to $\frac{1}{2}U$. **(5 marks)**

13 A lorry of mass $2000\,\text{kg}$ travels along a straight horizontal road. The engine of the lorry produces a constant driving force of magnitude $10\,000\,\text{N}$. At time t seconds, the speed of the lorry is $v\,\text{m s}^{-1}$. As the lorry moves, the total resistance to the motion of the lorry is of magnitude $(4000 + 500v)\,\text{N}$. The lorry starts from rest. Find:
  a v in terms of t **(6 marks)**
  b the terminal speed of the lorry. **(2 marks)**

14 At time $t = 0$, a particle of mass m is projected vertically upwards with speed U.
  The particle is subject to air resistance of magnitude $\dfrac{mgv}{k}$, where v is the speed of the particle at time t and k is a positive constant.
  a Show that the particle reaches its greatest height, H, above the point of projection at time, T,
  where $T = \dfrac{k}{g}\ln\dfrac{(k + U)}{k}$ **(6 marks)**
  b Find the greatest height, H, above the point of projection in terms of U and k. **(6 marks)**

15 At time $t = 0$, a particle of mass m is released from rest and falls vertically. At time t, the speed of the particle is v and the particle has fallen a distance x. The particle moves against air resistance, whose magnitude is modelled as being $mgkv^2$ where v is the speed of the particle at time t and k is a positive constant. Find:

a an expression for v in terms of x and k **(6 marks)**

b the terminal speed of the particle. **(1 mark)**

After a given time, the particle is observed to be falling with constant velocity.

c Comment on the model with reference to this observation. **(2 marks)**

Challenge

A particle P of mass m kg is acted on by a single force, F N, and moves in a straight line, passing a fixed point O at time $t = 0$. At the point when the displacement of the particle from O is x m, the force acts in the direction OP and has magnitude
$$F = 3x^2 - \sqrt[3]{x} \text{ N}$$

a Show that the work done by the force between times $t = a$ and $t = b$ is independent of the velocity of the particle at the point when it passes O.

b Find the work done by the force in the first 6 seconds of the motion of the particle.

5.2 Newton's law of gravitation

You can use Newton's law of gravitation to solve problems involving a particle moving away from (or towards) the Earth's surface.

Newton's law of gravitation states:

- **The force of attraction between two bodies of masses M_1 and M_2 is directly proportional to the product of their masses and inversely proportional to the square of the distance between them.**

This law is sometimes referred to as the inverse square law. It can be expressed mathematically by the following equation:

- $F = \dfrac{GM_1M_2}{d^2}$

   where G is a constant known as the constant of gravitation.

Watch out Newton's law of gravitation should be used when modelling large changes in distance relative to the sizes of the bodies, such as a rocket being launched into orbit. For small changes in height (such as when a ball is thrown into the air), gravity can be modelled as a constant force.

This force causes particles (and bodies) to fall to the Earth and the Moon to orbit the Earth.

Relationship between G and g

When a particle of mass m is resting on the surface of the Earth the force with which the Earth attracts the particle has magnitude mg and is directed towards the centre of the Earth.

Notation The numerical value of G was first determined by Henry Cavendish in 1798. In S.I. units, G is 6.67×10^{-11} kg^{-1} m^3 s^{-2}. Example 8 demonstrates the extremely small gravitational attraction between two everyday objects. You can ignore the gravitational force between small objects in your calculations.

A By modelling the Earth as a sphere of mass M and radius R and using Newton's law of gravitation:

$$F = \frac{GmM}{R^2} \quad \text{and} \quad F = mg$$

so

$$\frac{GmM}{R^2} = mg$$

and hence

$$G = \frac{gR^2}{M}$$

This relationship means you can answer questions involving gravity without using G explicitly.

When a particle is moving away from or towards the Earth the distance d between the two particles is changing. As the force of attraction between them is given by $F = \frac{GM_1M_2}{d^2}$, it follows that the force is a function of displacement and the methods of Section 5.1 must be used to solve problems.

Example 8

Two particles of masses 0.5 kg and 2.5 kg are 4 cm apart. Calculate the magnitude of the gravitational force between them.

$$F = \frac{GM_1M_2}{d^2}$$

$$= \frac{6.67 \times 10^{-11} \times 0.5 \times 2.5}{0.04^2}$$

$$= 5.210... \times 10^{-8}$$

The magnitude of the gravitational force between the particles is 5.21×10^{-8} N (3 s.f.).

Use $F = \frac{GM_1M_2}{d^2}$ with

$G = 6.67 \times 10^{-11}$, $M_1 = 0.5$, $M_2 = 2.5$ and $d = 0.04$.

This value is so small that it can be ignored in most calculations.

Example 9

Above the Earth's surface, the magnitude of the force on a particle due to the Earth's gravitational force is inversely proportional to the square of the distance of the particle from the centre of the Earth. The acceleration due to gravity on the surface of the Earth is g and the Earth can be modelled as a sphere of radius R. A particle P of mass m is a distance $(x - R)$, (where $x \geqslant R$), above the surface of the Earth.

a Prove that the magnitude of the gravitational force acting on P is $\frac{mgR^2}{x^2}$

A spacecraft S is fired vertically upwards from the surface of the Earth. When it is at a height $2R$ above the surface of the Earth its speed is $\frac{1}{2}\sqrt{gR}$. Assuming that air resistance can be ignored and the rocket's engine is turned off immediately after the rocket is fired,

b find, in terms of g and R, the speed with which S was fired.

a $F \propto \dfrac{1}{x^2}$ or $F = \dfrac{k}{x^2}$

So on the surface of the Earth $F = \dfrac{k}{R^2}$

On the surface of the Earth the magnitude of the force = mg.

$\therefore mg = \dfrac{k}{R^2} \Rightarrow k = mgR^2$

$\therefore$ the magnitude of the gravitational force is $\dfrac{mgR^2}{x^2}$

The force is the weight of the particle.

b
$$F = ma$$
$$\frac{mgR^2}{x^2} = -ma$$
$$v\frac{dv}{dx} = -\frac{gR^2}{x^2}$$
$$\int v\,dv = -gR^2 \int \frac{1}{x^2}\,dx$$
$$\tfrac{1}{2}v^2 = gR^2\frac{1}{x} + C$$

$$x = 3R, \ v = \tfrac{1}{2}\sqrt{gR}$$
$$\Rightarrow \tfrac{1}{2} \times \tfrac{1}{4}gR = gR^2 \times \frac{1}{3R} + C$$
$$C = -\tfrac{5}{24}gR$$
$$\tfrac{1}{2}v^2 = gR^2\frac{1}{x} - \tfrac{5}{24}gR$$
$$x = R \Rightarrow \tfrac{1}{2}v^2 = gR^2 \times \frac{1}{R} - \tfrac{5}{24}gR = \tfrac{19}{24}gR$$
$$\therefore S \text{ is fired with speed } \sqrt{\tfrac{19}{12}gR}.$$

Taking upwards as the positive direction, the force acts vertically downwards.

The force is a function of x so replace a with $v\frac{dv}{dx}$

Separate the variables and integrate. Don't forget to include the constant of integration.

Use the information in the question to obtain the value of C.

Finally make $x = R$ and obtain an expression for v as required.

Exercise 5B

P **1** Above the Earth's surface, the magnitude of the force on a particle due to the Earth's gravitational force is inversely proportional to the square of the distance of the particle from the centre of the Earth. The acceleration due to gravity on the surface of the Earth is g and the Earth can be modelled as a sphere of radius R. A particle P of mass m is a distance $(x - R)$ (where $x \geqslant R$) above the surface of the Earth. Prove that the magnitude of the gravitational force acting on P is $\dfrac{mgR^2}{x^2}$

P **2** The Earth can be modelled as a sphere of radius R. At a distance x (where $x \geqslant R$) from the centre of the Earth the magnitude of the acceleration due to the Earth's gravitational force is A. On the surface of the Earth, the magnitude of the acceleration due to the Earth's gravitational force is g. Prove that $A = \dfrac{gR^2}{x^2}$

Watch out In questions 3 to 6 you may assume either of the results proved in questions 1 and 2.

E/P **3** A spacecraft S is fired vertically upwards from the surface of the Earth. When it is at a height R where R is the radius of the Earth, above the surface of the Earth its speed is $\sqrt{gR}$.
Model the spacecraft as a particle and the Earth as a sphere of radius R and find, in terms of g and R, the speed with which S was fired. (You may assume that air resistance can be ignored and that the rocket's engine is turned off immediately after the rocket is fired.) **(7 marks)**

P **4** A rocket of mass m is fired vertically upwards from the surface of the Earth with initial speed U. The Earth is modelled as a sphere of radius R and the rocket as a particle. Find an expression for the speed of the rocket when it has travelled a distance X metres. (You may assume that air resistance can be ignored and that the rocket's engine is turned off immediately after the rocket is fired.)

A
P

5 A particle is fired vertically upwards from the Earth's surface. The initial speed of the particle is u where $u^2 = 3gR$ and R is the radius of the Earth. Find, in terms of g and R, the speed of the particle when it is at a height $4R$ above the Earth's surface. (You may assume that air resistance can be ignored.)

P **6** A particle is moving in a straight line towards the centre of the Earth, which is assumed to be a sphere of radius R. The particle starts from rest when its distance from the centre of the Earth is $3R$. Find the speed of the particle as it hits the surface of the Earth. (You may assume that air resistance can be ignored.)

E/P **7** A space shuttle S of mass m moves in a straight line towards the centre of the Earth. The Earth is modelled as a sphere of radius R and S is modelled as a particle. When S is at a distance x ($x \geqslant R$) from the centre of the Earth, the gravitational force exerted by the Earth on S is directed towards the centre of the Earth. The magnitude of this force is inversely proportional to x^2.

a Prove that the magnitude of the gravitational force on S is $\dfrac{mgR^2}{x^2}$

When S is at a height of $3R$ above the surface of the Earth, the speed of S is $\sqrt{2gR}$. Assuming that air resistance can be ignored,

b find, in terms of g and R, the speed of S as it hits the surface of the Earth. **(7 marks)**

Challenge

Given that $G = 6.67 \times 10^{-11} \text{ kg}^{-1} \text{ m}^3 \text{ s}^{-2}$, $g = 9.81 \text{ m s}^{-2}$ and that the radius of the Earth is 6.3781×10^6 m, estimate:

a the mass of the Earth

b the average density of the Earth.

5.3 Simple harmonic motion

You can solve problems about a particle which is moving in a straight line with simple harmonic motion.

- **Simple harmonic motion (S.H.M.) is motion in which the acceleration of a particle P is always towards a fixed point O on the line of motion of P and has magnitude proportional to the displacement of P from O.**

Notation The point O is called the **centre of oscillation**.

Online Explore simple harmonic motion using GeoGebra.

We write $\qquad \ddot{x} = -\omega^2 x$

This can be shown on a diagram:

The minus sign means that the acceleration is always directed towards O.

Links You usually use 'dot' notation when analysing simple harmonic motion.

$$\dot{x} = \frac{dx}{dt} \text{ and } \ddot{x} = \frac{d^2x}{dt^2}$$

← **Core Pure Book 2, Section 8.2**

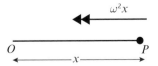

184

A As $\ddot{x}$ is a function of x, we use $\ddot{x} = v\dfrac{\mathrm{d}v}{\mathrm{d}x}$ to derive an expression for the velocity of P:

$$v\frac{\mathrm{d}v}{\mathrm{d}x} = -\omega^2 x$$

$$\int v\,\mathrm{d}v = \int -\omega^2 x\,\mathrm{d}x \quad\text{————————— Separate the variables and integrate.}$$

$$\tfrac{1}{2}v^2 = -\omega^2 \times \tfrac{1}{2}x^2 + C \quad\text{————————— } C \text{ is the constant of integration.}$$

The speed of P is the modulus of v or the modulus of $\dfrac{\mathrm{d}x}{\mathrm{d}t}$

This speed is zero when x has its maximum or minimum value.
Let the maximum displacement of P from O be a. This gives:

$$0 = -\omega^2 \times \tfrac{1}{2}a^2 + C$$

$$C = \tfrac{1}{2}\omega^2 a^2$$

Hence $\quad \tfrac{1}{2}v^2 = -\tfrac{1}{2}\omega^2 x^2 + \tfrac{1}{2}\omega^2 a^2$

or $\qquad \boldsymbol{v^2 = \omega^2(a^2 - x^2)}$

You can derive an expression for the displacement of P at time t by solving the differential equation $\ddot{x} = -\omega^2 x$ directly:

Links You could also write this equation as $\dfrac{\mathrm{d}^2 x}{\mathrm{d}t^2} + \omega^2 x = 0$. It is a second-order linear differential equation.

← **Core Pure Book 2, Section 7.2**

$$\ddot{x} + \omega^2 x = 0$$

The auxiliary equation is:

$$m^2 + \omega^2 = 0$$

$$m = \pm\omega\mathrm{i}$$

This has two imaginary roots, so the general solution is:

$$x = A\cos\omega t + B\sin\omega t$$

where A and B are arbitrary constants. ——————— This is the general solution when the auxiliary equation has two purely imaginary roots.

This can be rewritten in the form

$$\boldsymbol{x = a\sin(\omega t + \alpha)}$$

where a and α are arbitrary constants and $a > 0$. ——— By choosing α appropriately it is possible to make $a > 0$ for any possible boundary conditions.

So the motion of the particle is a sine function with maximum and minimum values $\pm a$ and period $\dfrac{2\pi}{\omega}$. The value a is called the **amplitude** of the motion, and $-a \leqslant x \leqslant a$ for all values of t.

We can use different values of α to investigate the motion further.

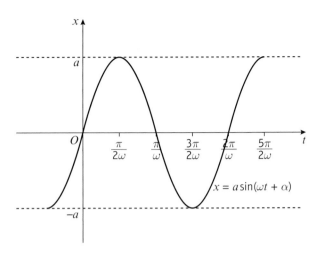

A Case 1: $\alpha = 0$

When $\alpha = 0$, the equation $x = a\sin(\omega t + \alpha)$ becomes $x = a\sin\omega t$.

Here is the graph of $x = a\sin\omega t$:

As the graph passes through the origin, we have $x = 0$ when $t = 0$. Thus $x = a\sin\omega t$ gives the displacement from the centre of oscillation of a particle moving with S.H.M. of amplitude a and period $\frac{2\pi}{\omega}$ which is at the centre of the oscillation and moving in the positive direction when $t = 0$.

Case 2: $\alpha = \dfrac{\pi}{2}$

When $\alpha = \frac{\pi}{2}$, the equation $x = a\sin(\omega t + \alpha)$ becomes $x = a\sin\left(\omega t + \frac{\pi}{2}\right)$.

The graph of $x = a\sin\left(\omega t + \frac{\pi}{2}\right)$ is:

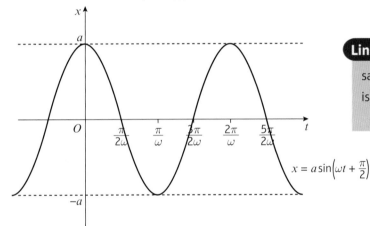

Links The graph of $x = a\sin\left(\omega t + \frac{\pi}{2}\right)$ is the same shape as the graph of $x = a\sin\omega t$ but is translated $\frac{\pi}{2\omega}$ to the left.

← Pure Year 2, Section 2.6

This is also the graph of $x = a\cos\omega t$. When $t = 0$ the particle's displacement from O is a.

Once again, the amplitude is a and the period is $\frac{2\pi}{\omega}$

Case 3: α is neither of the above.

When α takes some value other than 0 or $\frac{\pi}{2}$, the graph of $x = a\sin(\omega t + \alpha)$ is a translation of the graph of $x = a\sin\omega t$ through a distance $\frac{\alpha}{\omega}$ to the left:

The particle is neither at the centre nor at an extreme point of the oscillation when $t = 0$.

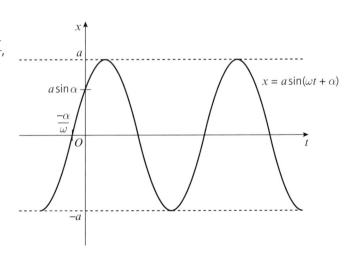

- **For S.H.M. of amplitude a defined by the equation $\ddot{x} = -\omega^2 x$,**
  - $v^2 = \omega^2(a^2 - x^2)$
  - **If P is at the centre of the oscillation when $t = 0$, use $x = a\sin\omega t$.**
  - **If P is at an end point of the oscillation when $t = 0$, use $x = a\cos\omega t$.**
  - **If P is at some other point when $t = 0$, use $x = a\sin(\omega t + \alpha)$.**
  - **The period of the oscillation is $T = \dfrac{2\pi}{\omega}$**

> **Note** These are standard results which may be quoted without proof in your exam.

Geometrical methods

You can also use a geometrical method to solve a simple harmonic motion problem.

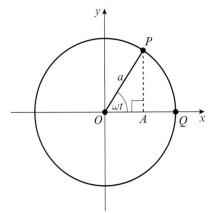

A particle P is moving round a circle of radius a, centre the origin O. The particle has a constant angular speed ω in an anticlockwise sense. The foot of the perpendicular from P to the x-axis is the point A.

> **Links** Angular speed, ω, is the rate at which the radius is turning, measured in rad s^{-1}. ← **Section 1.1**

The motion is timed from the instant when P is at the point Q on the x-axis. So t seconds later, $\angle POA = \omega t$ and $OA = x = a\cos\omega t$.

Hence $\dot{x} = -a\omega\sin\omega t$

and $\quad \ddot{x} = -a\omega^2\cos\omega t = -\omega^2 x$

This shows that the point A is moving along the x-axis with simple harmonic motion.

The amplitude is a and the period is $\dfrac{2\pi}{\omega}$

The circle associated with any particular simple harmonic motion is called the **reference circle**. Using the reference circle can be useful when calculating the time taken for a particle to move between two points of the oscillation, as shown in Example 14.

Example 10

A particle is moving along a straight line with S.H.M. The amplitude of the motion is 0.8 m. It passes through the centre of the oscillation O with speed 2 m s^{-1}. Calculate:

a the period of the oscillation

b the speed of the particle when it is 0.4 m from O

c the time the particle takes to travel 0.4 m from O.

a $v^2 = \omega^2(a^2 - x^2)$

$2^2 = \omega^2(0.8^2 - 0)$ ⟶ Use $v^2 = \omega^2(a^2 - x^2)$ with $v = 2$, $a = 0.8$ and $x = 0$.

$\omega = \frac{2}{0.8} = 2.5$ ⟶ Solve for ω.

period $= \frac{2\pi}{\omega}$

$= \frac{2\pi}{2.5} = \frac{4\pi}{5}$ ⟶ Use period $= \frac{2\pi}{\omega}$ with $\omega = 2.5$

The period is $\frac{4\pi}{5}$ s.

b $v^2 = \omega^2(a^2 - x^2)$

$v^2 = 2.5^2(0.8^2 - 0.4^2)$ ⟶ Use $v^2 = \omega^2(a^2 - x^2)$ with $\omega = 2.5$, $a = 0.8$ and $x = 0.4$ and solve for v.

$v = 1.732...$

The particle's speed is $1.73 \, \text{m s}^{-1}$ (3 s.f.).

c $x = a\sin\omega t$

$0.4 = 0.8\sin 2.5t$ ⟶ Take $t = 0$ at the centre and use $x = a\sin\omega t$ with $\omega = 2.5$, $a = 0.8$ and $x = 0.4$

$\sin 2.5t = 0.5$

$2.5t = \arcsin 0.5$

$t = \frac{1}{2.5}\arcsin 0.5 = 0.2094...$

The particle takes 0.209 s (3 s.f.).

Watch out Remember to have your calculator in radian mode.

Example 11

A particle P of mass 0.5 kg is moving along a straight line. At time t seconds, the distance of P from a fixed point O on the line is x metres. The force acting on P has magnitude $10x$ and acts in the direction PO.

a Show that P is moving with simple harmonic motion.

b Find the period of the motion.

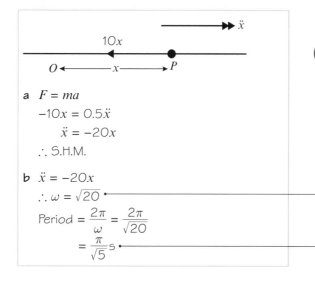

a $F = ma$

$-10x = 0.5\ddot{x}$

$\ddot{x} = -20x$

$\therefore$ S.H.M.

Problem-solving

If you need to show that a particle is moving with simple harmonic motion you must show that $\ddot{x} = -\omega^2 x$ and state the conclusion.

b $\ddot{x} = -20x$

$\therefore \omega = \sqrt{20}$ ⟶ Compare the equation for P found in **a** with $\ddot{x} = -\omega^2 x$ to find ω.

Period $= \frac{2\pi}{\omega} = \frac{2\pi}{\sqrt{20}}$

$= \frac{\pi}{\sqrt{5}}$ s ⟶ The answer can be left in its exact form or given correct to 3 s.f. (1.40 s).

A A particle is moving in a straight line with simple harmonic motion. Its maximum acceleration is $12\,\mathrm{m\,s^{-2}}$ and its maximum speed is $4\,\mathrm{m\,s^{-1}}$. Calculate:

a the period of the motion

b the amplitude of the motion

c the length of time, during one complete oscillation, that the particle is within $\frac{1}{3}\,\mathrm{m}$ of the centre of the oscillation.

a $v^2 = \omega(a^2 - x^2)$

$4^2 = \omega^2 a^2$

$\omega a = 4$ (1)

$\ddot{x} = -\omega^2 x$

$12 = \omega^2 a$ (2)

$\dfrac{\omega^2 a}{\omega a} = \dfrac{12}{4}$

$\omega = 3$

$\text{period} = \dfrac{2\pi}{\omega} = \dfrac{2\pi}{3}\,(= 2.094\ldots)$

The period is $\dfrac{2\pi}{3}$ (or $2.09\,\mathrm{s}$ (3 s.f.)).

b $\dfrac{(\omega a)^2}{\omega^2 a} = \dfrac{4^2}{12} = \dfrac{4}{3}$

$a = \dfrac{4}{3} = 1.333\ldots$

The amplitude is a $1.33\,\mathrm{m}$ (3 s.f.).

c Taking the initial displacement from O as 0.

$x = a\sin\omega t$ so $x = \dfrac{4}{3}\sin 3t$

Need to find t such that $-\dfrac{1}{3} \leqslant \dfrac{4}{3}\sin 3t \leqslant \dfrac{1}{3}$

$\dfrac{4}{3}\sin 3t = \dfrac{1}{3}$

$\sin 3t = \dfrac{1}{4} \Rightarrow t = 0.08423\ldots$

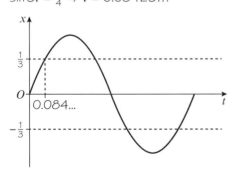

Total time within $\frac{1}{3}\,\mathrm{m} = 4 \times 0.08423\ldots$

$\qquad\qquad\qquad\qquad = 0.337\,\mathrm{s}$ (3 s.f.).

As $v^2 = \omega^2(a^2 - x^2)$, v is maximum when $x = 0$.

As $\ddot{x} = -\omega^2 x$, and hence $|\ddot{x}| = |-\omega^2 x| = \omega^2|x|$ acceleration is maximum when $|x| = a$.

Divide equation (2) by equation (1) to find ω.

Unless you are told the accuracy required, give the exact answer.

Square equation (1) and divide by equation (2) to find a.

The exact answer, $\frac{4}{3}$ or $1\frac{1}{3}$, is acceptable.

Problem-solving

The initial displacement does not affect the answer, so take $x = 0$ when $t = 0$ to simplify the expression for the displacement of the particle.

Choose the first positive value of t that satisfies the equation, and then draw a sketch to determine the total time that the particle is within $\frac{1}{3}\,\mathrm{m}$ of O. You could work out all the solutions of $\frac{4}{3}\sin 3t = \pm\frac{1}{3}$ in one complete oscillation, but it is easier to use symmetry.

Example 13

A A small rowing boat is floating on the surface of the sea, tied to a pier. The boat moves up and down in a vertical line and it can be modelled as a particle moving with simple harmonic motion. The boat takes 2 s to travel directly from its highest point, which is 3 m below the pier, to its lowest point. The maximum speed of the boat is 3 m s^{-1}. Calculate:

a the amplitude of the motion

b the time taken by the boat to rise from its lowest point to a point 5 m below the pier.

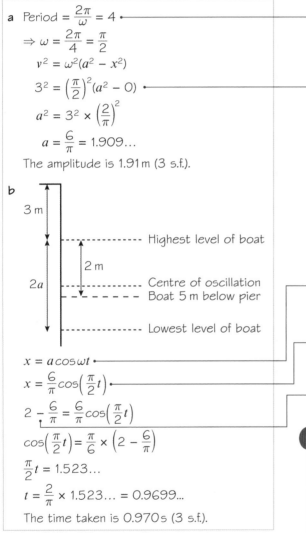

a Period $= \dfrac{2\pi}{\omega} = 4$

The time taken from the highest point to the lowest point is half the period.

$\Rightarrow \omega = \dfrac{2\pi}{4} = \dfrac{\pi}{2}$

$v^2 = \omega^2(a^2 - x^2)$

$3^2 = \left(\dfrac{\pi}{2}\right)^2(a^2 - 0)$

The maximum speed of the boat occurs when it passes the centre of oscillation, or when $x = 0$.

$a^2 = 3^2 \times \left(\dfrac{2}{\pi}\right)^2$

$a = \dfrac{6}{\pi} = 1.909\ldots$

The amplitude is 1.91 m (3 s.f.).

b

3 m .... Highest level of boat

2 m

2a .... Centre of oscillation
Boat 5 m below pier

.... Lowest level of boat

$x = a\cos\omega t$

$x = \dfrac{6}{\pi}\cos\left(\dfrac{\pi}{2}t\right)$

$2 - \dfrac{6}{\pi} = \dfrac{6}{\pi}\cos\left(\dfrac{\pi}{2}t\right)$

$\cos\left(\dfrac{\pi}{2}t\right) = \dfrac{\pi}{6} \times \left(2 - \dfrac{6}{\pi}\right)$

$\dfrac{\pi}{2}t = 1.523\ldots$

$t = \dfrac{2}{\pi} \times 1.523\ldots = 0.9699\ldots$

The time taken is 0.970 s (3 s.f.).

Time from the lowest point is needed so use $x = a\cos\omega t$.

Using the exact values of a and ω gives a more accurate answer.

x is measured from the centre of the oscillation. The diagram shows that $x = (2 - a)$ m.

Problem-solving

This analysis considers down as the positive direction, to simplify the working. If you want to consider up as the positive direction you could write the displacement as

$x = -\dfrac{6}{\pi}\cos\left(\dfrac{\pi}{2}t\right)$ and set $x = \dfrac{6}{\pi} - 2$

Example 14

A particle P is moving with simple harmonic motion along a straight line. The centre of the oscillation is the point O, the amplitude is 0.6 m and the period is 8 s. The points A and B are points on the line of P's motion and are on opposite sides of O. The distance OA is 0.5 m and OB is 0.1 m. Calculate the time taken by P to move directly from A to B.

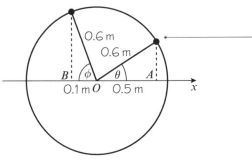

Draw a diagram showing a reference circle of radius 0.6 m. The diameter of this circle represents the line on which P is moving. Show points A and B on this diameter.

Online Explore calculations for simple harmonic motion with a reference circle using GeoGebra.

As P moves from A to B, the point on the circle moves round an arc of the circle which subtends an angle of $\pi - \theta - \phi$ at O.

$\theta = \arccos\left(\frac{0.5}{0.6}\right) = \arccos\left(\frac{5}{6}\right)$

$\phi = \arccos\left(\frac{0.1}{0.6}\right) = \arccos\left(\frac{1}{6}\right)$

Use the diagram to find θ and ϕ. Keep the exact forms.

$\text{Time} = \dfrac{\pi - \theta - \phi}{2\pi} \times 8$

$\quad\quad = \dfrac{\pi - \arccos\left(\frac{5}{6}\right) - \arccos\left(\frac{1}{6}\right)}{2\pi} \times 8$

$\quad\quad = 1.467\ldots$

The period is 8 s. This corresponds to an angle of 2π at the centre. Use proportion to find the time taken.

The time to travel from A to B is 1.47 s (3 s.f.).

Exercise 5C

1 A particle P is moving in a straight line with simple harmonic motion. The amplitude of the oscillation is 0.5 m and P passes through the centre of the oscillation O with speed $2\,\text{m s}^{-1}$. Calculate:

  a the period of the oscillation

  b the speed of P when $OP = 0.2\,\text{m}$.

2 A particle P is moving in a straight line with simple harmonic motion. The period is $\frac{\pi}{3}$ s and P's maximum speed is $6\,\text{m s}^{-1}$. The centre of the oscillation is O. Calculate:

  a the amplitude of the motion

  b the speed of P 0.3 s after passing through O.

3 A particle is moving in a straight line with simple harmonic motion. Its maximum speed is $10\,\text{m s}^{-1}$ and its maximum acceleration is $20\,\text{m s}^{-2}$. Calculate:

  a the amplitude of the motion

  b the period of the motion.

4 A particle is moving in a straight line with simple harmonic motion. The period of the motion is $\frac{3\pi}{5}$ s and the amplitude is 0.4 m. Calculate the maximum speed of the particle.

5 A particle is moving in a straight line with simple harmonic motion. Its maximum acceleration is $15\,\mathrm{m\,s^{-2}}$ and its maximum speed is $18\,\mathrm{m\,s^{-1}}$. Calculate the speed of the particle when it is 2.5 m from the centre of the oscillation.

6 A particle P is moving in a straight line with simple harmonic motion. The centre of the oscillation is O and the period is $\frac{\pi}{2}$ s. When $OP = 1.2\,\mathrm{m}$, P has speed $1.5\,\mathrm{m\,s^{-1}}$.

a Find the amplitude of the motion.

At time t seconds the displacement of P from O is x metres. When $t = 0$, P is passing through O.

b Find an expression for x in terms of t.

7 A particle is moving in a straight line with simple harmonic motion. The particle performs 6 complete oscillations per second and passes through the centre of the oscillation, O, with speed $5\,\mathrm{m\,s^{-1}}$. When P passes through the point A the magnitude of P's acceleration is $20\,\mathrm{m\,s^{-1}}$. Calculate:

a the amplitude of the motion

b the distance OA.

8 A particle P is moving on a straight line with simple harmonic motion between two points A and B. The midpoint of AB is O. When $OP = 0.6\,\mathrm{m}$, the speed of P is $3\,\mathrm{m\,s^{-1}}$ and when $OP = 0.2\,\mathrm{m}$ the speed of P is $6\,\mathrm{m\,s^{-1}}$. Find:

a the distance AB

b the period of the motion.

9 A particle is moving in a straight line with simple harmonic motion. When the particle is 1 m from the centre of the oscillation, O, its speed is $0.1\,\mathrm{m\,s^{-1}}$. The period of the motion is 2π seconds. Calculate:

a the maximum speed of the particle

b the speed of the particle when it is 0.4 m from O.

10 A piston of mass 1.2 kg is moving with simple harmonic motion inside a cylinder. The distance between the end points of the motion is 2.5 m and the piston is performing 30 complete oscillations per minute. Calculate the maximum value of the kinetic energy of the piston.

11 A marker buoy is moving in a vertical line with simple harmonic motion. The buoy rises and falls through a distance of 0.8 m and takes 2 s for each complete oscillation. Calculate:

a the maximum speed of the buoy

b the time taken for the buoy to fall a distance 0.6 m from its highest point.

12 Points O, A and B lie in that order in a straight line. A particle P is moving on the line with simple harmonic motion. The motion has period 2 s and amplitude 0.5 m. The point O is the centre of the oscillation, $OA = 0.2\,\mathrm{m}$ and $OB = 0.3\,\mathrm{m}$. Calculate the time taken by P to move directly from A to B. **(5 marks)**

A 13 A particle P is moving along the x-axis. At time t seconds the displacement, x metres, of P
E/P from the origin O is given by $x = 4\sin 2t$.

    a Prove that P is moving with simple harmonic motion. **(5 marks)**

    b Write down the amplitude and period of the motion. **(2 marks)**

    c Calculate the maximum speed of P. **(3 marks)**

    d Calculate the least value of t $(t > 0)$ for which P's speed is $4\,\mathrm{m\,s^{-1}}$. **(3 marks)**

    e Calculate the least value of t $(t > 0)$ for which $x = 2$. **(2 marks)**

E/P 14 A particle P is moving along the x-axis. At time t seconds the displacement, x metres, of P
from the origin O is given by $x = 3\sin\left(4t + \frac{1}{2}\right)$.

    a Prove that P is moving with simple harmonic motion. **(5 marks)**

    b Write down the amplitude and period of the motion. **(2 marks)**

    c Calculate the value of x when $t = 0$. **(2 marks)**

    d Calculate the value of t $(t > 0)$ the first time P passes through O. **(3 marks)**

E/P 15 On a certain day, low tide in a harbour is at 10 a.m. and the depth of the water is 5 m. High tide
on the same day is at 4.15 p.m. and the water is then 15 m deep. A ship which needs a depth of
water of 7 m needs to enter the harbour. Assuming that the water can be modelled as rising and
falling with simple harmonic motion, calculate:

    a the earliest time, to the nearest minute, after 10 a.m. at which the ship can enter the
harbour **(4 marks)**

    b the time by which the ship must leave. **(3 marks)**

E/P 16 Points A, O and B lie in that order in a straight line. A particle P is moving on the line with
simple harmonic motion with centre O. The period of the motion is 4 s and the amplitude
is 0.75 m. The distance OA is 0.4 m and AB is 0.9 m. Calculate the time taken by P to move
directly from B to A. **(5 marks)**

Challenge

A particle P is moving along the x-axis with simple harmonic motion.
The origin O is the centre of oscillation. When the displacements
from O are x_1 and x_2, the particle has speeds of v_1 and v_2 respectively.
Find the period of the motion in terms of x_1, x_2, v_1 and v_2.

5.4 Horizontal oscillation

You can investigate the motion of a particle which is attached to an elastic spring or string and is
oscillating in a horizontal line.

If an elastic spring has one end attached to a
fixed point A on a smooth horizontal surface a
particle P can be attached to the free end.
When P is pulled away from A and released P
will move towards A.

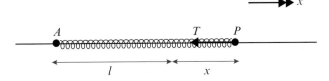

A Hooke's law: $T = \dfrac{\lambda x}{l}$

$$F = ma$$

$$-T = m\ddot{x}$$

$$m\ddot{x} = -\dfrac{\lambda x}{l}$$

$$\ddot{x} = -\dfrac{\lambda}{ml}x$$

Links λ is the modulus of elasticity of the spring and l is its natural length.   ← FM1, Chapter 3

λ, m and l are all positive constants, so the equation is of the form $\ddot{x} = -\omega^2 x$.

So P is moving with S.H.M.

The initial extension is the maximum value of x, so is the same as the amplitude.

When the particle is attached to an elastic **spring**, the particle will perform complete oscillations because there will always be a force acting – a tension when the spring is stretched and a thrust when the spring is compressed. The centre of the oscillation is where the tension is zero; that is the point when the spring has returned to its natural length.

When the particle is attached to an elastic **string**, the particle will move with S.H.M. only while the string is taut. Once the string becomes slack there is no tension and the particle continues to move with constant speed until the string becomes taut again.

- **For a particle moving on a smooth horizontal surface attached to one end of an elastic spring:**
  - **the particle will move with S.H.M.**
  - **the particle will perform complete oscillations.**

- **For a particle moving on a smooth horizontal surface attached to one end of an elastic string:**
  - **the particle will move with S.H.M. while the string is taut**
  - **the particle will move with constant speed while the string is slack.**

- **To solve problems involving elastic springs and strings:**
  - **use Hooke's law to find the tension**
  - **use $F = ma$ to obtain ω**
  - **use information given in the question to obtain the amplitude.**

Sometimes the particle is attached to two springs or strings which are stretched between two fixed points. When this happens you will need to find the tensions in both the springs or strings.

Example 15

A particle P of mass $0.6\,\text{kg}$ rests on a smooth horizontal floor attached to one end of a light elastic string of natural length $0.8\,\text{m}$ and modulus of elasticity $16\,\text{N}$. The other end of the string is fixed to a point A on the floor. The particle is pulled away from A until AP measures $1.2\,\text{m}$ and released.

a Show that, while the string remains taut, P moves with simple harmonic motion.

b Calculate the speed of P when the string returns to its natural length.

c Calculate the time that elapses between the point where the string becomes slack and the point where it next becomes taut.

d Calculate the time taken by the particle to return to its starting point for the first time.

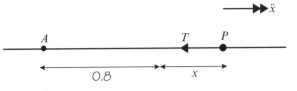

a $T = \dfrac{\lambda x}{l}$

$T = \dfrac{16x}{0.8} = 20x$

$F = ma$

$20x = -0.6\ddot{x}$

$\ddot{x} = -\dfrac{20}{0.6}x = -\dfrac{100}{3}x$

The particle is moving with S.H.M.

b $\ddot{x} = -\dfrac{100}{3}x$

$\omega^2 = \dfrac{100}{3}$

$v^2 = \omega^2(a^2 - x^2)$

$v^2 = \dfrac{100}{3} \times 0.4^2$

$v = 2.309\ldots$

At the natural length P has speed $2.31\,\text{m s}^{-1}$ (3 s.f.).

c The particle now moves a distance $1.6\,\text{m}$ at $2.309\ldots\text{m s}^{-1}$.

Time taken $= \dfrac{1.6}{2.309\ldots} = 0.6928\ldots$

The string is slack for $0.693\,\text{s}$ (3 s.f.).

d Period of the S.H.M. $= \dfrac{2\pi}{\omega} = 2\pi \times \sqrt{\dfrac{3}{100}} = 1.088\ldots$

Total time $= 1.088\ldots + (2 \times 0.6928\ldots) = 2.473\ldots$

The time taken is $2.47\,\text{s}$ (3 s.f.).

Use Hooke's law with $\lambda = 16$ and $l = 0.8$ to find the tension.

Use $F = ma$ with $F = T = 20x$ and $m = 0.6$. Remember that the positive direction is the direction of x increasing, and that the acceleration acts in the opposite direction.

The equation reduces to the form $\ddot{x} = -\omega^2 x$, so S.H.M. is proved.

Compare the equation obtained in **a** with $\ddot{x} = -\omega^2 x$ to find ω^2.

The amplitude is the same as the initial extension and at the natural length $x = 0$.

The particle moves at a constant speed while the string is slack.

The particle moves through a complete oscillation and two intervals of constant speed (when the string is slack).

Example 16

A particle P of mass $0.8\,\text{kg}$ is attached to the ends of two identical light elastic springs of natural length $1.6\,\text{m}$ and modulus of elasticity $16\,\text{N}$. The free ends of the springs are attached to two points A and B which are $4\,\text{m}$ apart on a smooth horizontal surface. The point C lies between A and B such that ABC is a straight line and $AC = 2.4\,\text{m}$. The particle is held at C and then released from rest.

a Show that the subsequent motion is simple harmonic motion.

b Find the period and amplitude of the motion.

c Calculate the maximum speed of P.

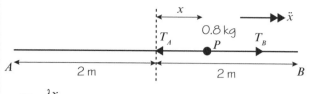

a $T = \dfrac{\lambda x}{l}$

$T_A = \dfrac{16(0.4 + x)}{1.6}$

$T_B = \dfrac{16(0.4 - x)}{1.6}$

$F = ma$

$T_B - T_A = 0.8\ddot{x}$

$\dfrac{16(0.4 - x)}{1.6} - \dfrac{16(0.4 + x)}{1.6} = 0.8\ddot{x}$

$-2 \times \dfrac{16x}{1.6} = 0.8\ddot{x}$

$\ddot{x} = -2 \times \dfrac{16x}{1.6 \times 0.8} = -25x$

The motion is S.H.M.

b $\ddot{x} = -25x$

$\omega^2 = 25, \omega = 5$

$\text{Period} = \dfrac{2\pi}{\omega} = \dfrac{2\pi}{5}\,\text{s}$

$\text{Amplitude} = 0.4\,\text{m}$

c $v^2 = \omega^2(a^2 - x^2)$

$v_{\text{max}}^2 = 25 \times 0.4^2$

$v_{\text{max}} = 5 \times 0.4 = 2$

P's maximum speed is $2\,\text{m s}^{-1}$.

Problem-solving

Use Hooke's law to find the tensions in each spring. Use your diagram to work out the extensions.

Use $F = ma$ to form an equation of motion for P. Reduce this to the form $\ddot{x} = -\omega^2 x$ to establish S.H.M.

Compare the equation found in **a** with $\ddot{x} = -\omega^2 x$ to find ω.

You can give an exact value or a 3 s.f. answer (1.26 s) for the period.

As the springs are identical the centre of the oscillation is at the midpoint of AB.

Exercise 5D

1 A particle P of mass $0.5\,\text{kg}$ is attached to one end of a light elastic spring of natural length $0.6\,\text{m}$ and modulus of elasticity $60\,\text{N}$. The other end of the spring is fixed to a point A on the smooth horizontal surface on which P rests. The particle is held at rest with $AP = 0.9\,\text{m}$ and then released.

a Show that P moves with simple harmonic motion.

b Find the period and amplitude of the motion.

c Calculate the maximum speed of P.

2 A particle P of mass $0.8\,\text{kg}$ is attached to one end of a light elastic string of natural length $1.6\,\text{m}$ and modulus of elasticity $20\,\text{N}$. The other end of the string is fixed to a point O on the smooth horizontal surface on which P rests. The particle is held at rest with $OP = 2.6\,\text{m}$ and then released.

a Show that, while the string is taut, P moves with simple harmonic motion.

b Calculate the time from the instant of release until P returns to its starting point for the first time.

3 A particle P of mass $0.4\,\text{kg}$ is attached to one end of a light elastic string of modulus of elasticity $24\,\text{N}$ and natural length $1.2\,\text{m}$. The other end of the string is fixed to a point A on the smooth horizontal table on which P rests. Initially P is at rest with $AP = 1\,\text{m}$. The particle receives an impulse of magnitude $1.8\,\text{N s}$ in the direction AP.

    a Show that, while the string is taut, P moves with simple harmonic motion.

    b Calculate the time that elapses between the moment P receives the impulse and the next time the string becomes slack.

    The particle comes to instantaneous rest for the first time at the point B.

    c Calculate the distance AB.

4 A particle P of mass $0.8\,\text{kg}$ is attached to one end of a light elastic spring of natural length $1.2\,\text{m}$ and modulus of elasticity $80\,\text{N}$. The other end of the spring is fixed to a point O on the smooth horizontal surface on which P rests. The particle is held at rest with $OP = 0.6\,\text{m}$ and then released.

    a Show that P moves with simple harmonic motion.

    b Find the period and amplitude of the motion.

    c Calculate the maximum speed of P.

5 A particle P of mass $0.6\,\text{kg}$ is attached to one end of a light elastic spring of modulus of elasticity $72\,\text{N}$ and natural length $1.2\,\text{m}$. The other end of the spring is fixed to a point A on the smooth horizontal table on which P rests. Initially P is at rest with $AP = 1.2\,\text{m}$. The particle receives an impulse of magnitude $3\,\text{N s}$ in the direction AP. Given that t seconds after the impulse the displacement of P from its initial position is x metres:

    a find an equation for x in terms of t

    b calculate the maximum magnitude of the acceleration of P.

6 A particle of mass $0.9\,\text{kg}$ rests on a smooth horizontal surface attached to one end of a light elastic string of natural length $1.5\,\text{m}$ and modulus of elasticity $24\,\text{N}$. The other end of the string is attached to a point on the surface. The particle is pulled so that the string measures $2\,\text{m}$ and released from rest.

    a State the amplitude of the resulting oscillation.

    b Calculate the speed of the particle when the string becomes slack.

    Before the string becomes taut again the particle hits a vertical surface which is at right angles to the particle's direction of motion. The coefficient of restitution between the particle and the vertical surface is $\frac{3}{5}$.

    c Calculate:

      i the period

      ii the amplitude

    of the oscillation which takes place when the string becomes taut once more.

7 A smooth cylinder is fixed with its axis horizontal. A piston of mass $2.5\,\text{kg}$ is inside the cylinder, attached to one end of the cylinder by a spring of modulus of elasticity $400\,\text{N}$ and natural length $50\,\text{cm}$. The piston is held at rest in the cylinder with the spring compressed to a length of $42\,\text{cm}$. The piston is then released. The spring can be modelled as a light elastic spring and the piston can be modelled as a particle.

    a Find the period of the resulting oscillations.

    b Find the maximum value of the kinetic energy of the piston.

8 A particle P of mass 0.5 kg is attached to one end of a light elastic string of natural length 0.4 m and modulus of elasticity 30 N. The other end of the string is attached to a point on the smooth horizontal surface on which P rests. The particle is pulled until the string measures 0.6 m and then released from rest.

  a Calculate the speed of P when the string becomes slack for the first time.

  The string breaks at the instant when it returns to its natural length for the first time. When P has travelled a distance 0.3 m from the point at which the string breaks the surface becomes rough. The coefficient of friction between P and the surface is 0.25. The particle comes to rest T seconds after it was released.

  b Find the value of T.

9 A particle P of mass 0.4 kg is attached to two identical light elastic springs each of natural length 1.2 m and modulus of elasticity 12 N. The free ends of the strings are attached to points A and B which are 4 m apart on a smooth horizontal surface. The point C lies between A and B with $AC = 1.4$ m and $CB = 2.6$ m. The particle is held at C and released from rest.

  a Show that P moves with simple harmonic motion. **(5 marks)**

  b Calculate the maximum value of the kinetic energy of P. **(3 marks)**

10 A particle P of mass m is attached to two identical light strings of natural length l and modulus of elasticity $3mg$. The free ends of the strings are attached to fixed points A and B which are $5l$ apart on a smooth horizontal surface. The particle is held at the point C, where $AC = l$ and A, B and C lie on a straight line, and is then released from rest.

  a Show that P moves with simple harmonic motion. **(3 marks)**

  b Find the period of the motion. **(2 marks)**

  c Write down the amplitude of the motion. **(4 marks)**

  d Find the speed of P when $AP = 3l$. **(2 marks)**

11 A light elastic string has natural length 2.5 m and modulus of elasticity 15 N. A particle P of mass 0.5 kg is attached to the string at the point K where K divides the unstretched string in the ratio $2:3$. The ends of the string are then attached to the points A and B which are 5 m apart on a smooth horizontal surface. The particle is then pulled along AB and held at rest in contact with the surface at the point C where $AC = 3$ m and ACB is a straight line. The particle is then released from rest.

  a Show that P moves with simple harmonic motion of period $\frac{\pi}{5}\sqrt{2}$. **(5 marks)**

  b Find the amplitude of the motion. **(4 marks)**

5.5 Vertical oscillation

You can investigate the motion of a particle which is attached to an elastic spring or string and is oscillating in a vertical line.

A particle which is hanging in equilibrium attached to one end of an elastic spring or string, the other end of which is fixed, can be pulled downwards and released. The particle will then oscillate in a vertical line about its equilibrium position.

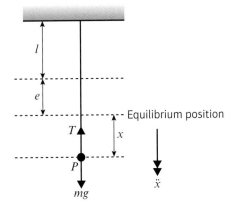

Taking downwards as the positive direction, when the particle is a distance x below its equilibrium position it has acceleration $\ddot{x}$.

At the equilibrium position, the tension in the spring or string is mg.

Using Hooke's law:

$$T = \frac{\lambda \times \text{extension}}{l}$$

$$mg = \frac{\lambda e}{l}$$

$$e = \frac{mgl}{\lambda}$$

λ is the modulus of elasticity and l is the natural length of the spring or string. e is the extension of the spring or string in the equilibrium position.

Now consider the particle at a distance x below its equilibrium position.

$$T = \frac{\lambda \times \text{extension}}{l}$$

$$T = \frac{\lambda(x + e)}{l}$$

$$T = \frac{\lambda\left(x + \dfrac{mgl}{\lambda}\right)}{l} = \frac{\lambda x}{l} + mg$$

The particle is a distance x below the equilibrium position, so the extension is $x + e$.

Using $F = ma$:

$$mg - T = m\ddot{x}$$

$$mg - \left(\frac{\lambda x}{l} + mg\right) = m\ddot{x}$$

$$-\frac{\lambda x}{l} = m\ddot{x}$$

$$\ddot{x} = -\frac{\lambda}{ml}x$$

λ, m and l are all positive constants, so the equation is of the form $\ddot{x} = -\omega^2 x$. It is the same result as obtained for a horizontal oscillation.

The particle is moving with S.H.M.

As in the case of horizontal oscillations, a particle attached to one end of a spring will perform complete oscillations. If the particle is attached to one end of an elastic string it will only move with S.H.M. while the string is taut. If the amplitude is greater than the extension at the equilibrium position the string will become slack before the particle reaches the upper end of the oscillation. Once the string becomes slack the oscillatory motion ceases and the particle moves freely under gravity until it falls back to the position where the string is once again taut.

■ **For a particle hanging in equilibrium attached to one end of an elastic spring and displaced vertically from its equilibrium position:**
  - **the particle will move with S.H.M.**
  - **the particle will perform complete oscillations**
  - **the centre of the oscillation will be the equilibrium position.**

■ **For a particle hanging in equilibrium attached to one end of an elastic string and displaced vertically from its equilibrium position:**
  - **the particle will move with S.H.M. while the string is taut**
  - **the particle will perform complete oscillations if the amplitude is no greater than the equilibrium extension**
  - **if the amplitude is greater than the equilibrium extension the particle will move freely under gravity while the string is slack.**

A particle can be attached to two springs or strings which are hanging side by side or stretched in a vertical line between two fixed points. The basic method of solution remains the same.

Example 17

A particle P of mass 1.2 kg is attached to one end of a light elastic spring of modulus of elasticity 60 N and natural length 60 cm. The other end of the spring is attached to a fixed point A on a ceiling. The particle hangs in equilibrium at the point B.

a Find the extension of the spring.

The particle is now raised vertically a distance 15 cm and released from rest.

b Prove that P will move with simple harmonic motion.

c Find the period and amplitude of the motion.

d Find the speed of P as it passes through B.

e Find the speed of P at the instant when the spring has returned to its natural length.

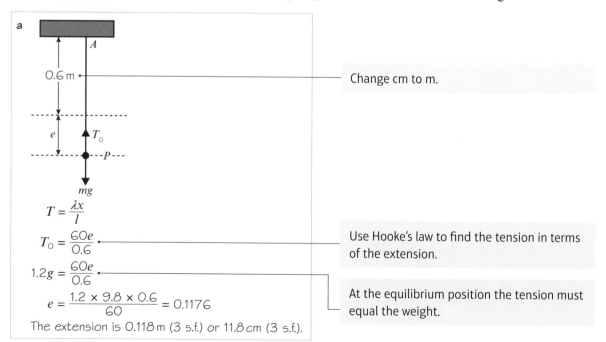

a

$T = \dfrac{\lambda x}{l}$

$T_0 = \dfrac{60e}{0.6}$ — Use Hooke's law to find the tension in terms of the extension.

$1.2g = \dfrac{60e}{0.6}$

$e = \dfrac{1.2 \times 9.8 \times 0.6}{60} = 0.1176$ — At the equilibrium position the tension must equal the weight.

The extension is 0.118 m (3 s.f.) or 11.8 cm (3 s.f.).

Change cm to m.

b

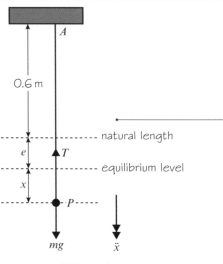

Online Explore the simple harmonic motion of a vertical spring using GeoGebra.

Draw a new diagram showing P at a distance x below the equilibrium level.

$$T = \frac{\lambda x}{l} = \frac{60(x + e)}{1.2}$$

Use Hooke's law once more. This time the extension is $x + e$.

$$F = ma$$

$$mg - T = m\ddot{x}$$

Watch out When you use $F = ma$ you must include the weight of the particle.

$$mg - \frac{60(x + e)}{0.6} = m\ddot{x}$$

$$\frac{60e}{0.6} - \frac{60(x + e)}{0.6} = 1.2\ddot{x}$$

$$\ddot{x} = -\frac{60x}{1.2 \times 0.6} = -\frac{250}{3}x$$

P moves with S.H.M.

Do not use an approximation for e. Instead, use your work from **a** to replace mg with the tension at the equilibrium level in terms of e.

When you simplify the equation e cancels.

c $\ddot{x} = -\frac{250}{3}x$

$$\omega^2 = \frac{250}{3}$$

$$\text{period} = \frac{2\pi}{\omega} = \frac{2\pi}{\sqrt{\frac{250}{3}}} = 0.6882\ldots$$

The period is $0.688\,\text{s}$ (3 s.f.).
The amplitude is $15\,\text{cm}$.

P was raised $15\,\text{cm}$ from its equilibrium level.

d $v^2 = \omega^2(a^2 - x^2)$

$$v_B^2 = \frac{250}{3}(0.15^2 - 0)$$

$x = 0$ at B.

$$v_B = \sqrt{\frac{250}{3}} \times 0.15 = 1.369\ldots$$

The speed at B is $1.37\,\text{ms}^{-1}$ (3 s.f.).

This is also the maximum speed of P.

e $v^2 = \omega^2(a^2 - x^2)$

$$v^2 = \frac{250}{3}(0.15^2 - (-0.1176)^2)$$

At the natural length $x = -e$. Use at least 4 s.f. in your approximation for e.

$$v = 0.8500\ldots$$

The speed at the natural length is $0.850\,\text{ms}^{-1}$ (3 s.f.).

A A particle P of mass $0.2\,\text{kg}$ is attached to one end of a light elastic string of natural length $0.6\,\text{m}$ and modulus of elasticity $8\,\text{N}$. The other end of the string is fixed to a point A on a ceiling. When the particle is hanging in equilibrium the length of the string is $L\,\text{m}$.

a Calculate the value of L.

The particle is held at A and released from rest. It first comes to instantaneous rest when the length of the string is $K\,\text{m}$.

b Use energy considerations to calculate the value of K.

c Show that while the string is taut, P is moving with simple harmonic motion.

The string becomes slack again for the first time T seconds after P was released from A.

d Calculate the value of T.

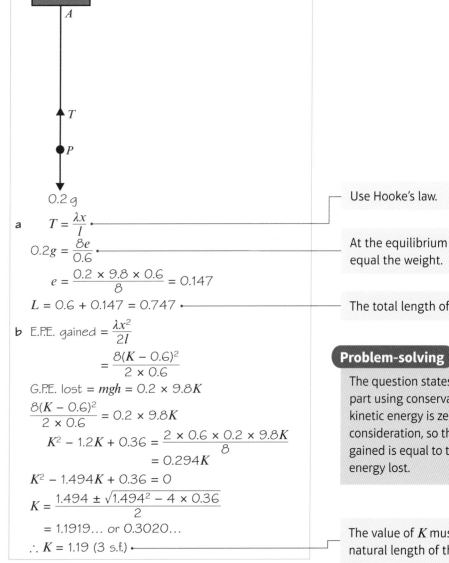

a
$$T = \frac{\lambda x}{l}$$

Use Hooke's law.

$$0.2g = \frac{8e}{0.6}$$

At the equilibrium position the tension must equal the weight.

$$e = \frac{0.2 \times 9.8 \times 0.6}{8} = 0.147$$

$$L = 0.6 + 0.147 = 0.747$$

The total length of the string is required.

b E.P.E. gained $= \dfrac{\lambda x^2}{2l}$

$$= \frac{8(K - 0.6)^2}{2 \times 0.6}$$

G.P.E. lost $= mgh = 0.2 \times 9.8K$

$$\frac{8(K - 0.6)^2}{2 \times 0.6} = 0.2 \times 9.8K$$

$$K^2 - 1.2K + 0.36 = \frac{2 \times 0.6 \times 0.2 \times 9.8K}{8}$$

$$= 0.294K$$

$$K^2 - 1.494K + 0.36 = 0$$

$$K = \frac{1.494 \pm \sqrt{1.494^2 - 4 \times 0.36}}{2}$$

$$= 1.1919\ldots \text{ or } 0.3020\ldots$$

$$\therefore K = 1.19 \ (3 \text{ s.f.})$$

Problem-solving

The question states that you must do this part using conservation of energy. The kinetic energy is zero at both points under consideration, so the elastic potential energy gained is equal to the gravitational potential energy lost. ← FM1, Section 3.4

The value of K must be greater than the natural length of the string.

c

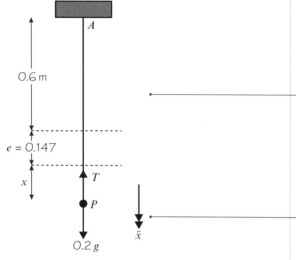

0.6 m

$e = 0.147$

x

T

P

$0.2g$

$\ddot{x}$

Draw a diagram which shows the natural length and the equilibrium level as well as the distance of P from the centre of the oscillation (x).

Remember that $\ddot{x}$ must be in the direction of increasing x.

$$T = \frac{\lambda \times \text{extension}}{l} = \frac{8(x + e)}{0.6}$$

$$F = ma$$

$$0.2g - \frac{8(x + e)}{0.6} = 0.2\ddot{x}$$

$$\frac{8e}{0.6} - \frac{8(x + e)}{0.6} = 0.2\ddot{x}$$

$$\ddot{x} = -\frac{8x}{0.6 \times 0.2} = -\frac{800}{12}x$$

$$\therefore \text{S.H.M.}$$

There is no need to use an approximation for e as $\frac{8e}{0.6}$ from part **a**.

Reduce the equation of motion to the form $\ddot{x} = -\omega^2 x$ to establish S.H.M.

d Time to fall 0.6 m from rest:

$$s = ut + \tfrac{1}{2}at^2$$

$$0.6 = 0 + \tfrac{1}{2} \times 9.8t^2$$

$$t^2 = \frac{0.6}{4.9}$$

$$t = 0.3499...$$

For S.H.M.:

$$\omega = \sqrt{\frac{800}{12}} = \sqrt{100 \times \frac{8}{12}} = 10\sqrt{\frac{2}{3}} = \frac{10\sqrt{6}}{3}$$

$$\text{amplitude} = K - L = 1.191 - 0.747 = 0.444$$

$$x = 0.444\cos\omega t$$

when $x = 0.147$, $0.147 = 0.444\cos\omega t$

$$\cos\omega t = \frac{0.147}{0.444}$$

$$t = \frac{1}{\omega}\arccos\left(\frac{0.147}{0.444}\right) = 0.1510...$$

$$\text{Period} = \frac{2\pi}{\omega} = 2\pi \times \frac{3}{10\sqrt{6}} = 0.7695...$$

Time for which string is taut

$$= 0.7695 - 2 \times 0.1510 = 0.4675...$$

Total time $= 0.4675... + 0.3499 = 0.8174...$

$$\therefore T = 0.817 \text{ (3 s.f.)}$$

Until the string is taut, P is falling freely under gravity.

Problem-solving

Because of the symmetry of S.H.M. there are several ways to obtain the time for which the string is taut. Whichever method you use you must show your working clearly.

Using $x = a\cos\omega t$ with the positive value of x when the string is at its natural length will give the time from the high point of the oscillation (if it were complete) to the point where the string becomes taut.

Subtracting twice the time just found from the period will give the time for which the string is taut in any one oscillation.

Finally, add the time taken while falling freely under gravity to the time for which the string is taut.

A Whenever a numerical value of g is required, take $g = 9.8\,\text{m s}^{-2}$.

E/P 1 A particle P of mass $0.75\,\text{kg}$ is hanging in equilibrium attached to one end of a light elastic spring of natural length $1.5\,\text{m}$ and modulus of elasticity $80\,\text{N}$. The other end of the spring is attached to a fixed point A vertically above P.

    a Calculate the length of the spring. **(3 marks)**

    The particle is pulled downwards and held at a point B which is vertically below A. The particle is then released from rest.

    b Show that P moves with simple harmonic motion. **(4 marks)**

    c Calculate the period of the oscillations. **(2 marks)**

    The particle passes through its equilibrium position with speed $2.5\,\text{m s}^{-1}$.

    d Calculate the amplitude of the oscillations. **(4 marks)**

E/P 2 A particle P of mass $0.5\,\text{kg}$ is attached to the free end of a light elastic spring of natural length $0.5\,\text{m}$ and modulus of elasticity $50\,\text{N}$. The other end of the spring is attached to a fixed point A and P hangs in equilibrium vertically below A.

    a Calculate the extension of the spring. **(3 marks)**

    The particle is now pulled vertically down a further $0.2\,\text{m}$ and released from rest.

    b Calculate the period of the resulting oscillations. **(2 marks)**

    c Calculate the maximum speed of the particle. **(2 marks)**

P 3 A particle P of mass $2\,\text{kg}$ is hanging in equilibrium attached to the free end of a light elastic spring of natural length $1.5\,\text{m}$ and modulus of elasticity $\lambda\,\text{N}$. The other end of the spring is fixed to a point A vertically above P. The particle receives an impulse of magnitude $3\,\text{N s}$ in the direction AP.

    a Find the speed of P immediately after the impact.

    b Show that P moves with simple harmonic motion.

    The period of the oscillations is $\dfrac{\pi}{2}\,\text{s}$.

    c Find the value of λ.

    d Find the amplitude of the oscillations.

E/P 4 A piston of mass $2\,\text{kg}$ moves inside a smooth cylinder which is fixed with its axis vertical. The piston is attached to the base of the cylinder by a spring of natural length $12\,\text{cm}$ and modulus of elasticity $500\,\text{N}$. The piston is released from rest at a point where the spring is compressed to a length of $8\,\text{cm}$. Assuming that the spring can be modelled as a light elastic spring and the piston as a particle, calculate:

    a the period of the resulting oscillations **(5 marks)**

    b the maximum speed of the piston. **(2 marks)**

5 A light elastic string of natural length 40 cm has one end A attached to a fixed point. A particle P of mass 0.4 kg is attached to the free end of the string and hangs freely in equilibrium vertically below A. The distance AP is 45 cm.

  a Find the modulus of elasticity of the string. **(3 marks)**

The particle is now pulled vertically downwards until AP measures 52 cm and then released from rest.

  b Show that, while the string is taut, P moves with simple harmonic motion. **(4 marks)**

  c Find the period and amplitude of the motion. **(3 marks)**

  d Find the greatest speed of P during the motion. **(2 marks)**

  e Find the time taken by P to rise 11 cm from the point of release. **(3 marks)**

6 A particle P of mass 0.4 kg is attached to one end of a light elastic string of natural length 0.5 m and modulus of elasticity 10 N. The other end of the string is attached to a fixed point A and P is initially hanging freely in equilibrium vertically below A. The particle is then pulled vertically downwards a further 0.2 m and released from rest.

  a Calculate the time from release until the string becomes slack for the first time. **(4 marks)**

  b Calculate the time between the string first becoming slack and the next time it becomes taut. **(4 marks)**

7 A particle P of mass 1.5 kg is hanging freely attached to one end of a light elastic string of natural length 1 m and modulus of elasticity 40 N. The other end of the string is attached to a fixed point A on a ceiling. The particle is pulled vertically downwards until AP is 1.8 m and released from rest. When P has risen a distance 0.4 m the string is cut.

  a Calculate the greatest height P reaches above its equilibrium position. **(4 marks)**

  b Calculate the time taken from release to reach that greatest height. **(3 marks)**

8 A particle P of mass 1.5 kg is attached to the midpoint of a light elastic string of natural length 1.2 m and modulus of elasticity 15 N. The ends of the string are fixed to the points A and B where A is vertically above B and $AB = 2.8$ m.

  a Given that P is in equilibrium calculate the length AP. **(3 marks)**

The particle is now pulled downwards a distance 0.15 m from its equilibrium position and released from rest.

  b Prove that P moves with simple harmonic motion. **(4 marks)**

T seconds after being released P is 0.1 m above its equilibrium position.

  c Find the value of T. **(3 marks)**

9 A rock climber of mass 70 kg is attached to one end of a rope. He falls from a ledge which is 8 m vertically below the point to which the other end of the rope is fixed. The climber falls vertically without hitting the rock face. Assuming that the climber can be modelled as a particle and the rope as a light elastic string of natural length 16 m and modulus of elasticity 40 000 N, calculate:

  a the climber's speed at the instant when the rope becomes taut **(3 marks)**

  b the maximum distance of the climber below the ledge **(3 marks)**

  c the time from falling from the ledge to reaching his lowest point. **(2 marks)**

Mixed exercise 5

Whenever a numerical value of g is required, take $g = 9.8\,\text{m}\,\text{s}^{-2}$.

(E/P) **1** A particle P of mass 0.6 kg moves along the positive x-axis under the action of a single force which is directed towards the origin O and has magnitude $\dfrac{k}{(x+2)^2}$ N where $OP = x$ metres and k is a constant. Initially P is moving away from O. At $x = 2$ the speed of P is $8\,\text{m}\,\text{s}^{-1}$ and at $x = 10$ the speed of P is $2\,\text{m}\,\text{s}^{-1}$.

   a Find the value of k. **(6 marks)**

   The particle first comes to instantaneous rest at the point B.

   b Find the distance OB. **(4 marks)**

(E) **2** A particle P of mass 0.5 kg is moving along the x-axis, in the positive x direction. At time t seconds (where $t > 0$) the resultant force acting on P has magnitude $\dfrac{5}{\sqrt{(3t+4)}}$ N and is directed towards the origin O. When $t = 0$, P is moving through O with speed $12\,\text{m}\,\text{s}^{-1}$.

   a Find an expression for the velocity of P at time t seconds. **(5 marks)**

   b Find the distance of P from O when P is instantaneously at rest. **(5 marks)**

(E/P) **3** A particle of mass m moves in a straight line on a smooth horizontal plane in a medium which exerts a resistance of magnitude mkv^2, where v is the speed of the particle and k is a positive constant. At time $t = 0$ the particle has speed U.

   Find, in terms of k and U, the time at which the particle's speed is $\frac{3}{4}U$. **(5 marks)**

(E/P) **4** A small pebble of mass m is placed in a viscous liquid and sinks vertically from rest through the liquid. When the speed of the particle is v the magnitude of the resistance due to the liquid is modelled as mkv^2, where k is a positive constant.

   Find the speed of the pebble after it has fallen a distance D through the liquid. **(5 marks)**

(E/P) **5** A car of mass 1000 kg is driven by an engine which generates a constant power of 12 kW. The only resistance to the car's motion is air resistance of magnitude $10v^2$ N, where $v\,\text{m}\,\text{s}^{-1}$ is the speed of the car.

   Find the distance travelled as the car's speed increases from $5\,\text{m}\,\text{s}^{-1}$ to $10\,\text{m}\,\text{s}^{-1}$. **(8 marks)**

A

E/P **6** A bullet B, of mass m kg, is fired vertically downwards into a block of wood W which is fixed in the ground. The bullet enters W with speed U m s^{-1} and W offers a resistance of magnitude $m(14.8 + 5bv^2)$ N, where v m s^{-1} is the speed of B and b is a positive constant. The path of B in W remains vertical until B comes to rest after travelling a distance d metres into W.

Find d in terms of b and U. **(8 marks)**

E/P **7** A particle of mass m is projected vertically upwards, with speed V, in a medium which exerts a resisting force of magnitude $\dfrac{mgv^2}{c^2}$, where v is the speed of the particle and c is a positive constant.

a Show that the greatest height attained above the point of projection is

$$\frac{c^2}{2g}\ln\left(1 + \frac{V^2}{c^2}\right)$$ **(8 marks)**

b Find an expression, in terms of V, c and g, for the time to reach this height. **(3 marks)**

E/P **8** A particle is projected vertically upwards with speed U in a medium in which the resistance is proportional to the square of the speed. Given that U is also the speed for which the resistance offered by the medium is equal to the weight of the particle, show that:

a the time of ascent is $\dfrac{\pi U}{4g}$ **(6 marks)**

b the distance ascended is $\dfrac{U^2}{2g}\ln 2$. **(6 marks)**

E/P **9** At time t, a particle P, of mass m, moving in a straight line has speed v. The only force acting is a resistance of magnitude $mk(V_0^2 + 2v^2)$, where k is a positive constant and V_0 is the speed of P when $t = 0$.

a Show that, as v reduces from V_0 to $\frac{1}{2}V_0$, P travels a distance $\dfrac{\ln 2}{4k}$ **(6 marks)**

b Express the time P takes to cover this distance in the form $\dfrac{\lambda}{kV_0}$, giving the value of λ to two decimal places. **(6 marks)**

E/P **10** A car of mass m is moving along a straight horizontal road. When displacement of the car from a fixed point O is x, its speed is v. The resistance to the motion of the car has magnitude $\dfrac{mkv^2}{3}$, where k is a positive constant. The engine of the car is working at a constant rate P.

a Show that $3mv^2\dfrac{\mathrm{d}v}{\mathrm{d}x} = 3P - mkv^3$. **(6 marks)**

When $t = 0$, the speed of the car is half of its limiting speed.

b Find x in terms of m, k, P and v. **(6 marks)**

E/P **11** A spacecraft S of mass m is moving in a straight line towards the centre of the Earth. When the distance of S from the centre of the Earth is x metres, the force exerted by the Earth on S has magnitude $\dfrac{k}{x^2}$, where k is a constant, and is directed towards the centre of the Earth.

a By modelling the Earth as a sphere of radius R and S as a particle, show that $k = mgR^2$. **(2 marks)**

The spacecraft starts from rest when $x = 5R$.

b Assuming that air resistance can be ignored, find the speed of S as it crashes onto the Earth's surface. **(7 marks)**

A **12** A particle P is moving with simple harmonic motion between two points A and B which are
0.4 m apart on a horizontal line. The midpoint of AB is O. At time $t = 0$, P passes through O,
E moving towards A, with speed u m s^{-1}. The next time P passes through O is when $t = 2.5$ s.

    a Find the value of u.   **(4 marks)**

    b Find the speed of P when $t = 3$ s.   **(2 marks)**

    c Find the distance of P from A when $t = 3$ s.   **(5 marks)**

E **13** A particle P of mass 1.2 kg moves along the x-axis. At time $t = 0$, P passes through the origin
O, moving in the positive x direction. At time t seconds, the velocity of P is v m s^{-1} and
$OP = x$ metres. The resultant force acting on P has magnitude $6(2.5 - x)$ N and acts in the
positive x direction. The maximum speed of P is 8 m s^{-1}.

    a Find the value of x when the speed of P is 8 m s^{-1}.   **(5 marks)**

    b Find an expression for v^2 in terms of x.   **(5 marks)**

E/P **14** A particle P moves along the x-axis in such a way that at time t seconds its distance x metres
from the origin O is given by $x = 3\sin\left(\dfrac{\pi t}{4}\right)$.

    a Prove that P moves with simple harmonic motion.   **(4 marks)**

    b Write down the amplitude and the period of the motion.   **(3 marks)**

    c Find the maximum speed of P.   **(2 marks)**

    The points A and B are on the same side of O with $OA = 1.2$ m and $OB = 2$ m.

    d Find the time taken by P to travel directly from A to B.   **(4 marks)**

E/P **15** A particle P moves on the x-axis with simple harmonic motion such that its centre of
oscillation is the origin, O. When P is a distance 0.09 m from O, its speed is 0.3 m s^{-1} and the
magnitude of its acceleration is 1.5 m s^{-2}.

    a Find the period of the motion.   **(3 marks)**

    The amplitude of the motion is a metres. Find:

    b the value of a **(3 marks)**

    c the total time, within one complete oscillation, for which the distance OP is greater
than $\dfrac{a}{2}$ metres.   **(5 marks)**

E/P **16** A particle P of mass 0.6 kg is attached to one end of a light elastic spring of natural length
2.5 m and modulus of elasticity 25 N. The other end of the spring is attached to a fixed point A
on the smooth horizontal table on which P lies. The particle is held at the point B where
$AB = 4$ m and released from rest.

    a Prove that P moves with simple harmonic motion.   **(4 marks)**

    b Find the period and amplitude of the motion.   **(3 marks)**

    c Find the time taken for P to move 2 m from B.   **(2 marks)**

E/P **17** A particle P of mass 0.4 kg is attached to the midpoint of a light elastic string of natural length
1.2 m and modulus of elasticity 2.5 N. The ends of the string are attached to points A and B on
a smooth horizontal table where $AB = 2$ m. The particle P is released from rest at the point C
on the table, where A, C and B lie in a straight line and $AC = 0.7$ m.

    a Show that P moves with simple harmonic motion.   **(4 marks)**

    b Find the period of the motion.   **(3 marks)**

The point D lies between A and B and $AD = 0.85$ m.

c Find the time taken by P to reach D for the first time. **(4 marks)**

E/P **18** A and B are two points on a smooth horizontal floor, where $AB = 12$ m.

A particle P has mass 0.4 kg. One end of a light elastic spring, of natural length 5 m and modulus of elasticity 20 N, is attached to P and the other end is attached to A. The ends of another light elastic spring, of natural length 3 m and modulus of elasticity 18 N, are attached to P and B.

a Find the extensions in the two springs when the particle is at rest in equilibrium.   **(5 marks)**

Initially P is at rest in equilibrium. It is then set in motion and starts to move towards B. In the subsequent motion P does not reach A or B.

b Show that P oscillates with simple harmonic motion about the equilibrium position.

(4 marks)

c Given that P stays within 0.4 m of the equilibrium position for $\frac{1}{3}$ of the time within each complete oscillation, find the initial speed of P.   **(7 marks)**

E/P **19** A particle P of mass 0.5 kg is attached to one end of a light elastic string of natural length 1.2 m and modulus of elasticity λ N. The other end of the string is attached to a fixed point A. The particle is hanging in equilibrium at the point O, which is 1.4 m vertically below A.

a Find the value of λ.   **(3 marks)**

The particle is now displaced to a point B, 1.75 m vertically below A, and released from rest.

b Prove that while the string is taut P moves with simple harmonic motion.   **(4 marks)**

c Find the period of the simple harmonic motion.   **(4 marks)**

d Calculate the speed of P at the first instant when the string becomes slack.   **(4 marks)**

e Find the greatest height reached by P above O.   **(4 marks)**

E/P **20** A particle P of mass m is attached to the midpoint of a light elastic string of natural length $4l$ and modulus of elasticity $5mg$. One end of the string is attached to a fixed point A and the other end to a fixed point B, where A and B lie on a smooth horizontal surface and $AB = 6l$. The particle is held at the point C where A, C and B are collinear and $AC = \frac{9l}{4}$, and released from rest.

a Prove that P moves with simple harmonic motion.   **(4 marks)**

Find, in terms of g and l:

b the period of the motion   **(2 marks)**

c the maximum speed of P.   **(2 marks)**

Challenge

The motion of a space shuttle which is launched from a point O on the surface of the Earth can be modelled as a particle of mass m moving in a straight line, subject to the universal law of gravitation using $F = -\dfrac{mMG}{(R + x)^2}$ where:

M is the mass of the Earth

m is the mass of the space shuttle

R is the radius of the Earth

x is the height of the space shuttle above the Earth

G is the universal constant of gravitation.

a Given a space shuttle is launched with initial velocity u m s^{-1}, show that the maximum height, H, above the Earth that the spaceship reaches can be expressed as $H = \dfrac{Ru^2}{\left(\dfrac{2MG}{R} - u^2\right)}$

The minimum velocity required to project the space shuttle into space is called the **escape velocity**. This is the value of u for which H tends to infinity.

b Use

$$M = 5.98 \times 10^{24}$$
$$R = 6.4 \times 10^6$$
$$G = 6.7 \times 10^{-11}$$

to work out the escape velocity for the space shuttle correct to 3 significant figures.

Summary of key points

1 In forming an equation of motion, forces that tend to decrease the displacement are negative and forces that tend to increase the displacement are positive.

2 Newton's law of gravitation states that the force of attraction between two bodies of mass M_1 and M_2 is directly proportional to the product of their masses and inversely proportional to the square of the distance between them.

- $F = \dfrac{GM_1M_2}{d^2}$ where G is a constant known as the constant of gravitation.

3 Simple harmonic motion (S.H.M.) is motion in which the acceleration of a particle is always towards a fixed point O on the line of motion of P, and has magnitude proportional to the displacement of P from O.

4 For S.H.M. of amplitude a defined by the equation $\ddot{x} = -\omega^2 x$,

- $v^2 = \omega^2(a^2 - x^2)$
- If P is at the centre of the oscillation when $t = 0$, use $x = a\sin\omega t$.
- If P is at an end point of the oscillation when $t = 0$, use $x = a\cos\omega t$.
- If P is at some other point when $t = 0$, use $x = a\sin(\omega t + \alpha)$.
- The period of the oscillation is $T = \dfrac{2\pi}{\omega}$

5 For a particle moving on a smooth horizontal surface attached to one end of an elastic spring:

- the particle will move with S.H.M.
- the particle will perform complete oscillations.

6 For a particle moving on a smooth horizontal surface attached to one end of an elastic string:

- the particle will move with S.H.M. while the string is taut
- the particle will move with constant speed while the string is slack.

7 To solve problems involving elastic springs and strings:

- use Hooke's law to find the tension
- use $F = ma$ to obtain ω
- use information given in the question to obtain the amplitude.

8 For a particle hanging in equilibrium attached to one end of an elastic spring and displaced vertically from its equilibrium position:

- the particle will move with S.H.M.
- the particle will perform complete oscillations
- the centre of the oscillation will be the equilibrium position.

9 For a particle hanging in equilibrium attached to one end of an elastic string and displaced vertically from its equilibrium position:

- the particle will move with S.H.M. while the string is taut
- the particle will perform complete oscillations if the amplitude is no greater than the equilibrium extension
- if the amplitude is greater than the equilibrium extension the particle will move freely under gravity while the string is slack.

Review exercise

(E) **1** A particle P moves in a straight line. At time t seconds, the acceleration of P is e^{2t} m s^{-2}, where $t \geqslant 0$. When $t = 0$, P is at rest. Show that the speed, v m s^{-1}, of P at time t seconds is given by

$$v = \tfrac{1}{2}(e^{2t} - 1) \qquad \textbf{(6)}$$

← **Section 4.1**

(E) **2** A particle P moves along the x-axis in the positive direction. At time t seconds, the velocity of P is v m s^{-1} and its acceleration is $\tfrac{1}{2}e^{-\frac{1}{6}t}$ m s^{-2}. When $t = 0$ the speed of P is 10 m s^{-1}.

  a Express v in terms of t. **(6)**

  b Find, to 3 significant figures, the speed of P when $t = 3$. **(2)**

  c Find the limiting value of v. **(1)**

← **Section 4.1**

(E) **3** A particle P moves along the x-axis. At time t seconds the velocity of P is v m s^{-1} and its acceleration is $2\sin\tfrac{1}{2}t$ m s^{-2}, both measured in the direction Ox. Given that $v = 4$ when $t = 0$,

  a find v in terms of t **(6)**

  b calculate the distance travelled by P between the times $t = 0$ and $t = \dfrac{\pi}{2}$ **(7)**

← **Section 4.1**

(E) **4** A particle P moves along the x-axis. At time t seconds its acceleration is $-4e^{-2t}$ m s^{-2} in the direction of x increasing. When $t = 0$, P is at the origin O and is moving with speed 1 m s^{-1} in the direction of x increasing.

  a Find an expression for the velocity of P at time t. **(6)**

  b Find the distance of P from O when P comes to instantaneous rest. **(7)**

← **Section 4.1**

(E/P) **5** At time $t = 0$, a particle P is at the origin O and is moving with speed 18 m s^{-1} along the x-axis in the positive x direction. At time t seconds ($t > 0$), the acceleration of P has magnitude

$$\dfrac{3}{\sqrt{t+4}}\ \text{m s}^{-2}\ \text{and is directed towards } O.$$

  a Show that, at time t seconds, the velocity of P is $(30 - 6\sqrt{t+4})$ m s^{-1}. **(6)**

  b Find the distance of P from O when P comes to instantaneous rest. **(7)**

← **Section 4.1**

(E/P) **6** A particle moving in a straight line starts from rest at a point O at time $t = 0$. At time t seconds, the velocity v m s^{-1} is given by

$$v = \begin{cases} 3t(t-4), & 0 \leqslant t \leqslant 5 \\ 75t^{-1}, & 5 < t \leqslant 10 \end{cases}$$

  a Sketch a velocity–time graph for the particle for $0 \leqslant t \leqslant 10$. **(3)**

  b Find the set of values of t for which the acceleration of the particle is positive. **(2)**

  c Show that the total distance travelled by the particle in the interval $0 \leqslant t \leqslant 5$ is 39 m. **(5)**

  d Find, to 3 significant figures, the value of t at which the particle returns to O. **(3)**

← **Section 4.1**

(E/P) 7 A car is travelling along a straight horizontal road. As it passes a point O on the road, the engine is switched off. At time t seconds after the car has passed O, it is at a point P, where $OP = x$ metres, and its velocity is $v\,\mathrm{m\,s^{-1}}$. The motion of the car is modelled by

$$v = \frac{1}{p + qt}$$

where p and q are positive constants.

a Show that, with this model, the retardation of the car is proportional to the square of the speed. **(3)**

When $t = 0$, the retardation of the car is $0.75\,\mathrm{m\,s^{-2}}$ and $v = 20$. Using the model, find:

b the value of p and the value of q **(2)**

c x in terms of t. **(5)**

← Section 4.1

(A)(E) 8 A particle P moves along the x-axis in such a way that when its displacement from the origin O is $x\,\mathrm{m}$, its velocity is $v\,\mathrm{m\,s^{-1}}$ and its acceleration is $4x\,\mathrm{m\,s^{-2}}$. When $x = 2$, $v = 4$.

Show that $v^2 = 4x^2$. **(4)**

← Section 4.2

(E) 9 A particle P moves on the positive x-axis. When $OP = x$ metres, where O is the origin, the acceleration of P is directed away from O and has magnitude $\left(1 - \dfrac{4}{x^2}\right)\mathrm{m\,s^{-2}}$. When $OP = x$ metres, the velocity of P is $v\,\mathrm{m\,s^{-1}}$. Given that when $x = 1$, $v = 3\sqrt{2}$, show that when $x = \frac{3}{2}$, $v^2 = \frac{49}{3}$ **(5)**

← Section 4.2

(E) 10 A particle P is moving in a straight line. When P is at a distance x metres from a fixed point O on the line, the acceleration of P is $(5 + 3\sin 3x)\,\mathrm{m\,s^{-2}}$ in the direction OP. Given that P passes through O with speed $4\,\mathrm{m\,s^{-1}}$, find the speed of P at $x = 6$. Give your answer to 3 significant figures. **(5)**

← Section 4.2

(A)(E) 11 A particle P is moving along the positive x-axis in the direction of x increasing. When $OP = x$ metres, the velocity of P is $v\,\mathrm{m\,s^{-1}}$ and the acceleration of P is $\dfrac{4k^2}{(x + 1)^2}\,\mathrm{m\,s^{-2}}$, where k is a positive constant. At $x = 1$, $v = 0$.

a Find v^2 in terms of x and k. **(5)**

b Deduce that v cannot exceed $2k$. **(2)**

← Section 4.2

(E) 12 A particle P moves along the x-axis. At time $t = 0$, P passes through the origin O, moving in the positive x direction. At time t seconds, the velocity of P is $v\,\mathrm{m\,s^{-1}}$ and $OP = x$ metres. The acceleration of P is $\frac{1}{12}(30 - x)\,\mathrm{m\,s^{-2}}$, measured in the positive x direction.

a Give a reason why the maximum speed of P occurs when $x = 30$. **(2)**

Given that the maximum speed of P is $10\,\mathrm{m\,s^{-1}}$,

b find an expression for v^2 in terms of x. **(5)**

← Section 4.2

(E/P) 13 A particle P starts at rest and moves in a straight line. The acceleration of P initially has magnitude $20\,\mathrm{m\,s^{-2}}$ and, in a first model of the motion of P, it is assumed that this acceleration remains constant.

a For this model, find the distance moved by P while accelerating from rest to a speed of $6\,\mathrm{m\,s^{-1}}$. **(4)**

The acceleration of P when it is x metres from its initial position is $a\,\mathrm{m\,s^{-2}}$ and it is then established that $a = 12$ when $x = 2$. A refined model is proposed in which $a = p - qx$, where p and q are constants.

b Show that, under the refined model, $p = 20$ and $q = 4$. **(5)**

c Hence find, for the refined model, the distance moved by P in first attaining a speed of $6\,\mathrm{m\,s^{-1}}$. **(4)**

← Section 4.2

A **E** **14** A particle P moves along the positive x-axis. When $OP = x$ metres, the velocity of P is $v\,\text{m}\,\text{s}^{-1}$ and its acceleration is

$$\frac{72}{(2x + 1)^2}\,\text{m}\,\text{s}^{-2}$$ in the direction of

x increasing. Initially $x = 1$ and P is moving toward O with speed $6\,\text{m}\,\text{s}^{-1}$. Find:

a v^2 in terms of x **(5)**

b the minimum distance of P from O. **(3)**

← Section 4.2

E/P **15** A particle moves on the positive x-axis. The particle is moving towards the origin O when it passes through the point A, where $x = 2c$, with speed $\sqrt{\dfrac{k}{c}}$, where k is a constant. Given that the particle experiences an acceleration of $\dfrac{k}{2x^2} + \dfrac{k}{4c^2}$ acting in the direction of x increasing,

a show that it comes instantaneously to rest at a point B, where $x = c$. **(5)**

As soon as the particle reaches B the acceleration changes to $\dfrac{k}{2x^2} - \dfrac{k}{4c^2}$ in a direction away from O.

b Show that the particle next comes instantaneously to rest at A. **(5)**

← Section 4.2

E **16** A particle moves along the x axis in the direction of x increasing. When the speed of the particle is $v\,\text{m}\,\text{s}^{-1}$, the acceleration is $-\dfrac{v^2}{10}\,\text{m}\,\text{s}^{-2}$. Initially the particle is at the origin, O, and is moving with a speed of $12\,\text{m}\,\text{s}^{-1}$. At time T seconds, the particle is at the point A with a velocity of $6\,\text{m}\,\text{s}^{-1}$. Find:

a the value of T **(6)**

b the distance OA. **(5)**

← Section 4.3

E **17** A particle P moves along the positive x-axis. At time t seconds, the acceleration of the particle is $-(k + v)$ where $v\,\text{m}\,\text{s}^{-1}$ is the velocity of the particle and k is a positive constant. When $t = 0$, P is at O and $v = U$. The particle comes to rest at the point A. Find:

a the distance OA **(6)**

b the time P takes to travel from O to A. **(5)**

← Section 4.3

E **18** A van starts from rest and moves along a straight horizontal road. At time t seconds, the acceleration of the van is $\dfrac{169 - v^2}{100}\,\text{m}\,\text{s}^{-2}$ where $v\,\text{m}\,\text{s}^{-1}$ is the velocity of the van.

a Find v in terms of t **(6)**

b Show that the speed of the van cannot exceed $13\,\text{m}\,\text{s}^{-1}$. **(2)**

← Section 4.3

E/P **19** A motorbike travels along a straight horizontal road. At time t seconds, the speed of the motorbike is $v\,\text{m}\,\text{s}^{-1}$ and the acceleration is $\left(6 - \dfrac{v}{5}\right)\,\text{m}\,\text{s}^{-2}$. The motorbike starts from rest. Find:

a v in terms of t **(6)**

b the terminal speed of the motorbike. **(3)**

← Section 4.3

A **E** **20** A particle, P, is moving along the x axis Ox in the direction of x increasing. At time t seconds, the velocity of P is $v\,\text{m}\,\text{s}^{-1}$ and the acceleration is $-(a^2 + v^2)\,\text{m}\,\text{s}^{-2}$ where a is a constant. At time $t = 0$, P is at O and its speed is $20\,\text{m}\,\text{s}^{-1}$. At time $t = T$, the particle is at point A and its velocity is $12\,\text{m}\,\text{s}^{-1}$. Find:

a T in terms of a **(5)**

b the distance OA. **(6)**

← Section 4.3

A **E** **21** A particle P of mass 0.2 kg moves away from the origin along the positive x-axis. It moves under the action of a force directed away from the origin O of magnitude $\frac{5}{x+1}$ N, where $OP = x$ m. Given that the speed of P is 5 m s^{-1} when $x = 0$, find the value of x, to 3 significant figures, when the speed of P is 15 m s^{-1}. **(8)**

← Section 5.1

E **22** A particle P of mass 2.5 kg moves along the positive x-axis. It moves away from a fixed origin O, under the action of a force directed away from O. When $OP = x$ metres, the magnitude of the force is $2e^{-0.1x}$ N and the speed of P is v m s^{-1}. When $x = 0$, $v = 2$. Find:

a v^2 in terms of x **(4)**

b the value of x when $v = 4$. **(2)**

c Give a reason why the speed of P does not exceed $\sqrt{20}$ m s^{-1}. **(2)**

← Section 5.1

E/P **23** A toy car of mass 0.2 kg is travelling in a straight line on a horizontal floor. The car is modelled as a particle. At time $t = 0$ the car passes through a fixed point O. After t seconds the speed of the car is v m s^{-1} and the car is at a point P with $OP = x$ metres. The resultant force on the car is modelled as $\frac{1}{10}x(4 - 3x)$ N in the direction OP. The car comes to instantaneous rest when $x = 6$. Find:

a an expression for v^2 in terms of x **(4)**

b the initial speed of the car. **(2)**

← Section 5.1

E **24** A particle P of mass 0.6 kg is moving along the positive x-axis under the action of a force which is directed away from the origin O. At time t seconds, the force has magnitude $3e^{-0.5t}$ N. When $t = 0$, the particle P is at O and moving with speed 2 m s^{-1} in the direction of x increasing.

A Find:

a the value of t when the speed is 8 m s^{-1} **(7)**

b the distance of P from O when $t = 2$. **(5)**

← Section 5.1

E **25** A car of mass 800 kg moves along a horizontal straight road. At time t seconds, the resultant force on the car has magnitude $\frac{48\,000}{(t+2)^2}$ N, acting in the direction of motion of the car. When $t = 0$, the car is at rest.

a Show that the speed of the car approaches a limiting value as t increases and find this value. **(7)**

b Find the distance moved by the car in the first 6 s of its motion. **(5)**

← Section 5.1

E/P **26** A particle P of mass $\frac{1}{3}$ kg moves along the positive x-axis under the action of a single force. The force is directed towards the origin O and has magnitude $\frac{k}{(x+1)^2}$ N, where $OP = x$ metres and k is a constant. Initially P is moving away from O. At $x = 1$ the speed of P is 4 m s^{-1}, and at $x = 8$ the speed of P is $\sqrt{2}$ m s^{-1}. Find:

a the value of k **(7)**

b the distance of P from O when P first comes to instantaneous rest. **(5)**

← Section 5.1

E/P **27** A particle P of mass 0.7 kg falls vertically from rest. A resisting force of magnitude $2.1v$ N acts on P as it falls where v m s^{-1} is the velocity of P after t seconds. Find:

a v in terms of t **(8)**

b the distance that P falls in the first two seconds. **(5)**

← Section 5.1

A **E/P** **28** A particle of mass m kg is projected vertically upwards with velocity U m s^{-1}. The motion of the particle is subject to air resistance of magnitude $\dfrac{mv}{k}$ N where v m s^{-1} is the speed of the particle at time t seconds and k is a positive constant.

Find, in terms of U, g and k, the maximum height above O reached by the particle. **(8)**

← Section 5.1

E/P **29** Above the Earth's surface, the magnitude of the force on a particle due to the Earth's gravity is inversely proportional to the square of the distance of the particle from the centre of the Earth. Assuming that the Earth is a sphere of radius R, and taking g as the acceleration due to gravity at the surface of the Earth,

  a prove that the magnitude of the gravitational force on a particle of mass m when it is a distance x (where $x \geqslant R$) from the centre of the Earth is $\dfrac{mgR^2}{x^2}$ **(3)**

A particle is fired vertically upwards from the surface of the Earth with initial speed u, where $u^2 = \frac{3}{2}gR$. Ignoring air resistance,

  b find, in terms of g and R, the speed of the particle when it is at a height $2R$ above the surface of the Earth. **(7)**

← Section 5.2

E/P **30** A rocket is fired vertically upwards with speed U from a point on the Earth's surface. The rocket is modelled as a particle P of constant mass m, and the Earth as a fixed sphere of radius R. At a distance x from the centre of the Earth, the speed of P is v. The only force acting on P is directed towards the centre of the Earth and has magnitude $\dfrac{cm}{x^2}$, where c is a constant.

  a Show that $v^2 = U^2 + 2c\left(\dfrac{1}{x} - \dfrac{1}{R}\right)$ **(7)**

A The kinetic energy of P at $x = 2R$ is half of the kinetic energy at $x = R$.

  b Find c in terms of U and R. **(3)**

← Section 5.2

E/P **31** A projectile P is fired vertically upwards from a point on the Earth's surface. When P is at a distance x from the centre of the Earth its speed is v. Its acceleration is directed towards the centre of the Earth and has magnitude $\dfrac{k}{x^2}$, where k is a constant. The Earth is assumed to be a sphere of radius R.

  a Show that the motion of P may be modelled by the differential equation
  $$v\frac{\mathrm{d}v}{\mathrm{d}x} = -\frac{gR^2}{x^2}$$ **(3)**

The initial speed of P is U, where $U^2 < 2gR$. The greatest distance of P from the centre of the Earth is X.

  b Find X in terms of U, R and g. **(7)**

← Section 5.2

E **32** A particle P moves in a straight line with simple harmonic motion about a fixed centre O with period 2 s. At time t seconds the speed of P is v m s^{-1}. When $t = 0$, $v = 0$ and P is at a point A where $OA = 0.25$ m.

Find the smallest positive value of t for which $AP = 0.375$ m. **(5)**

← Section 5.3

E **33** A particle P of mass 0.2 kg oscillates with simple harmonic motion between the points A and B, coming to rest at both points. The distance AB is 0.2 m, and P completes 5 oscillations every second.

  a Find, to 3 significant figures, the maximum resultant force exerted on P. **(3)**

A When the particle is at A, it is struck a blow in the direction BA. The particle now oscillates with simple harmonic motion with the same frequency as previously but twice the amplitude.

  b Find, to 3 significant figures, the speed of the particle immediately after it has been struck. **(3)**

← **Section 5.3**

E/P **34** A piston P in a machine moves in a straight line with simple harmonic motion about a fixed centre O. The period of the oscillations is π s. When P is 0.5 m from O, its speed is 2.4 m s^{-1}. Find:

  a the amplitude of the motion **(4)**

  b the maximum speed of P during its motion **(2)**

  c the maximum magnitude of the acceleration of P during the motion **(2)**

  d the total time, in seconds to 2 decimal places, in each complete oscillation for which the speed of P is greater than 2.4 m s^{-1}. **(5)**

← **Section 5.3**

E/P **35**

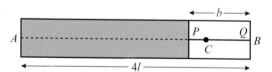

In a game at a fair, a small target C moves horizontally with simple harmonic motion between the points A and B, where $AB = 4l$. The target moves inside a box and takes 3 s to travel from A to B. A player has to shoot at C, but C is only visible to the player when it passes a window PQ where $PQ = b$. The window is initially placed with Q at the point shown in the figure above. The target takes 0.75 s to travel from Q to P.

  a Show that $b = (2 - \sqrt{2})l$. **(3)**

  b Find the speed of C as it passes P. **(3)**

A

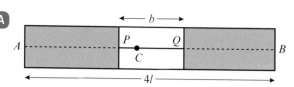

For advanced players, the window PQ is moved to the centre of AB so that $AP = QB$, as shown in the second figure above.

  c Find the time, in seconds to 2 decimal places, taken for C to pass from Q to P in this new position. **(4)**

← **Section 5.3**

E **36** The points O, A, B and C lie in a straight line, in that order, with $OA = 0.6$ m, $OB = 0.8$ m and $OC = 1.2$ m. A particle P, moving in a straight line, has speed $\left(\frac{3}{10}\sqrt{3}\right)$ m s^{-1} at A, $\left(\frac{1}{5}\sqrt{5}\right)$ m s^{-1} at B and is instantaneously at rest at C.

  a Show that this information is consistent with P performing simple harmonic motion with centre O. **(4)**

Given that P is performing simple harmonic motion with centre O,

  b show that the speed of P at O is 0.6 m s^{-1}. **(2)**

  c Find the magnitude of the acceleration of P as it passes A. **(4)**

  d Find, to 3 significant figures, the time taken for P to move directly from A to B. **(4)**

← **Section 5.3**

E/P **37** The rise and fall of the water level in a harbour is modelled as simple harmonic motion. On a particular day the maximum and minimum depths of the water in the harbour are 10 m and 4 m and these occur at 1100 hours and 1700 hours respectively.

  a Find the speed, in m h^{-1}, at which the water level in the harbour is falling at 1600 hours on this particular day. **(7)**

b Find the total time, between 1100 hours and 2300 hours on this particular day, for which the depth in the harbour is less than 5.5 m. **(5)**

← Section 5.3

E/P 38 A piston in a machine is modelled as a particle of mass 0.2 kg attached to one end A of a light elastic spring, of natural length 0.6 m and modulus of elasticity 48 N. The other end B of the spring is fixed and the piston is free to move in a horizontal tube which is assumed to be smooth. The piston is released from rest when $AB = 0.9$ m.

a Prove that the motion of the piston is simple harmonic with period $\frac{\pi}{10}$ s. **(4)**

b Find the maximum speed of the piston. **(2)**

c Find, in terms of π, the length of time during each oscillation for which the length of the spring is less than 0.75 m. **(5)**

← Section 5.4

E/P 39 A particle P of mass 0.8 kg is attached to one end A of a light elastic spring OA, of natural length 60 cm and modulus of elasticity 12 N. The spring is placed on a smooth table and the end O is fixed. The particle is pulled away from O to a point B, where $OB = 85$ cm, and is released from rest.

a Prove that the motion of P is simple harmonic motion with period $\frac{2\pi}{5}$ s. **(4)**

b Find the greatest magnitude of the acceleration of P during the motion. **(2)**

Two seconds after being released from rest, P passes through the point C.

c Find, to 2 significant figures, the speed of P as it passes through C. **(2)**

d State the direction in which P is moving 2 s after being released. **(2)**

← Section 5.4

A 40 A particle P of mass 0.3 kg is attached
E/P to one end of a light elastic spring. The other end of the spring is attached to a fixed point O on a smooth horizontal table. The spring has natural length 2 m and modulus of elasticity 21.6 N. The particle P is placed on the table at a point A, where $OA = 2$ m. The particle P is now pulled away from O to the point B, where OAB is a straight line with $OB = 3.5$ m. It is then released from rest.

a Prove that P moves with simple harmonic motion of period $\frac{\pi}{3}$ s. **(4)**

b Find the speed of P when it reaches A. **(2)**

The point C is the midpoint of AB.

c Find, in terms of π, the time taken for P to reach C for the first time. **(4)**

Later in the motion, P collides with a particle Q of mass 0.2 kg which is at rest at A.

After impact, P and Q coalesce to form a single particle R.

d Show that R also moves with simple harmonic motion and find the amplitude of this motion. **(4)**

← Section 5.4

E/P 41 A light elastic string of natural length l has one end attached to a fixed point A. A particle P of mass m is attached to the other end of the string and hangs in equilibrium at the point O, where $AO = \frac{5}{4}l$.

a Find the modulus of elasticity of the string. **(3)**

The particle P is then pulled down and released from rest. At time t the length of the string is $\frac{5l}{4} + x$.

b Prove that, while the string is taut,

$$\frac{\mathrm{d}^2 x}{\mathrm{d}t^2} = -\frac{4gx}{l}$$ **(5)**

A

When P is released, $AP = \frac{7}{4}l$. The point B is a distance l vertically below A.

c Find the speed of P at B. **(2)**

d Describe briefly the motion of P after it has passed through B for the first time until it next passes through O. **(2)**

← **Section 5.5**

E/P **42** A light elastic string, of natural length $4a$ and modulus of elasticity $8mg$, has one end attached to a fixed point A. A particle P of mass m is attached to the other end of the string and hangs in equilibrium at the point O.

a Find the distance AO. **(4)**

The particle is now pulled down to a point C vertically below O, where $OC = d$. It is released from rest. In the subsequent motion the string does not become slack.

b Show that P moves with simple harmonic motion of period $\pi\sqrt{\dfrac{2a}{g}}$ **(5)**

The greatest speed of P during this motion is $\frac{1}{2}\sqrt{ga}$.

c Find d in terms of a. **(2)**

Instead of being pulled down a distance d, the particle is pulled down a distance a. Without further calculation,

d describe briefly the subsequent motion of P. **(2)**

← **Section 5.5**

Challenge

A

1 A particle moves in a straight line with an initial velocity of $u\,\mathrm{m\,s^{-1}}$. When the particle is moving with a velocity of $v\,\mathrm{m\,s^{-1}}$ the acceleration is $-e^{2kv}$ where k is a constant.

a Show that the time taken for the particle to come to rest is $\dfrac{1}{2k}\left(\dfrac{e^{2ku}-1}{e^{2ku}}\right)$

b Find the distance the particle moves before coming to rest.

← **Section 4.3**

2 A particle P travels on the x-axis, passing the origin at time $t = 0$ with velocity $-k\,\mathrm{m\,s^{-1}}$, where k is a positive constant. At time t the particle is a distance $x\,\mathrm{m}$ from the origin and its acceleration, $a\,\mathrm{m\,s^{-2}}$, is given by $a = 8x\dfrac{\mathrm{d}x}{\mathrm{d}t}$

Show that the distance of the particle from the origin never exceeds $\frac{1}{2}\sqrt{k}$ metres.

← **Section 4.2**

3 A particle of mass m is projected vertically upwards at time $t = 0$, with speed U. The particle moves against air resistance of magnitude $mgkv^2$, where v is the speed of the particle at time t and k is a constant. Find in terms of U, g and k the height of the particle at the first point where its speed is half of its initial speed.

← **Section 5.1**

Exam-style practice
Further Mathematics
AS Level
Further Mechanics 2

Time: 50 minutes
You must have: Mathematical Formulae and Statistical Tables, Calculator

1 A particle P moves along the positive x axis. At time t seconds, the acceleration of the particle is $\dfrac{144 - v^2}{48}$ where $v\,\text{m s}^{-1}$ is the velocity of the particle. The particle starts from rest.

   a Find v in terms of t. **(6)**

   b Show that the speed of the particle cannot exceed $12\,\text{m s}^{-1}$. **(2)**

2
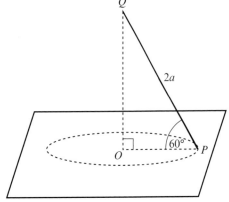

A particle P of mass m is attached to one end of a light inextensible string of length $2a$. The other end of the string is fixed to a point Q which is vertically above the point O on a smooth horizontal table. The particle P remains in contact with the surface of the table and moves in a circle with centre O and with angular speed $\sqrt{\dfrac{kg}{4a}}$, where k is a constant.

Throughout the motion the string remains taut and $\angle QPO = 60°$, as shown in the figure above.

   a Show that the tension in the string is $\dfrac{mkg}{2}$ **(3)**

   b Find, in terms of m, g and k, the normal reaction between P and the table. **(3)**

   c Deduce the range of possible values of k in order for the particle to remain in contact with the table. **(2)**

The angular speed of P is changed to $\sqrt{\dfrac{3g}{a}}$. The particle P now moves in a horizontal circle above the table. The centre of this circle is X.

   d Show that $QX : QO = 1 : 3\sqrt{3}$ **(7)**

3

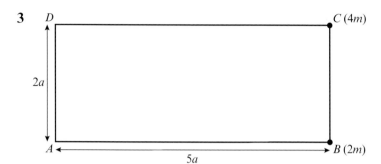

The figure above shows four uniform rods joined to form a rigid rectangular framework $ABCD$, where $AB = CD = 5a$, and $BC = AD = 2a$. The mass of each rod is $\frac{m}{a}$ per unit length.

Particles, of masses $2m$ and $4m$, are attached to the framework at points B and C respectively.

a Find the distance of the centre of mass of the loaded framework from:

  i AB

  ii AD **(8)**

A new uniform rod, PQ, of mass $10m$ is added to the loaded framework to join AB and CD. P lies on AB and Q lies on DC such that PQ is parallel to AD. Given that the centre of mass of the new framework is a distance $2.5a$ from AD,

b find the distance AP. **(5)**

The loaded framework is freely suspended from D and hangs in equilibrium.

c Find, correct to three significant figures, the angle which DC makes with the vertical. **(3)**

d State how in your calculations you have used the assumption that the rods are uniform. **(1)**

Exam-style practice

Further Mathematics
A Level
Further Mechanics 2

Time: 1 hour and 30 minutes
You must have: Mathematical Formulae and Statistical Tables, Calculator

1 In a harbour, sea level at low tide is 10 m below the level of the sea at high tide. At low tide the depth of the water in the harbour is 8 m. On a particular day, low tide occurs at 1 p.m. and the next low tide occurs at 1.30 a.m. A ship can remain in the harbour safely when the depth of water is at least 12 m. The sea level is modelled as rising and falling with simple harmonic motion.

   a Write down the

     i period

     ii amplitude of the motion.   **(2)**

   A 'safe mooring height' marker is attached to the harbour wall at a depth of 12 m.

   b Find the speed, in metres per hour, at which the water level is rising when it passes this marker.   **(4)**

   c Find the total length of time between two consecutive low tides for which the water in the harbour is at a safe mooring depth.   **(4)**

2 A uniform rectangular piece of card $ABCD$ has $AB = 2a$ and $AD = a$. Corner C is folded down to meet side AB as shown in the diagram.

   a Find the distance of the centre of mass of the lamina from

     i AD ii AB **(7)**

   The lamina is freely suspended by a string attached to the point A and hangs at rest.

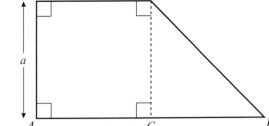

   b Find, to the nearest degree, the angle between DE and the vertical.   **(4)**

3 A particle P of mass m moves along the positive x-axis. At time t seconds, the acceleration of the particle is $-2(k^2 + v^2)\,\mathrm{m\,s^{-2}}$ where $v\,\mathrm{m\,s^{-1}}$ is the velocity of the particle and k is a positive constant. When $t = 0$, P is at O and $v = 2U$. The particle passes through the point A with velocity U.

   a Find the distance OA.   **(6)**

   b Show that the time P takes to travel from O to A is $\dfrac{1}{2k}\left(\arctan\dfrac{2U}{k} - \arctan\dfrac{U}{k}\right)$ **(5)**

4 A circular track is banked at an angle α to the horizontal, where $\tan \alpha = \frac{3}{4}$. A car moves round the track at constant speed in a horizontal circle of radius 306 m. The car is modelled as a particle and the track is modelled as being smooth and any non-gravitational resistance to motion is ignored.

a Find the speed of the car. **(6)**

The model is refined to account for the roughness of the track, and the coefficient of friction between the car and the track is taken to be 0.5. Any non-gravitational resistances to motion are still ignored.

b Given that the car does not skid either up or down the banked surface of the track, find the range of possible speeds of the car. **(8)**

5 The region R is bounded by the curve with equation $y = \sqrt{\cos 2x}$, the positive x-axis and the line $x = \frac{\pi}{4}$.

The unit of length on both axes is the metre.
A uniform solid S is formed by rotating R through 2π about the x-axis.

a Show that the volume of S is $\frac{\pi}{2}$ m³. **(4)**

b Find, in terms of π, the x-coordinate of the centre of mass of S. You must use algebraic integration to obtain your answer. **(7)**

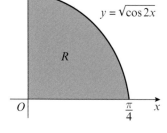

The solid S is placed on an inclined plane, rough enough to prevent slipping, with its circular face on the plane. The plane is slowly tilted until the solid S is about to topple.

c Find the inclination of the plane to the horizontal when the solid S is about to topple. **(4)**

6 A particle P of mass m is attached to one end of a light inextensible string of length a. The other end of the string is attached to a fixed point O. The particle is held with the string taut so that OP makes an angle of 60° with the downward vertical, OQ. The particle is then projected with speed $\sqrt{3ga}$ in a direction which is perpendicular to OP so that the particle moves in the vertical plane OPQ.

Show that P does not make complete circles about O and find the angle that OP makes with the upward vertical when the string first goes slack. **(14)**

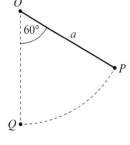

Answers

CHAPTER 1
Prior knowledge check
1 a 21.9 N (3 s.f.)   b 3.05 kg (3 s.f.)
2 a 10.6 m (3 s.f.)   b 58.6 J (3 s.f.)

Exercise 1A
1 a $0.524\,\mathrm{rad\,s^{-1}}$ (3 s.f.)   b $12.6\,\mathrm{rad\,s^{-1}}$ (3 s.f.)
  c $38.2\,\mathrm{rev\,min^{-1}}$ (3 s.f.) d $1720\,\mathrm{rev\,h^{-1}}$ (3 s.f.)
2 a $80\,\mathrm{m\,s^{-1}}$ b $83.8\,\mathrm{m\,s^{-1}}$ (3 s.f.)
3 a $8\,\mathrm{rad\,s^{-1}}$ b $76.4\,\mathrm{rev\,min^{-1}}$ (3 s.f.)
4 a $2\,\mathrm{m\,s^{-1}}$ b $2.09\,\mathrm{m\,s^{-1}}$ (3 s.f.)
5 a 44.9 s (3 s.f.)   b $0.14\,\mathrm{rad\,s^{-1}}$ (3 s.f.)
6 a $0.628\,\mathrm{rad\,s^{-1}}$ (3 s.f.) b $0.0754\,\mathrm{m\,s^{-1}}$ (3 s.f.)
  c $0.0503\,\mathrm{m\,s^{-1}}$ (3 s.f.)
7 a $0.279\,\mathrm{rad\,s^{-1}}$ (3 s.f.) b 39.8 m (3 s.f.)
8 $3.14\,\mathrm{m\,s^{-1}}$ (3 s.f.), $5.24\,\mathrm{m\,s^{-1}}$ (3 s.f.)
9 a $0.242\,\mathrm{rad\,s^{-1}}$ (3 s.f.) b $0.362\,\mathrm{m\,s^{-1}}$ (3 s.f.)
10 $0.056\,\mathrm{rad\,s^{-1}}$ (3 s.f.)
11 a $0.000145\,\mathrm{rad\,s^{-1}}$ (3 s.f.), $0.00175\,\mathrm{rad\,s^{-1}}$ (3 s.f.)
   b $1.45 \times 10^{-5}\,\mathrm{m\,s^{-1}}$ (3 s.f.), $2.62 \times 10^{-4}\,\mathrm{m\,s^{-1}}$ (3 s.f.)
12 $62.8\,\mathrm{m\,s^{-1}}$ (3 s.f.)
13 a $4.71\,\mathrm{rad\,s^{-1}}$ (3 s.f.)   b 2.55 cm (3 s.f.)
14 $29\,900\,\mathrm{m\,s^{-1}}$ (3 s.f.)
15 $r > 5$

Challenge
$\dfrac{\pi}{19}\,\mathrm{rad\,s^{-1}}$

Exercise 1B
1 $4\,\mathrm{m\,s^{-2}}$
2 $20.8\,\mathrm{m\,s^{-2}}$ (3 s.f.)
3 a $5\,\mathrm{rad\,s^{-1}}$ b $15\,\mathrm{m\,s^{-1}}$
4 a $12.9\,\mathrm{rad\,s^{-1}}$ (3 s.f.) b $7.75\,\mathrm{m\,s^{-1}}$ (3 s.f.)
5 $2.14\,\mathrm{m\,s^{-2}}$ (3 s.f.)
6 $0.283\,\mathrm{rad\,s^{-1}}$ (3 s.f.)
7 0.72 N
8 48.6 N
9 a 0.588 N (3 s.f.)   b 4.5 N
10 a $0.24\,\mathrm{m\,s^{-1}}$ b 0.0072 N
11 0.0294 (3 s.f.)
12 $3.13\,\mathrm{rad\,s^{-1}}$ (3 s.f.)
13 0.157 (3 s.f.)
14 $0.233\,\mathrm{rad\,s^{-1}}$ (3 s.f.)
15 a 320 N (3 s.f.)   b 0.000153 (3 s.f.)
16 a $2.42\,\mathrm{rad\,s^{-1}}$ (3 s.f.)
   b No, because it is the minimum possible value for W.
      If the speed or the coefficient of friction reduced at
      all, the people would slip down the cylinder.
17 $1.4\,\mathrm{m\,s^{-1}}$
18 a $\mu > \dfrac{v^2}{gR}$
   b Model assumes that the tyres all experience the
      same friction.
19 0.322 m (3 s.f.)
20 $\omega \leqslant \dfrac{\sqrt{197}}{10}$

Challenge
a $y = \dfrac{q}{p^2}x^2$

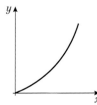

b Acceleration is $2q$ in the positive y-direction. Speed at
   the origin is p.
c $y = R - \sqrt{R^2 - x^2}$
d $R = \dfrac{p^2}{2q}$
e $2q$
f The acceleration of P and Q are equal.

Exercise 1C
1 18.4 N (3 s.f.), $4.52\,\mathrm{rad\,s^{-1}}$ (3 s.f.)
2 10.3 N (3 s.f.), $4.43\,\mathrm{rad\,s^{-1}}$ (3 s.f.)
3 23.7 N (3 s.f.), 60° (nearest degree)
4 73.5 N, 0.6 m
5 $T = ml\omega^2$
6 Let the tension in the string be T. The angle between
   the string and the vertical is θ, and the radius of the
   circle is r.
   R($\rightarrow$): $T\sin\theta = mr\omega^2$
   R($\rightarrow$): $T\cos\theta = mg$
   Dividing the horizontal component by the vertical
   component
   $\tan\theta = \dfrac{mr\omega^2}{mg}$
   $\dfrac{r}{x} = \dfrac{r\omega^2}{g}$ so, $\omega^2 x = g$
7 $\omega = \sqrt{\dfrac{g}{3}}$
8 $d = 5\,\mathrm{cm}$
9 $\omega = 1.8\,\mathrm{rad\,s^{-1}}$, $v = 5.4\,\mathrm{m\,s^{-1}}$
10 $18.1\,\mathrm{rad\,s^{-1}}$ (3 s.f.)
11 9.5°
12 $22.8\,\mathrm{rad\,s^{-1}}$ (3 s.f.)
13 $9.44\,\mathrm{rad\,s^{-1}}$ (3 s.f.)
14 a R($\rightarrow$): $R\cos\alpha = mg$
      R($\uparrow$): $R\sin\alpha = \dfrac{mv^2}{r}$
      Dividing the horizontal component by the vertical
      component to eliminate R:
      $\tan\alpha = \dfrac{mv^2}{rmg} = \dfrac{v^2}{rg}$
      So, $v^2 = rg\tan\alpha = \sqrt{rg\tan\alpha}$
   b This model assumes there is no friction between the
      vehicle and the road.
15 $2.22\,\mathrm{m\,s^{-1}}$ (3 s.f.)
16 0.4
17 42° (nearest degree)
18 $29.5\,\mathrm{m\,s^{-1}}$ (3 s.f.), $9.94\,\mathrm{m\,s^{-1}}$ (3 s.f.)

19 a 20° (nearest degree), 20 800 N (3 s.f.);
68° (nearest degree), 52 700 N (3 s.f.)

b To turn in a shorter time the aircraft will need to decrease the radius of the circular arc in which it turns. Thus the angle to the horizontal and lift force must both increase.

20 R(↑): $T\sin\theta + R\sin\theta = mg \Rightarrow T + R = \dfrac{mg}{\sin\theta}$

R(→): $T\cos\theta - R\cos\theta = ml\cos\theta\,\omega^2 \Rightarrow T - R = ml\omega^2$

Eliminating R:

$2T = \dfrac{mg}{\sin\theta} + ml\omega^2$

$T = \frac{1}{2}m(l\omega^2 - g\,\mathrm{cosec}\,\theta)$

21 2.20 rad s⁻¹ (3 s.f.)

Exercise 1D

1 a 3.13 m s⁻¹ (3 s.f.) **b** 17.6 N (3 s.f.)
2 a 3.43 m s⁻¹ (3 s.f.) **b** 19.6 N
3 a 2.97 m s⁻¹ (3 s.f.) **b** 15.7 N (3 s.f.)
4 a 2.21 m s⁻¹ (3 s.f.) **b** 5.88 N (3 s.f.)
5 a 8.52 m s⁻¹ (3 s.f.) **b** 46.9 N (3 s.f.)
6 a $v = \sqrt{u^2 - 1.4g(1 - \cos\theta)}$ **b** $u \geqslant \sqrt{2.8g}$
7 a $T = 4.5g\cos\theta + \dfrac{3u^2}{4} - 3g$ **b** $u > \sqrt{10g}$
8 a 0.27 **b** 0.26
9 a 0.30 **b** 0.28

c The particle will continue in a parabolic arc (projectile motion) in the negative x-direction, initially increasing in y before decreasing in y.

10 a 6.26 m s⁻¹ (3 s.f.), 25.3 N (3 s.f.)
b 5.6 m s⁻¹, 18.8 N (3 s.f.)
11 a 9.66 m s⁻¹ (3 s.f.) **b** g m s⁻²
c 45.8 N (3 s.f.)
12 39.6° (3 s.f.) to the upward vertical, $v = 2.74$ m s⁻¹ (3 s.f.)
13 $\sqrt{8gr}$
14 a $\sqrt{\dfrac{16gr}{5}}$ **b** $\dfrac{11mg}{5}$
15 a $T = mg(1 + 3\sin\theta)$ **b** 19.5° (3 s.f.)
16 a At point S: G.P.E. = $0.4 \times g \times 3.8 = 1.52g$, K.E. = 0
At point P: G.P.E. = $0.4 \times g \times 4\sin\theta = 1.6g\sin\theta$,
K.E. = $\frac{1}{2} \times 0.4 \times v^2 = 0.2v^2$
By conservation of energy: $1.52g = 1.6g\sin\theta + 0.2v^2$
$\Rightarrow 0.2v^2 = 1.52g - 1.6g\sin\theta$
$\Rightarrow v^2 = 7.6g - 8g\sin\theta$
$\Rightarrow v = \sqrt{7.6g - 8g\sin\theta}$
b 3.8 m
c In reality there will be frictional forces acting on the handle so the height will be less than 3.8 m.

Exercise 1E

1 a $mg + 3mg\cos\theta$ **b** $\dfrac{4l}{3}$ **c** $\dfrac{40l}{27}$
2 a $9g\cos\theta - 6g$ **b** 48° (nearest degree)
c 6.7 m
3 a $\dfrac{9rg}{4} - 2rg\cos\theta$ **b** $\dfrac{3}{4}$ **c** $\sqrt{\dfrac{3rg}{4}}$
d $\dfrac{3\sqrt{rg}}{2}$ **e** 64° (nearest degree)
4 a 48° (nearest degree)
b $\sqrt{8g}$, 74° (nearest degree)

5 a $\dfrac{a}{4}$ **b** $\sqrt{\dfrac{ga}{4}}$
c $\sqrt{\dfrac{9ga}{4}}$, 64° (nearest degree)
6 a 49° (nearest degree)
b The particle will fall through a parabolic arc (projectile motion) towards the surface in the positive x-direction.
7 a 10.3 m s⁻¹
b At R: $\frac{1}{2} \times 2 \times v^2 = 2g(12 - 5\cos70° - 7\cos40°)$
So, $v^2 = 96.58$
Resolving towards B: $mg\cos\theta - R = \dfrac{mv^2}{7}$
$R = 2g\cos40° - \dfrac{2v^2}{7} = -12.6$
$R < 0$
This is impossible, so the particle must have lost contact with the chute.
c In reality, energy is lost due to friction between the laundry bags and the chute.
8 a $\sqrt{\dfrac{17ga}{3}}$
b Energy would be lost due to the frictional force acting on the marble, requiring a larger initial speed for the marble to leave the bowl.

Mixed exercise 1

1 R(→): $R\cos\theta = mg$

R(↑): $R\sin\theta = \dfrac{mv^2}{r} = \dfrac{2mu^2}{3a}$

Dividing the horizontal component by the vertical component gives $\tan\theta = \dfrac{2u^2}{3ag}$

From the problem's geometry, $\tan\theta = \dfrac{\left(\frac{3a}{2}\right)}{\left(\frac{\sqrt{7}a}{2}\right)} = \dfrac{3}{\sqrt{7}}$

So, $\dfrac{3}{\sqrt{7}} = \dfrac{2u^2}{3ag}$

$9ag = 2\sqrt{7}\,u^2$

2 a $\dfrac{3mg}{2}$ **b** $\dfrac{mg}{2}$
3 a $\dfrac{13mg}{5}$ **b** $\sqrt{60gl}$
4 108 m (3 s.f.)
5 a T is the tension in string AP. S is the tension in string BP.
The triangle is equilateral.
R(→): $T\cos60° = mg + S\cos60°$
$T - S = 2mg$
R(↑): $T\cos30° + S\cos30° = mr\omega^2$
$(T + S)\cos30° = m(l\cos30°)\omega^2$
$T + S = ml\omega^2$
Eliminating S to find T:
$2T = 2mg + ml\omega^2$
$T = \dfrac{m}{2}(2g + l\omega^2)$
b $\dfrac{m}{2}(l\omega^2 - 2g)$
c Both strings taut $\Rightarrow l\omega^2 - 2g > 0$, $\omega^2 > \dfrac{2g}{l}$
6 a Let T be the tension in the string.
R(↑): $T\cos45° = mg$
$T = \sqrt{2}\,mg$
b $\omega = \sqrt{\dfrac{\sqrt{2}g}{l}}$

7 a $6.48\,\text{N}$ **b** $25°$ (nearest degree)

8 a $4mg$ **b** $3\pi\sqrt{\dfrac{r}{g}}$

9 a
$$\frac{mv^2}{r} = \mu R = \mu mg$$
$$\frac{v^2}{rg} = \mu$$
$$\frac{21^2}{100 \times 9.8} = \mu$$
$$\mu = 0.45$$

  b $\dfrac{35}{136}$

10 a $\dfrac{\sqrt{3}\,m}{4}(r\omega^2 + 2g)$

  b Maximum speed gives the shortest time. At the maximum speed with the rod still on the surface of the sphere, $R = 0$. Radius of the circle is $\dfrac{\sqrt{3}\,r}{2}$.

When $R = 0$, $T\cos\alpha = mg \Rightarrow T = \dfrac{mg}{\cos\alpha} = \dfrac{2mg}{\sqrt{3}}$

$$T\sin\alpha = m \times \frac{\sqrt{3}\,r}{2} = \omega^2$$

So
$$\frac{2mg}{\sqrt{3}} \times \frac{1}{2} = m \times \frac{\sqrt{3}\,r}{2} = \omega^2$$
$$\omega^2 = \sqrt{\frac{2g}{3r}}$$

Time for one revolution $= \dfrac{2\pi}{\omega} = \pi\sqrt{\dfrac{4 \times 3r}{2g}} = \pi\sqrt{\dfrac{6r}{g}}$

  c i The minimum period decreases.
     ii The minimum period increases.

11 a Let F be the force due to friction.
Let R be the normal reaction force.
R($\rightarrow$): $R = mg$
R($\uparrow$): $F = mr\omega^2$
If P does not slip, then $\mu R \geqslant mr\omega^2$
$$\frac{3}{7}mg \geqslant \frac{3a}{5}m\omega^2 \Rightarrow \omega^2 \leqslant \frac{5g}{7a}$$

  b $\dfrac{5g}{42a} \leqslant \omega^2 \leqslant \dfrac{65g}{42a}$

12 a $\frac{4}{3}ga + 2ga\sin\theta$ **b** $mg\left(\frac{4}{3} + 3\sin\theta\right)$

  c $206°$

  d $v = 0$ before the particle reaches the top of the circle.

13 $\sqrt{2.6g}$

14 a $\dfrac{ga}{2}(5 - 4\cos\theta)$

  b Resolving towards the centre O:
$$mg\cos\theta - R = \frac{mv^2}{r}$$
$$mg\cos\theta - R = \frac{mg}{2}(5 - 4\cos\theta)$$
Substituting $\cos\theta = 0.9$
$$R = 0.9mg - \frac{mg}{2}(5 - 3.6)$$
$$R = 0.2mg$$
$R > 0 \Rightarrow P$ is still on the hemisphere

  c i $\dfrac{5}{6}$ **ii** $\sqrt{\dfrac{5ga}{6}}$

  d $\sqrt{\dfrac{5ga}{2}}$ **e** $61°$ (nearest degree)

15 a $u^2 - \dfrac{16}{5}gr$ **b** $\dfrac{mg}{5}$

c $\sqrt{\dfrac{19gr}{5}}$ **d** $\sqrt{\dfrac{73gr}{15}}$

16 a $\dfrac{u^2 + 2ag}{3ag}$ **b** $34°$ (nearest degree)

Challenge

a At point (x, x^2), $\dfrac{dy}{dx} = 2x$

  R($\uparrow$): $R\cos\theta = mg$ (1)
  R($\rightarrow$): $R\sin\theta = mx\omega^2$ (2)

  (2) ÷ (1): $\tan\theta = \dfrac{x\omega^2}{g}$ (3)

$$\tan\theta = \frac{dy}{dx} = 2x$$
$$\Rightarrow 2x = \frac{x\omega^2}{g} \Rightarrow 2g = \omega^2 \Rightarrow \omega = \sqrt{2g}$$
Hence ω is independent of the vertical height.

b From (3), $\omega^2 = \dfrac{g\tan\theta}{x}$

For ω to be independent of $x \Rightarrow \dfrac{g\tan\theta}{x} = k$ for constant k
$\Rightarrow \tan\theta = ax$ for constant a
$$\frac{dy}{dx} = \tan\theta = ax \Rightarrow y = \tfrac{1}{2}ax^2 + b$$
Hence $f(x) = px^2 + q$ for for constants p and q

CHAPTER 2
Prior knowledge check

1 $x = 3, y = 2$
2 $3.6\,\text{m}$
3 $144\,\text{cm}^2$

Exercise 2A

1 $(3.2, 0)$
2 $(0, 2.5)$
3 $(1.1, 0)$
4 $2\frac{1}{3}\,\text{m}$
5 $m = 6$
6 $0.7\,\text{kg}$
7 6.5
8 $(0, -2)$
9 $m_1 = 2, m_2 = 3$
10 $1 \times (2m + 5) = ((m - 1) \times -1) + ((5 - m) \times 1) + (2 \times m)$
                     $+ ((m + 1) \times 0)$
             $2m + 5 = 6$
                $2m = 1$
                $m = 0.5\,\text{kg}$

Challenge
$(2 \times PQ) + \left(3 \times \left(PQ + \frac{3}{2}PQ\right)\right) = (6 \times PG)$
$\frac{19}{2}PQ = 6PG$
$19PQ = 12PG$
$PQ : PG$ is $12 : 19$

Exercise 2B

1 $(3, 2)$
2 $(0.5, -0.75)$
3 $(4.6, 4.2)$
4 $3\mathbf{i} + 2.5\mathbf{j}$
5 $(2.1, 0.3)$
6 a 1 **b** 1.5
7 $p = 1, q = -2$
8 $(1, 3)$

9 $3\frac{3}{4}$ cm from AB and $4\frac{1}{3}$ cm from AD.

10 a 3g   **b** 3.2 cm

Challenge

0.2 kg

Exercise 2C

1 a (2, 3)   **b** (3, 4)

   c $\left(\frac{1}{3}, 1\right)$ **d** $\left(\frac{8a}{3}, 3a\right)$

2 Centre of mass is on the axis of symmetry at a distance $\frac{16}{3\pi}$ cm from the centre.

3 $a = 3, b = 3$

4 a Distance a from AB, Distance $\frac{2a}{3}$ from BC.

   b Distance $\frac{a}{3}$ from BC, Distance $\frac{4a}{3}$ from AB.

   c On the line of symmetry, $\frac{4a}{3}$ from the line AB.

   d $\left(3a, \frac{2a}{3}\right)$ with A as the origin and AC as the x-axis.

5 (2, 7) and (−13, 4)

6 a B has coordinates $\left(\frac{18}{5}, \frac{41}{5}\right)$, D has coordinates $\left(\frac{12}{5}, -\frac{1}{5}\right)$

   b (3, 4)

7 a (3, 5), (3,−3)

   b $\left(3, \frac{7}{3}\right)$, $\left(3, -\frac{1}{3}\right)$

8 $\bar{y} = \frac{y}{3} = y - \frac{4\sqrt{3}}{3}$

$\frac{2}{3}y = \frac{4\sqrt{3}}{3}$

$y = 2\sqrt{3}$

Centre of mass is at $x = 2$

So, using Pythagoras:

$AB^2 = BC^2 = \sqrt{2^2 + (2\sqrt{3})^2}$

$AB = BC = 4$

So the triangle is equilateral.

Exercise 2D

1 a $\left(\frac{5}{2}, \frac{13}{14}\right)$ **b** (1.7, 2.6)   **c** $\left(\frac{113}{30}, \frac{49}{30}\right)$ **d** $\left(\frac{7}{3}, 2\right)$

   e $\left(\frac{32}{9}, \frac{53}{18}\right)$ **f** $\left(2, \frac{63}{17}\right)$ **g** $\left(\frac{25}{8}, 3\right)$

2 $\frac{2}{9}a$

3 $2.89a$ (3 s.f.)

4 a 10 cm

   b 12 cm horizontally to the right of A and 4.5 cm vertically above A.

5 a 4.5 cm horizontally to the right of O, and 3.5 cm vertically above O.

   b 5.83 cm horizontally to right of O, and 4.32 cm vertically above O.

6 a $12\binom{1}{3} + 24\binom{5}{4} + 16\binom{9}{2} = 52\binom{\bar{x}}{\bar{y}}$

$\binom{\bar{x}}{\bar{y}} = \binom{\frac{69}{13}}{\frac{41}{13}}$

$\left(\frac{69}{13}, \frac{41}{13}\right)$

   b Since $\bar{y} = \frac{41}{13}$ for original plate, holes must be symmetrically placed about the line $y = \frac{41}{13}$

   c $\frac{70}{9}$

7 $x = a$

Challenge

Area of hexagon $ABCDEF = \frac{3\sqrt{3}}{2}x^2$

Using Pythagoras, find the height of the midpoint:

$x^2 = \left(\frac{x}{2}\right)^2 + h^2 \Rightarrow h = \frac{\sqrt{3}}{2}$

So the centre of mass of $ABCDEF$ is $\begin{pmatrix}\frac{x}{2}\\\frac{\sqrt{3}x}{2}\end{pmatrix}$

Area of triangle to be removed $= \frac{1}{2} \times \sqrt{3}x \times \frac{x}{2} = \frac{\sqrt{3}}{4}x^2$

So the centre of mass of DEF is $\begin{pmatrix}x + \frac{x}{6}\\\frac{\sqrt{3}x}{2}\end{pmatrix}$

$\bar{y} = \frac{\sqrt{3}x}{2}$ due to symmetry

$\frac{3\sqrt{3}}{2}x^2\begin{pmatrix}\frac{x}{2}\\\frac{\sqrt{3}x}{2}\end{pmatrix} - \frac{\sqrt{3}}{4}\begin{pmatrix}\frac{7x}{6}\\\frac{\sqrt{3}x}{2}\end{pmatrix} = \frac{5\sqrt{3}}{4}\begin{pmatrix}\bar{x}\\\frac{\sqrt{3}x}{2}\end{pmatrix}$

$\bar{x} = \frac{11x}{30}$, so $N\left(\frac{11x}{30}, \frac{\sqrt{3}x}{2}\right)$

$MN = \frac{x}{2} - \frac{11x}{30} = \frac{2}{15}x$

Exercise 2E

1 a $\left(\frac{43}{16}, \frac{21}{16}\right)$ **b** $\left(\frac{53}{18}, \frac{20}{9}\right)$

   c (2.67, 2.26) (3 s.f.)   **d** (3.73, 3.00) (3 s.f.)

2 Centre of mass is on line of symmetry through O, and a distance of $\frac{9(\sqrt{3} + 2)}{6 + \pi}$ from O.

3 $\frac{11}{7}a$

4 a 5.83 cm (3 s.f.)

   b i 817 g (3 s.f.)   **ii** 4.41 cm (3 s.f.)

5 $\frac{12}{5}$ m horizontally to the right of A and $\frac{19}{10}$ m vertically below A.

6 Centre of mass is on the line of symmetry at a distance of $\frac{3}{2\pi}$ below the line AB.

7 a 1.75 m

   b $(20 \times 1.75) - (1 \times 0.5) = 19\bar{y} \Rightarrow \bar{y} = \frac{69}{38}$

$\triangle y = \frac{69}{38} - 1.75 = \frac{5}{76}$

Challenge

$\frac{1}{5}\sqrt{5}$ cm

Exercise 2F

1 a 20.4° (3 s.f.)   **b** 24.4° (3 s.f.).   **c** 56.8° (3 s.f.)

2 63.0° (3 s.f.)

3 80.5° (3 s.f.)

4 33.1° (3 s.f.)

5 81.0° (3 s.f.)

6 a $\frac{26}{7}$ cm

   b $\frac{18}{7}$ cm

   c $\alpha = 22.2°$ (3 s.f.)

7 67.1° (3 s.f.)

8 65.3° (3 s.f.)

9 a 3.5 m

   b At A: 5gN; At B: 7gN

   c 149°

10 $\frac{8 - \pi}{\pi}$

227

11 a For a rectangular lamina: Area $= 2a \times 3a = 6a^2$

Centre of mass $\begin{pmatrix} a \\ \frac{3a}{2} \end{pmatrix}$

For the quarter circle: Area $= \frac{1}{4} \times \pi \times a^2 = \frac{\pi a^2}{4}$

Centre of mass:

$d = \dfrac{2r\sin\alpha}{3\alpha} = \dfrac{2a\sin\left(\frac{\pi}{4}\right)}{3 \times \frac{\pi}{4}} = \dfrac{4\sqrt{2}a}{3\pi}$

This distance is the hypotenuse of a right-angled triangle, with the other sides equal.

$d^2 = 2l^2 \Rightarrow \left(\dfrac{4\sqrt{2}a}{3\pi}\right)^2 = 2l^2 \Rightarrow l^2 = \dfrac{16a^2}{9\pi^2} \Rightarrow l = \dfrac{4a}{3\pi}$

So, centre of mass of the quarter circle $\begin{pmatrix} \left(2 - \frac{4}{3\pi}\right)a \\ \frac{4a}{3\pi} \end{pmatrix}$

Centre of mass of the lamina:

$6a^2\begin{pmatrix} a \\ \frac{3a}{2} \end{pmatrix} - \dfrac{\pi a^2}{4}\begin{pmatrix} \left(2 - \frac{4}{3\pi}\right)a \\ \frac{4a}{3\pi} \end{pmatrix} = \left(6 - \dfrac{\pi}{4}\right)a^2\begin{pmatrix} \bar{x} \\ \bar{y} \end{pmatrix}$

$\left(6 - \dfrac{\pi}{4}\right)a^2\bar{x} = 6a^3 - \dfrac{\pi a^3}{2} + \dfrac{a^3}{3} \Rightarrow \bar{x} = \dfrac{2a(38 - 3\pi)}{3(24 - \pi)}$

$\left(6 - \dfrac{\pi}{4}\right)a^2\bar{y} = 9a^3 - \dfrac{a^3}{3} \Rightarrow \bar{y} = \dfrac{104a}{3(24 - \pi)}$

b $\dfrac{34W}{3(24 - \pi)}$ and $\dfrac{W(38 - 3\pi)}{3(24 - \pi)}$

c $\arctan\left(\dfrac{68}{112 - 9\pi}\right)$ (39.1° (3 s.f.))

Challenge
0.0343M (3 s.f.)

Exercise 2G
1 21.8° (3 s.f.)
2 104° (3 s.f.)
3 a $x = 8\,\text{cm}$ **b** 0.2
4 30.6° (3 s.f.)
5 28.5° (3 s.f.)
6 19.4°
7 a $T_1 = \frac{5}{8}W,\ T_2 = \frac{3}{8}W$ **b** 78.7° (3 s.f.)
8 a $T_1 = \frac{7}{18}W,\ T_2 = \frac{11}{9}W$ **b** 25.5° (3 s.f.)

Challenge
$\theta = 26.6°$ (1 d.p.)

Exercise 2H
1 $\frac{7}{3}$ cm horizontally to the right of AD and 2 cm vertically below AB.
2 2.4 cm vertically below AB and 5.6 cm to the right of (original) AD.
3 20.1° (3 s.f.)
4 $\left(\frac{35}{11}, \frac{12}{11}\right)$
5 (2.89, 1.31) (3 s.f.)
6 50.2° (3 s.f.)
7 a 0.07 m
  b C, as the centre of mass of the composite lamina is between O and C.
  c At B: $17.2g$; At C: $2.8g$
8 a 61.2° (3 s.f.)   **b** 27.6° (3 s.f.)

Challenge
a Let width $= 2$, then height $= \dfrac{1}{\tan 22.5°} = \dfrac{1}{\sqrt{2} - 1} = \sqrt{2} + 1$

b On central axis, a distance $0.54x$ from the bottom edge of the paper.

Mixed exercise 2
1 a 0.413 m (3 s.f.)   **b** 12° (nearest degree)
  c 0.275 (3 s.f.)
2 $\theta = 36.9°$ (3 s.f.)
3 $\left(-\frac{1}{7}, \frac{3}{2}\right)$
4 a $\dfrac{13a}{9}$ **b** $\dfrac{4a}{9}$ **c** 45°   **d** $m = \dfrac{5M}{9}$
5 a i $\dfrac{4a}{5}$ **ii** $\dfrac{a}{2}$ **b** $\theta = 58°$
6 a $1.7a$ **b** $1.1a$
7 59.1° (3 s.f.)
8 a 39.0° (3 s.f.)   **b** 7.8° (3 s.f.)
9 $AB = 0.25M,\ BC = \frac{2}{3}M,\ CD = 0.5M,\ AD = M$

$0.25M\begin{pmatrix} 3 \\ 3 \end{pmatrix} + \dfrac{2}{3}M\begin{pmatrix} 5 \\ 5 \end{pmatrix} + 0.5M\begin{pmatrix} 8 \\ 3 \end{pmatrix} + M\begin{pmatrix} 6 \\ 1 \end{pmatrix} = \dfrac{29}{12}M\begin{pmatrix} \bar{x} \\ \bar{y} \end{pmatrix}$

$\begin{pmatrix} \bar{x} \\ \bar{y} \end{pmatrix} = \begin{pmatrix} \frac{169}{29} \\ \frac{79}{29} \end{pmatrix}$

$\tan\theta = \dfrac{\left(\frac{169}{29} - 3\right)}{\left(5 - \frac{79}{29}\right)} = \dfrac{\left(\frac{82}{29}\right)}{\left(\frac{66}{29}\right)} = \dfrac{41}{33}$

$\theta = \arctan\left(\dfrac{41}{33}\right)$

10 a $T_1 = 1.2Mg$ and $T_2 = 0.8Mg$ **b** 45°
11 a $T_1 = Mg$ and $T_2 = 2Mg$ **b** 47.3° (3 s.f.)

Challenge
13.6° (3 s.f.)

CHAPTER 3
Prior knowledge check
1 $T_1 = \dfrac{\sqrt{2}}{\sqrt{3} + 1}mg,\ T_2 = \dfrac{\sqrt{3} - 1}{\sqrt{3} + 1}mg$

2 $A = 8$ and $B = 4$

3 $\frac{262}{91}$

Exercise 3A
1 $\left(\frac{2}{3}, 2\right)$
2 (1.5, 3.6)
3 (2.4, 0.75)
4 $\left(\frac{14}{25}, \frac{23}{35}\right)$
5 $\left(\frac{3}{5}a, 0\right)$
6 $\left(\frac{\pi}{2}, \frac{\pi}{8}\right)$
7 $\left(\dfrac{1 - \ln 2}{\ln 2}, \dfrac{1}{4\ln 2}\right)$
8 $\left(\dfrac{4\sqrt{2}r}{3\pi}, 0\right)$
9 $\left(\frac{16}{15}, \frac{64}{21}\right)$
10 $\left(2\frac{54}{55}, 1\frac{5}{11}\right)$
11 (1.01, 0) (3 s.f.)
12 (1.34, 0.206) (3 s.f.)

Challenge
$(0, 0), (4, 0), (4, 2), (4, -2), \left(\dfrac{19}{10}, \sqrt{\dfrac{19}{10}}\right)$ and $\left(\dfrac{19}{10}, -\sqrt{\dfrac{19}{10}}\right)$

Exercise 3B

1 $(2, 0)$ **2** $(1, 0)$

3 $(\pi, 0)$ **4** $(0, -3)$

5 $\left(\frac{5}{6}, 0\right)$

6 $\left(2\frac{2}{3}, 0\right)$

7 $\left(\frac{35}{48}, 0\right)$

8 $(1.65, 0)$ (3 s.f.)

9 $\left(\frac{1}{2}\frac{(e^2 + 1)}{(e^2 - 1)}, 0\right)$

10 $(0, 1.34)$ (3 s.f.)

11 $(1.01, 0)$ (3 s.f.)

12 1.46 (3 s.f)

13 $3.04\,\text{cm}$

14 $\frac{r}{3}$ above the base

15 $\frac{5}{6}$ cm above the base

16 The arc of the circle $x^2 + y^2 = a^2$, $a - h \leqslant x \leqslant a$ is rotated about the x-axis.

$$\bar{x} = \frac{\pi \int_{a-h}^{a}(a^2 - x^2)x\,dx}{\pi \int_{a-h}^{h}(a^2 - x^2)dx} = \frac{\left[\frac{1}{2}a^2x^2 - \frac{1}{4}x^4\right]_{a-h}^{a}}{\left[a^2x - \frac{1}{2}x^3\right]_{a-h}^{a}}$$

$$= \frac{\frac{1}{4}a^4 - \frac{1}{2}a^2(a-h)^2 + \frac{1}{4}(a-h)^4}{\frac{2}{3}a^3 - a^2(a-h) + \frac{1}{3}(a-h)^2}$$

$$= \frac{\frac{1}{4}(a^2 - (a-h)^2)^2}{\frac{1}{3}(2a^2 - 3a^2(a-h) + (a-h)^3)}$$

$$= \frac{3}{4}\frac{(2ah - h^2)^2}{3h^2a - h^3}$$

$$= \frac{3}{4}\frac{(2a - h)^2}{(3a - h)}$$

∴ Distance of centre of mass from base of cap (i.e. $x = a - h$) is

$$\frac{3(2a - h)^2}{4(3a - h)} - (a - h) = \frac{3(2a - h)^2 - 4(3a - h)(a - h)}{4(3a - h)}$$

$$= \frac{4ah - h^2}{4(3a - h)}$$

$$= \frac{h(4a - h)}{4(3a - h)}$$

17 a $\bar{y} = 3$ **b** $4.18\,\text{cm}$

18 $k = \frac{7}{5}$

19 a $\frac{23h}{44}$

 b On the common axis of symmetry of the cone and cylinder, a distance $\frac{23h}{44}$ below the top plane face of the cylinder and a distance $\frac{r}{11}$ to the left of the vertical axis of the cylinder.

20 a $4.07\,\text{cm}$ (3 s.f.)   **b** $4.17\,\text{cm}$ (3 s.f.)

Challenge

$\frac{R}{2}$ from the plane face

Exercise 3C

1 The centre of mass lies on the axis of symmetry at a point $0.42\,\text{cm}$ below the base of the cone.

2 Centre of mass is on the axis of symmetry at a distance $5.33\,\text{cm}$ away from base of the cylinder.

---

3 a A square based pyramid has base area A and height h, the centre of mass is on the axis of symmetry.

Volume $= \frac{1}{3}Ah$

Mass $= \frac{1}{3}Ah\rho$

Take a slice of thickness δx at a distance x_i from vertex. The base of the slice is an enlargement of the base of the pyramid with scale factor $\frac{x_i}{h}$.

Ratio of areas is $\left(\frac{x_i}{h}\right)^2$

Area of base of slice is $\frac{x_i^2}{h^2}A$

Mass of slice $m_i = \delta x$

$$\lim_{\delta x \to 0}\sum_{x=0}^{h}m_i x_i = \int_{0}^{h}\frac{x^3}{h^2}A\rho\,dx = \frac{1}{4}h^2A\rho$$

$$\bar{x} = \frac{\sum m_i x_i}{M} = \frac{\frac{1}{4}h^2A\rho}{\frac{1}{3}Ah\rho} = \frac{3}{4}h$$

The centre of mass lies on the line of symmetry at a distance $\frac{3}{4}h$ from the vertex, or $\frac{h}{4}$ from base.

 b $3.39\,\text{cm}$ (3 s.f.) below O

4 a $\frac{h}{4}$ where h is the height of the tetrahedron.

 b $\frac{3}{8}\sqrt{6}$ m below O

5 a $1.96\,\text{cm}$ (3 s.f.)   **b** $0.48\,\text{cm}$ below O (3 s.f.)

6 a $11.4\,\text{kg}$

 b $0.589\,\text{m}$ (3 s.f.) from its base.

 c Mass of post will decrease as cross-section of cylinder is less than 1. Position of centre of mass will be the same.

7 a

Shape	Mass	Vertical distance from O to CoM
Cylinder	$32\pi r^3$	r
Hemisphere	$\frac{2\pi r^3}{3}$	$\frac{3r}{8}$
Composite body	$32\pi r^3 - \frac{2\pi r^3}{3} = \frac{94\pi r^3}{3}$	$\bar{x}$

$$32\pi r^3(r) - \frac{2\pi r^3}{3}\left(\frac{3r}{8}\right) = \frac{94\pi r^3}{3}(\bar{x})$$

$$\frac{127\pi r^4}{4} = \frac{94\pi r^3}{3}\bar{x} \Rightarrow \bar{x} = \frac{381r}{376}$$

 b i $\frac{387r}{392}$ ii $\frac{r}{49}$

8 a $\frac{81}{19}$ m   **b** $4.76\,\text{m}$ (3 s.f.)

 c $0.727\,\text{m}$ (3 s.f.)   **d** $3.45\,\text{m}$ (3 s.f.)

9 a $824\,\text{kg}$ (3 s.f.)   **b** $8.73\,\text{m}$

10 a $\frac{3}{2}$ **b** $\frac{6}{7}$ m

Challenge

$a = -2.5$ and $b = 3$

Exercise 3D

1 $\theta = 68°$ (to the nearest degree)

2 $\theta = 40°$ (to the nearest degree)

3 $\theta = 63°$ (to the nearest degree)

4 a 2.78 cm from large circular face

 b $\alpha = 51°$ (to the nearest degree)

5 $T_1 = 13.3\,\text{N}$, $T_2 = 6.34\,\text{N}$

6 a $\frac{\sqrt{2}}{3}$ m   **b** $T_1 = 38.3\,\text{N}$, $T_2 = 42.9\,\text{N}$

7 a At P reaction is $\dfrac{195g}{4}$N and at Q reaction is $\dfrac{445g}{4}$N.

  b 390 kg

8 a Mass of elemental disc $= \rho\pi r^2 h = 100\pi\,\mathrm{e}^{0.1h}\,\delta h$

Total mass $= 100\pi \int_0^{30} \mathrm{e}^{0.1h}\,\mathrm{d}h = 100\pi\,[10\mathrm{e}^{0.1h}]_0^{30}$

$= 100\pi(10\,\mathrm{e}^3 - 10) = 1000\pi(\mathrm{e}^3 - 1)$

$1000\pi(\mathrm{e}^3 - 1)\bar{y} = 100\pi \int_0^{30} h\mathrm{e}^{0.1h}\,\mathrm{d}h \Rightarrow \bar{y} = \dfrac{\displaystyle\int_0^{30} h\mathrm{e}^{0.1h}\,\mathrm{d}h}{10(\mathrm{e}^3 - 1)}$

$\int_0^{30} h\mathrm{e}^{0.1h}\,\mathrm{d}h = [10h\mathrm{e}^{0.1h}]_0^{30} - \int_0^{30} 10\mathrm{e}^{0.1h}\,\mathrm{d}h$

$= [10h\mathrm{e}^{0.1h} - 100\mathrm{e}^{0.1h}]_0^{30} = 100(2\mathrm{e}^3 + 1)$

$\therefore \bar{y} = \dfrac{100(2\mathrm{e}^3 + 1)}{10(\mathrm{e}^3 - 1)} = \dfrac{10(2\mathrm{e}^3 + 1)}{\mathrm{e}^3 - 1}$

  b $40°$

9 a $T_1 = 1.07 \times 10^{-3}\,\mathrm{N}$, $T_2 = 1.21 \times 10^{-3}\,\mathrm{N}$

  b $46.7°$ (3 s.f.)

10 a In equilibrium the centre of mass ^{G} lies below the point of suspension S. Let distance $SG = x$.
O is the centre of the base of the cone and V is its vertex.
A is the point on the base connected to the string and B is the point on the line SG a distance r from G.
$\tan\theta = \dfrac{x}{3r}$ (from triangle VSG)

Also $\tan\theta = \dfrac{x - r}{r}$ (from triangle ABS)

$\dfrac{x}{3r} = \dfrac{x - r}{r}$

$x = 3x - 3r$

$2x = 3r$

$x = \dfrac{3r}{2}$

$\tan\theta = \dfrac{1}{2}$

  b $\dfrac{\sqrt{5}\,mg}{2}$

11 $22.8°$ (3 s.f.)

Exercise 3E

1 4.77 cm (3 s.f.)

2 a $31°$ (to the nearest degree)   **b** $\mu = \dfrac{3}{5}$

3 a $30°$ **b** 35 cm (2 s.f.)

4 a $\dfrac{2Mg}{2 + \sqrt{3}}$ **b** $\dfrac{1}{2}$

5 a Let the height of the small cone be h.
Using similar triangles:

$\dfrac{h}{h + 2r} = \dfrac{r}{2r}$

$2h = h + 2r$

$h = 2r$

Shape	Mass	Mass ratio	Distance of CoM from centre axis
Large cone	$\rho\frac{1}{3}\pi(2r)^2 4r$	8	r
Small cone	$\rho\frac{1}{3}\pi r^2 \times 2r$	1	$2r + \dfrac{2r}{4} = \dfrac{5r}{2}$
Frustum	$\rho\frac{1}{3}\pi \times 14r^3$	7	$\bar{x}$

Take moments about centre axis:

$8r - \dfrac{5r}{2} = 7\bar{x}$

$\bar{x} = \dfrac{11r}{14}$

  b i Yes   **ii** No

  c As the angle of the slope is $40°$, limiting friction would imply $\mu = \tan 40°$.
No slipping implies that $\mu \geqslant 0.839$ (3 s.f.)

6 a Consider the cube in equilibrium, on the point of toppling, so the reaction force acts through the bottom corner A.
$R(\rightarrow): P - F = 0 \Rightarrow F = P$
$R(\uparrow): R - W = 0 \Rightarrow R = W$
Moments about A: $P \times 4a = W \times 3a$
$\Rightarrow P = \dfrac{3}{4}W$
If equilibrium is broken by toppling:
$P = \dfrac{3}{4}W \Rightarrow F = \dfrac{3}{4}W$
But $F < \mu R$
$\Rightarrow \dfrac{3}{4}W < \mu W \Rightarrow \mu > \dfrac{3}{4}$ is the condition for toppling.
If $\mu < \dfrac{3}{4}$ then the cube will be on the point of slipping when $F = \mu R$.

  b $2a$

7 a $25.6°$ **b** $0 < k < 6$

8 $16.1°$

9 a

Shape	Mass	Mass ratio	Distance of CoM from O
Hemisphere	$\frac{2}{3}\pi\rho r^3$	$2r$	$h + \dfrac{3}{8}r$
Cylinder	$\pi\rho r^2 h$	$3h$	$\dfrac{h}{2}$
Composite solid	$\pi\rho r^2\left(\frac{2}{3}r + h\right)$	$2r + 3h$	$\bar{x}$

Moments: $2r\left(h + \dfrac{3}{8}r\right) + 3h \times \dfrac{h}{2} = (2r + 3h)\bar{x}$

$\Rightarrow 2rh + \dfrac{3}{4}r^2 + \dfrac{3}{2}h^2 = (2r + 3h)\bar{x}$

$8rh + 3r^2 + 6h^2 = 4(2r + 3h)\bar{x}$

$\Rightarrow \bar{x} = \dfrac{6h^2 + 8hr + 3r^2}{4(3h + 2r)}$

  b $\alpha = 29°$ (to the nearest degree)

  c $\mu > \dfrac{44}{81}$

10 a

Shape	Mass	Mass ratio	Distance of CoM from O
Large cone	$\frac{1}{3}\pi\rho(2r^2)2h$	8	$\dfrac{2h}{4}$
Small cone	$\frac{1}{3}\pi\rho r^2 h$	1	$h + \dfrac{h}{4}$
Frustum	$\frac{1}{3}\pi\rho(8r^2h - r^2h)$	7	$\bar{x}$

The centre of the base is the point O.
The radius of the smaller cone is obtained by similar triangles.

Moments: $8 \times \dfrac{2h}{4} - 1 \times \dfrac{5h}{4} = 7\bar{x}$

$\dfrac{11h}{4} = 7\bar{x} \Rightarrow \bar{x} = \dfrac{11}{28}h$

  b $\theta = 38°$ (to the nearest degree)

11 a $P > \mu Mg$ **b** $P > \dfrac{3}{8}Mg$

  c i Slide   **ii** Topple   **iii** Slide and topple

12 a $P = \dfrac{9g}{40}$

  b $P = \dfrac{3g}{10}$

  c $\dfrac{9g}{40} < \dfrac{3g}{10}$

Online Full worked solutions are available in SolutionBank.

Challenge

a

Shape	Mass	Mass ratio	Distance of CoM from O
Cone	$\frac{1}{3}\pi\rho r^2 h$	h	$\frac{h}{4}$
Hemisphere	$\frac{2}{3}\pi\rho r^3$	$2r$	$\frac{-3r}{8}$
Toy	$\frac{1}{3}\pi\rho(r^2h + 2r^3)$	$h + 2r$	$\bar{x}$

Moments: $(h + 2r)\bar{x} = h \times \frac{h}{4} + 2r\left(\frac{-3r}{8}\right)$

$(h + 2r)\bar{x} = \frac{h^2}{4} - \frac{3r^2}{4}$

$\bar{x} = \frac{h^2 - 3r^2}{4(h + 2r)}$

b i Fall over

  ii Return to vertical position

  iii Remain in new position

Mixed exercise 3

1 a $V = \int \pi y^2\,dx = \pi\int_0^4 4x\,dx$

   $= \pi[2x^2]_0^4$

   $= 32\pi$

  b $\bar{x} = \frac{8}{3}$

2 a $V = \int \pi y^2\,dx = \pi\int_1^2 \frac{1}{x^2}\,dx$

   $= \pi\left[\frac{-1}{x}\right]_1^2$

   $= \frac{\pi}{2}m^2$

  b 39 cm to the nearest cm

3 34 cm

4 $\frac{9r}{10}$

5 a 0.71 (2 s.f.)

  b The centre of mass of the body is at C which is always directly above the contact point.

6 a $\bar{y} = \frac{\rho\int\frac{1}{2}y^2\,dx}{\rho\int y\,dx} = \frac{\frac{1}{2}\int_0^4 \frac{x^2}{16}(16 - 8x + x^2)\,dx}{\frac{1}{4}\int_0^4 4x - x^2\,dx}$

   $= 2\frac{\left[\frac{1}{3}x^3 - \frac{1}{8}x^4 + \frac{1}{80}x^5\right]_0^4}{\left[2x^2 - \frac{1}{3}x^3\right]_0^4} = \frac{2}{5}$

  b $\theta = 79°$ (to the nearest degree)

7 a Take the diameter as the y-axis and the midpoint of the diameter as the origin.

   Then $M\bar{x} = \rho\int 2yx\,dx$ where

   $M = \frac{1}{2}\rho\pi(2a)^2$ and where $x^2 + y^2 = (2a)^2$

   $2\rho\pi a^2\bar{x} = \rho\int_0^{2a} 2x\sqrt{4a^2 - x^2}\,dx$

   $= \frac{-2\rho}{3}[(4a^2 - x^2)^{\frac{3}{2}}]_0^{2a}$

   $2\rho\pi a^2\bar{x} = \frac{2\rho}{3} \times 8a^3$

   $\bar{x} = \frac{16}{3}a^3 \div 2\pi a^2$

   $= \frac{8a}{3\pi}$

b

Shape	Mass	Mass ratio	Distance of CoM from AB
Large semicircle	$2\pi\rho a^2$	4	$\frac{8a}{3\pi}$
Semicircle diameter AD	$\frac{1}{2}\pi\rho a^2$	1	$\frac{4a}{3\pi}$
Semicircle diameter OB	$\frac{1}{2}\pi\rho a^2$	1	$\frac{4a}{3\pi}$
Remainder	$\pi\rho a^2$	2	$\bar{x}$

Moments: $4 \times \frac{8a}{3\pi} - 1 \times \frac{4a}{3\pi} - 1 \times \frac{4a}{3\pi} = 2\bar{x}$

$\frac{24a}{3\pi} = 2\bar{x}$

$\bar{x} = \frac{4a}{\pi}$

c The distance from OC is a

   The distance from OB is $\frac{2a}{\pi}$

d 78° (to the nearest degree)

8 a

Shape	Mass	Mass ratio	Distance of CoM from O
Cylinder	$\pi\rho r^2 h$	h	$\frac{-h}{2}$
Hemisphere	$\frac{2}{3}\pi\rho(3r)^3$	$18r$	$\frac{3}{8}(3r)$
Mushroom	$\pi\rho r^2(h + 18r)$	$h + 18r$	0

Moments: $-h \times \frac{h}{2} + 18r \times \frac{3}{8} \times 3r = 0$

$\frac{h^2}{2} = \frac{81r^2}{4}$

$h = r\sqrt{\frac{81}{2}}$

b $\theta = 9°$ (to the nearest degree)

9 a $V = \pi\int_0^a 4ax\,dx$

   $= \pi[2ax^2]_0^a$

   $= 2\pi a^3$

  b $\bar{x} = \frac{\pi\int_0^a 4ax^2\,dx}{2\pi a^3}$

   $= \pi\frac{\left[\frac{4ax^3}{3}\right]_0^a}{2\pi a^3}$

   $= \frac{2}{3}a$

  c $\rho_1 : \rho_2 = 6 : 1$

  d As centre of mass is at centre of hemisphere this will always be above the point of contact with the plane. (Tangent–radius property).

10 a

Shape	Mass	Mass ratio	Distance of CoM from AB
Cylinder	$\pi\rho(2r)^2 \times 3r$	$12r$	$\frac{3r}{2}$
Cone	$\frac{1}{3}\pi\rho r^2 \times h$	$\frac{1}{3}h$	$\frac{1}{4}h$
Remainder	$\pi\rho\left(12r^3 - \frac{1}{3}r^2h\right)$	$12r - \frac{1}{3}h$	$\bar{x}$

Moments: $\left(12r - \frac{1}{3}h\right)\bar{x} = 12r \times \frac{3r}{2} - \frac{h}{3} \times \frac{h}{4}$

$$\left(12r - \frac{1}{3}h\right)\bar{x} = 18r^2 - \frac{h^2}{12}$$

$$\bar{x} = \frac{18r^2 - \frac{h^2}{12}}{12r - \frac{h}{3}}$$

$$\bar{x} = \frac{216r^2 - h^2}{4(36r - h)}$$

  b $\theta = 38°$ (to the nearest degree)

11 a   0.265 m (3 s.f.)   **b** $0.478mg$ (3 s.f.)

12 a   1.11 (3 s.f.)   **b**   14.4° (3 s.f.)

13 a $\int_0^9 (1000 + 400x^{\frac{1}{2}})dx = \left[1000x + \frac{800}{3}x^{\frac{3}{2}}\right]_0^9$

                  $= (9000 + 7200) - (0)$

                  $= 16\,200\,\text{kg}$

  b   4.9 m

14 a   40 cm

  b i   e.g. Suitable if uniform across cross-section, or suitable as height $\gg$ diameter, or unsuitable as may be non-uniform across cross-section.

    ii   Unsuitable as rod has no width so will never be stable.

  c   5.71° (3 s.f.)

15 $T_A = 48g\,\text{N}$, $T_B = 120g\,\text{N}$

16   60 m

Challenge

a   Mass of elemental disc $= \rho\pi(h - x)^2\,\delta x$

     $= \pi(x + 1)(h^2 - 2hx + x^2)\delta x$

     $= \pi(h^2x - 2hx^2 + x^3 + h^2 - 2hx + x^2)\delta x$

     Mass $= \pi\int_0^h h^2x - 2hx^2 + x^3 + h^2 - 2hx + x^2\,dx$

        $= \pi\left[\frac{1}{2}h^2x^2 - \frac{2}{3}hx^3 + \frac{1}{4}x^4 + h^2x - hx^2 + \frac{1}{3}x^3\right]_0^h$

        $= \pi\left(\left(\frac{1}{2}h^4 - \frac{2}{3}h^4 + \frac{1}{4}h^4 + h^3 - h^3 + \frac{1}{3}h^3\right) - (0)\right)$

        $= \pi\left(\frac{1}{12}h^4 + \frac{1}{3}h^3\right)$

        $= \frac{1}{12}\pi h^3(h + 4)$

b   5 m

c $\bar{y} = \dfrac{\frac{1}{60}\pi h^4(2h + 5)}{\frac{1}{12}\pi h^3(h + 4)}$

     If $\bar{y} = kh$ for some constant k

     $\Rightarrow k = \dfrac{\frac{1}{60}\pi h^3(2h + 5)}{\frac{1}{12}\pi h^3(h + 4)} = \dfrac{2h + 5}{5(h + 4)}$

     $= \dfrac{2(h + 4) - 3}{5(h + 4)} = \dfrac{2}{5} - \dfrac{3}{5(h + 4)}$

     As $h \to \infty$, $k \to \frac{2}{5}$

     Hence as h varies the height of the centre of mass of the cone above its base cannot exceed $\frac{2}{5}h$.

Review exercise 1

1 $3.67\,\text{m s}^{-1}$ (3 s.f.)

2 a $T = 5.5\,\text{N}$ (2 s.f.)   **b** $\theta = 26°$ (nearest degree)

3 a   R($\uparrow$): $T\cos 60° - mg = 0 \Rightarrow T = \dfrac{mg}{\cos 60°} = 2mg$

  b $\omega = \sqrt{\dfrac{2g}{L}}$

4   190 m (2 s.f.)

5 $24\,\text{m s}^{-1}$ (2 s.f.)

6 a $\dfrac{5mg}{4}$ **b** $v = \sqrt{6gl}$

  c   The tensions could not be assumed to have the same magnitude.

7 a $\alpha = 70.5°$ (3 s.f.)   **b** $l = \dfrac{3}{2k}$

8 a $2mg$ **b** $2\pi\sqrt{\dfrac{r}{2g}}$

9 a $\tan 60° = \dfrac{r}{\frac{h}{r}} \Rightarrow r = \dfrac{h}{2}\tan 60° = \dfrac{\sqrt{3}}{2}h$

  b   Tension in $AP = mg + \frac{1}{2}mh\omega^2$ and tension in $BP = \frac{1}{2}mh\omega^2 - mg$

  c   Tension in $BP > 0 \Rightarrow \omega > \sqrt{\dfrac{2g}{h}}$

     As $T = \dfrac{2\pi}{\omega}$, $T < 2\pi\sqrt{\dfrac{h}{2g}} \Rightarrow T < \pi\sqrt{\dfrac{2h}{g}}$

10 a   R($\uparrow$): $T\cos\theta - mg = 0 \Rightarrow T = \dfrac{mg}{\cos\theta}$

     R($\leftarrow$): $T + T\sin\theta = mr\omega^2$

     $r = h\tan\theta \Rightarrow \dfrac{mg}{\cos\theta}(1 + \sin\theta) = mh\dfrac{\sin\theta}{\cos\theta}\omega^2$

     $\Rightarrow \omega^2 = \dfrac{g}{h}\left(\dfrac{1 + \sin\theta}{\sin\theta}\right)$

  b $\omega^2 = \dfrac{g}{h}\left(\dfrac{1}{\sin\theta} + 1\right)$ and $\sin\theta < 1$ so $\dfrac{1}{\sin\theta} > 1$

     $\Rightarrow \omega^2 > \dfrac{g}{h} \times 2 \Rightarrow \omega > \sqrt{\dfrac{2g}{h}}$

  c $T = \dfrac{2\sqrt{3}}{3}mg$ or $1.15mg$

11 $2a$

12 a   R($\leftarrow$): $T\cos 30° = m(2a\cos 30°)\dfrac{kg}{3a} \Rightarrow T = \dfrac{2kmg}{3}$

  b $mg\left(1 - \dfrac{k}{3}\right)$ **c** $k < 3$

  d $PX = 2a\cos\theta$

     R($\leftarrow$): $T'\cos\theta = m \times 2a\cos\theta \times \dfrac{2g}{a} \Rightarrow T' = 4mg$

     R($\uparrow$): $T'\sin\theta - mg = 0 \Rightarrow \sin\theta = \dfrac{mg}{T'} = \dfrac{mg}{4mg} = \dfrac{1}{4}$

     $AX = 2a\sin\theta = \frac{1}{2}a, AO = 2a\sin 30 = a$

     So $AX = \frac{1}{2}AO$

13 a   R($\uparrow$): $T\cos\theta - S\cos\theta - mg = 0$

     $\Rightarrow T - S = \dfrac{mg}{\cos\theta} = \dfrac{4mg}{3}$ (1)

     R($\leftarrow$): $T\sin\theta + S\sin\theta = mr\omega^2 = ml\sin\theta\omega^2$

     $\Rightarrow T + S = ml\omega^2$ (2)

     Solving (1) and (2) simultaneously

     $\Rightarrow T = \frac{1}{6}m(3l\omega^2 + 4g)$

  b $\frac{1}{6}m(3l\omega^2 - 4g)$ **c**   As $S \geq 0$, $\omega^2 \geq \dfrac{4g}{3l}$

14 a   R($\uparrow$): $R - mg = 0 \Rightarrow R = mg$

     R($\leftarrow$): $F = mr\omega^2 = m\left(\frac{4}{3}a\right)\omega^2$

     As P remains at rest $F \leq \mu R$

     $\Rightarrow m\left(\frac{4}{3}a\right)\omega^2 \leq \frac{3}{5}mg \Rightarrow \omega^2 \leq \dfrac{9g}{20a}$

  b $\omega^2_{\max} = \dfrac{19g}{20a}$ and $\omega^2_{\min} = \dfrac{g}{20a}$

15 a i   2.5 gN   **ii** $\arccos\left(\frac{4}{5}\right)$ or 36.9°

  b $T\sin\theta = \dfrac{mv^2}{r} \Rightarrow \left(\dfrac{5g}{2}\right)\left(\dfrac{3}{5}\right) = \dfrac{2v^2}{\left(\frac{3}{4}x\right)}$

     $\Rightarrow v^2 = \dfrac{9gx}{16} \Rightarrow v = \frac{3}{4}\sqrt{gx}$

16 a $u = \sqrt{\dfrac{gl}{2}}$ **b** $\dfrac{5l}{6}$

17 a   Use conservation of energy:

     $\frac{1}{2}m(u^2 - v^2) = mgl(1 - \cos\theta)$

     $\Rightarrow v^2 = u^2 - 2gl(1 - \cos\theta) = 3gl - 2gl + 2gl\cos\theta$

     $v^2 = gl + 2gl\cos\theta$

Online   Full worked solutions are available in SolutionBank.

Resolve along the string:

$$T - mg\cos\theta = \frac{mv^2}{l} = \frac{mgl + 2mgl\cos\theta}{l}$$

$$\Rightarrow T = mg(1 + 3\cos\theta)$$

b $v = \sqrt{\dfrac{gl}{3}}$ **c** $\dfrac{40l}{27}$

18 a $\sqrt{\dfrac{2gl}{3}}$

b Resolve along the string: $T - mg\cos\theta = \dfrac{mv^2}{l}$ (1)

Conservation of energy:

$\frac{1}{2}mu^2 - \frac{1}{2}mv^2 = mgl(1 - \cos\theta)$ (2)

Solving (1) and (2) simultaneously: $T = \dfrac{mg}{3}(9\cos\theta - 4)$

c $\dfrac{2mg}{3} \leqslant T \leqslant \dfrac{5mg}{3}$

19 a $\frac{3}{4}$ or 0.75

b Resolve along the radius OB: $mg\cos\theta = \dfrac{mv^2}{l} = \dfrac{mv^2}{0.8}$

So $v^2 = 0.8g\cos\theta = 0.6g = 5.88$

c $u = 1.4$

20 a $\frac{1}{2}mv^2 = mg(a\cos\alpha - a\cos\theta) \Rightarrow v^2 = 2ga(\cos\alpha - \cos\theta)$

b $\theta = 60°$ or $\frac{\pi}{3}$ radians   **c** $\sqrt{\dfrac{7ga}{2}}$

21 a $\frac{3}{2}mg$

b $\frac{1}{2}m \times \dfrac{7ga}{2} - \frac{1}{2}mV^2 = mga(1 + \cos\theta)$ (1)

Resolving along radius: $mg\cos\theta = \dfrac{mV^2}{a}$ (2)

Eliminate V^2 from equations (1) and (2):

$\Rightarrow \cos\theta = \frac{1}{2} \Rightarrow \theta = 60°$

c $\sqrt{\dfrac{6a}{g}}$

22 a $v = 8$ **b** $m = 6$ **c** 1300 N (2 s.f.)

23 a 30 m s⁻¹ (2 s.f.)   **b** 1900 N

c 28 m s⁻¹ (2 s.f.)   **d** 560 N (2 s.f.)

e Lower speed at $C \Rightarrow$ the normal reaction is reduced.

24 a $v^2 = ag(1 - 2\cos\theta)$

b R(↗) along radius:

$T + mg\cos\theta = \dfrac{mv^2}{a} \Rightarrow T = (1 - 3\cos\theta)mg$

c $\frac{4}{3}a$ **d** $\frac{4}{27}a$ or $0.148a$

25 a $\frac{3}{2}ga + 2ga\sin\theta$ **b** $\dfrac{3mg}{2}(1 + 2\sin\theta)$

c Put $T = 0$, then $\sin\theta = -\frac{1}{2}$ so $\theta = 210°$

d No. $v = 0$ when $\theta = 229°$ (3 s.f.), so P changes direction before it reaches the top of the circle.

e Consider motion at start and at A:

no change in P.E. $\Rightarrow$ no change in K.E.

so $v = u = \sqrt{\dfrac{3ga}{2}}$

f 73.2° (3 s.f.)

26 a $\frac{2}{3} + \dfrac{u^2}{3ag}$ **b** 34° (nearest degree)

27 a $u^2 - 3ga$ **b** $\dfrac{5mg}{2}$

c $\sqrt{\dfrac{7ga}{2}}$ **d** $u = \sqrt{5ag}$

28 a Taking moments about the x-axis:

$2(8 + \lambda) = 3 \times 4 + 5 \times 0 + 4\lambda$ so $\lambda = 2$

b $k - 1.1$

29 $x = y = \frac{1}{2}$

30 $(3\mathbf{i} + 2.5\mathbf{j})$ m

31 a The total mass is $2M + M + kM = (3 + k)M$

M(Oy): $(3 + k)M \times 3 = 2M \times 6 + M \times 0 + kM \times 2$

$\Rightarrow 9 + 3k = 12 + 2k \Rightarrow k = 3$

b $c = -\frac{1}{3}$

32 a 10.7 cm (3 s.f.)

b 25° (nearest degree)

33 a The area of rectangle $ABDE$ is $6a \times 8a = 48a^2$

The area of $\triangle BCD$ is $\frac{1}{2} \times 6a \times 4a = 12a^2$

The area of lamina $ABCDE$ is $48a^2 + 12a^2 = 60a^2$

Shape	Lamina	Rectangle	Triangle
Mass ratios	$60a^2$	$48a^2$	$12a^2$
Displacement from X	GX	$4a$	$-\frac{4}{3}a$

M(X): $60a^2 \times GX = 48a^2 \times 4a + 12a^2 \times \left(-\frac{4}{3}a\right)$

$= 192a^3 - 16a^3 = 176a^3$

$GX = \dfrac{176a^3}{60a^2} = \frac{44}{15}a$

b $\frac{11}{15}$

34 a $\dfrac{6l}{5}$ **b** l **c** 51° (nearest degree)

35 a $\frac{19}{15}a$ **b** $\frac{7}{45}M$

36 a 6 cm   **b** 22.6° (1 d.p.)

37 a 6.86 cm (2 d.p.)   **b** 32.1° (1 d.p.)

38 a 25 cm   **b** $\frac{3}{11}m$

39 a $\left(\frac{9}{2}, \frac{2}{3}\right)$

b The centre of mass of the lamina is $(3, 0)$

Let the centre of mass of the combined system of the lamina and the three particles be at the point G.

The total mass of the system is $12m + km = (12 + k)m$

Shape	Combined system	Particles	Lamina
Mass	$(12 + k)m$	$12m$	km
Distances (x)	4	$\frac{9}{2}$	3
Distances (y)	λ	$\frac{2}{3}$	0

M(Oy): $(12 + k)m \times 4 = 12m \times \frac{9}{2} + km \times 3$

$\Rightarrow 48 + 4k = 54 + 3k \Rightarrow k = 6$

c $\frac{4}{9}$ **d** 83.7° (1 d.p.)

40 a Let the distances of the centre of mass of L, say G, from AD and AB be $\bar{x}$ and $\bar{y}$ respectively.

The mass of L is $3m + 4m + m + 2m = 10m$

Shape	L	$ABCD$	A	B	C
Mass	$10m$	$3m$	$4m$	m	$2m$
Distances (x)	$\bar{x}$	$2.5a$	0	$5a$	$5a$
Distances (y)	$\bar{y}$	a	0	0	$2a$

b $0.7a$ **c** 20° (nearest degree)

d M(O): $P \times 2a = 10mg \times (2.5a - \bar{x}) = 10mg \times 0.25a$

$\Rightarrow P = \dfrac{2.5mga}{2a} = \frac{5}{4}mg$

e $\dfrac{5\sqrt{65}}{4}mg$

41 a 3   **b** 2   **c** 37° (nearest degree)

42 a **i** $\frac{5}{2}a$ **ii** $\frac{4}{3}a$

b 15° (nearest degree)

43 a **i** $\dfrac{5l}{12}$ **ii** $\dfrac{l}{3}$

b 39° (nearest degree)

44 a d **b** d **c** 63° (nearest degree)

45 a 3.28 cm (3 s.f.)   **b** 0.211 (3 s.f.)

46 a 43.8 cm (3 s.f.)
  b $T_1 = 0.452$ N (3 s.f.), $T_2 = 0.548$ N (3 s.f.)
  c $k = 7.45$ (3 s.f.)

47 $\frac{8}{3}$ from O

48 a The centre of mass lies on the x-axis from symmetry.
An elemental strip of area is $2y\delta x$
The boundary of the semicircle has equation
$x^2 + y^2 = a^2, 0 \le x \le a$
$$\bar{x} = \frac{\rho \int 2x(a^2 - x^2)^{\frac{1}{2}}dx}{\rho \times \frac{\pi a^2}{2}} = \frac{2}{\pi a^2}\left[-\frac{2}{3}(a^2 - x^2)^{\frac{3}{2}}\right]_0^a = \frac{4a}{3\pi}$$

b

Shape	Mass	Mass ratio	Distance of CoM from O
Semicircle radius a	$\frac{1}{2}\pi\rho a^2$	a^2	$\frac{4a}{3\pi}$
Semicircle radius b	$\frac{1}{2}\pi\rho b^2$	b^2	$\frac{4b}{3\pi}$
Remainder	$\frac{1}{2}\pi\rho(a^2 - b^2)$	$a^2 - b^2$	$\bar{x}$

Moments about $O : a^2 \times \frac{4a}{3\pi} - b^2 \times \frac{4b}{3\pi} = (a^2 - b^2)\bar{x}$

so $\bar{x} = \frac{4}{3\pi}\frac{(a^3 - b^3)}{(a^2 - b^2)} = \frac{4}{3\pi}\frac{(a - b)(a^2 + ab + b^2)}{(a - b)(a + b)}$

$\quad = \frac{4(a^2 + ab + b^2)}{3\pi(a + b)}$

c $\frac{2a}{\pi}$

49 a Let the equation of the line AC be $y = c - mx$
$$\bar{y} = \frac{\frac{1}{2}\int_0^{\frac{c}{m}}(c - mx)^2 dx}{\int_0^{\frac{c}{m}}c - mx\,dx} = \frac{\frac{1}{2}\left[-\frac{1}{3m}(c - mx)^3\right]_0^{\frac{c}{m}}}{\left[-\frac{1}{2m}(c - mx)^2\right]_0^{\frac{c}{m}}}$$

$\quad = \frac{c^3}{6m} \div \frac{c^2}{2m} = \frac{1}{3}c$

b, c

Shape	Mass	Position of CoM
Rectangle	$2a^2\rho$	$\left(\frac{a}{2}, a\right)$
Triangle	$a^2\rho$	$\left(\frac{4a}{3}, \frac{2a}{3}\right)$
Lamina	$3a^2\rho$	$(\bar{x}, \bar{y})$

Taking moments:
$$2a^2\rho\begin{pmatrix}\frac{a}{2} \\ a\end{pmatrix} + a^2\rho\begin{pmatrix}\frac{4a}{3} \\ \frac{2a}{3}\end{pmatrix} = 3a^2\rho\begin{pmatrix}\bar{x} \\ \bar{y}\end{pmatrix}$$

$$\begin{pmatrix}a + \frac{4a}{3} \\ 2a + \frac{2a}{3}\end{pmatrix} = 3\begin{pmatrix}\bar{x} \\ \bar{y}\end{pmatrix} \Rightarrow \bar{x} = \frac{7}{9}a, \bar{y} = \frac{8}{9}a$$

50 a $\bar{x} = \frac{\int \pi x y^2 dx}{\int \pi y^2 dx} = \frac{\frac{1}{4}\int_1^2 x^{-3} dx}{\frac{1}{4}\int_1^2 x^{-4} dx} = \frac{\left[-\frac{1}{2}x^{-2}\right]_1^2}{\left[-\frac{1}{3}x^{-3}\right]_1^2} = \frac{9}{7}$

The centre of mass is $\left(\frac{9}{7} - 1\right) = \frac{2}{7}$m from the larger plan face.

b $\frac{29}{30}$ m or 0.967 m

51 a $\bar{x} = \frac{\int_0^R \pi x(R^2 - x^2) dx}{\frac{2}{3}\pi R^3} = \frac{\left[\pi R^2\frac{x^2}{2} - \frac{\pi x^4}{4}\right]_0^R}{\frac{2}{3}\pi R^3} = \frac{\frac{1}{4}\pi R^4}{\frac{2}{3}\pi R^3}$

$\quad = \frac{3}{8}R$

b

Shape	Mass	Mass ratio	Position of CoM
Cone	$\frac{1}{3}\pi\rho a^3 k$	k	$\frac{3}{4}ka$
Hemisphere	$\frac{2}{3}\pi\rho a^3$	2	$ka + \frac{3}{8}a$
Top	$\frac{1}{3}\pi\rho a^3(k + 2)$	$k + 2$	$\bar{x}$

Taking moments about V:
$k\left(\frac{3}{4}ka\right) + 2\left(ka + \frac{3}{8}a\right) = (k + 2)\bar{x} \Rightarrow \bar{x} = \frac{(3k^2 + 8k + 3)a}{4(k + 2)}$

c $k = \sqrt{3}$

52 a

Shape	Mass	Mass ratio	Distance of CoM from O
Large hemisphere	$\frac{2}{3}\pi\rho a^3$	8	$\frac{3}{8}a$
Small hemisphere	$\frac{2}{3}\pi\rho\left(\frac{a}{2}\right)^3$	1	$\frac{3}{16}a$
Remainder	$\frac{2}{3}\pi\rho\frac{7a^3}{8}$	7	$\bar{x}$

Taking moments about O: $8 \times \frac{3}{8}a - 1 \times \frac{3}{16}a = 7\bar{x}$

$\quad\quad\quad \Rightarrow \bar{x} = \frac{45a}{112}$

b $k = \frac{2}{7}$

53 a $\frac{8\pi}{5}$

b $\bar{x} = \frac{\pi\int_0^2 \frac{1}{4}x(x - 2)^4 dx}{\frac{8\pi}{5}} = \frac{5}{8\pi} \times \frac{\pi}{4}\int_{-2}^0(u + 2)u^4 du$

$\quad = \frac{5}{32}\int_{-2}^0 u^5 + 2u^4 du = \frac{5}{32}\left[\frac{1}{6}u^6 + \frac{2}{5}u^5\right]_{-2}^0$

$\quad = \frac{5}{32}\left(0 - \left(-\frac{64}{30}\right)\right) = \frac{1}{3}$

c $\frac{59W}{12}$

54 a

Shape	Mass	Mass ratio	Distance of CoM from O
Cylinder	$\pi\rho r^2 h$	1	$\frac{h}{2}$
Cone	$\frac{1}{3}\pi\rho r^2\left(\frac{h}{2}\right)$	$\frac{1}{6}$	$h - \frac{1}{4}\left(\frac{h}{2}\right)$
Ornament	$\frac{5}{6}\pi\rho r^2 h$	$\frac{5}{6}$	$\bar{x}$

Taking moments about O, centre of plane base:
$1 \times \frac{h}{2} - \frac{1}{6} \times \frac{7h}{8} = \frac{5}{6}\bar{x} \Rightarrow \frac{5}{6}\bar{x} = \frac{17h}{48} \Rightarrow \bar{x} = \frac{17h}{40}$

b $\alpha = 66.5°$ (1 d.p.)

55 a The centre of mass lies on the y-axis from symmetry. An elemental strip of area is $2x\,\delta x$
$$\bar{y} = \frac{\int_0^h y2x\,dy}{\int_0^h 2x\,dy} = \frac{\int_0^h y(h - y)\,dy}{\int_0^h(h - y)\,dy}$$

$\quad = \frac{\left[\frac{y^2h}{2} - \frac{y^3}{3}\right]_0^h}{\left[yh - \frac{y^2}{2}\right]_0^h} = \frac{h}{3}$

Online Full worked solutions are available in SolutionBank.

b

Shape	Mass	Distance of CoM from A
$\triangle ABC$	$12\rho a^2$	$2a$
$\triangle BDC$	$4\rho a^2$	$2a + \dfrac{2a}{3}$
Remainder	$8\rho a^2$	$\bar{x}$

Taking moments about A:

$$12\rho a^2 \times 2a - 4\rho a^2\left(\frac{8a}{3}\right) = 8\rho a^2 \bar{x} \Rightarrow \frac{40a}{3} = 8\bar{x}$$

$$\Rightarrow \bar{x} = \frac{5a}{3}$$

c Taking moments about C:

$$Mg \times 8a\sin\theta = Mg\left(\frac{4a}{3}\cos\theta - 4a\sin\theta\right)$$

$$\Rightarrow 12Mga\sin\theta = \frac{4}{3}Mga\cos\theta$$

$$\Rightarrow \tan\theta = \frac{1}{9}$$

So CB makes an angle $\arctan\left(\frac{1}{9}\right)$ with the vertical.

56 a

Shape	Mass	Distance of CoM from O
Circular disc	$\pi\rho a^2$	0
Hemispherical bowl	$2\pi\rho a^2$	$\frac{1}{2}a$
Closed container	$3\pi\rho a^2$	$\bar{x}$

Taking moments about O:

$$0 + 2\pi\rho a^2 \times \frac{a}{2} = 3\pi\rho a^2 \bar{x} \Rightarrow \bar{x} = \frac{a}{3}$$

b $56°$ (nearest degree)

57 $\theta = 53.1°$ (1 d.p.)

58 a

Shape	Mass	Distance of CoM from C
H	$8M$	$\dfrac{3a}{8}$
K	M	$-\dfrac{3a}{16}$
S	$9M$	$\bar{x}$

Taking moments about C:

$$8M \times \frac{3a}{8} - M \times \frac{3a}{16} = 9M\bar{x} \Rightarrow 3a - \frac{3a}{16} = 9\bar{x}$$

$$\Rightarrow \bar{x} = \frac{45a}{16} \div 9 = \frac{5a}{16}$$

b $\dfrac{16}{45}$

59 a

$$\bar{x} = \frac{\pi\displaystyle\int_0^h x\left(r^2 - \frac{2r^2 x}{h} + \frac{r^2 x^2}{h^2}\right)dx}{\frac{1}{3}\pi r^2 h}$$

$$= \frac{3}{r^2 h}\int_0^h xr^2 - \frac{2r^2 x^2}{h} + \frac{r^2 x^3}{h^2}\,dx$$

$$= \frac{3}{r^2 h}\left[\frac{x^2 r^2}{2} - \frac{2r^2 x^3}{3h} + \frac{r^2 x^4}{4h^2}\right]_0^h = \frac{3}{r^2 h} \times \frac{h^2 r^2}{12} = \frac{h}{4}$$

b

Shape	Mass	Mass ratio	Position of CoM
Large cone	$\frac{1}{3}\pi\rho r^3 H$	H	$\dfrac{H}{4}$
Small cone	$\frac{1}{3}\pi\rho r^3 h$	h	$\dfrac{h}{4}$
Remainder	$\frac{1}{3}\pi\rho r^2(H-h)$	$H-h$	$\bar{x}$

$$H \times \frac{H}{4} - h \times \frac{h}{4} = (H - h)\bar{x} \Rightarrow \bar{x} = \frac{1(H^2 - h^2)}{4(H - h)} = \frac{1}{4}(H + h)$$

Distance from the vertex $= H - \frac{1}{4}(H + h) = \frac{1}{4}(3H - h)$

c $\dfrac{H + h}{3H - h}$

60 a

Shape	Mass	Mass ratio	Position of CoM
Hemisphere	$4\pi\rho r^3$	$4r$	$\dfrac{5r}{8}$
Cylinder	$\pi\rho r^2 h$	h	$\dfrac{h}{2} + r$
Toy	$\pi\rho r^2(4r + h)$	$4r + h$	$\bar{x}$

Taking moments about O:

$$4r \times \frac{5r}{8} + h\left(\frac{h}{2} + r\right) = (4r + h)\bar{x}$$

$$\Rightarrow \bar{x} = \frac{\frac{5r^2}{2} + \frac{h^2}{2} + rh}{4r + h} \text{ so } d = \frac{h^2 + 2rh + 5r^2}{2(h + 4r)}$$

b $h = \sqrt{3}\,r$

61 a

Shape	Mass	Distance of CoM from AB
Hemisphere	M	$\dfrac{3r}{8}$
Cone	m	$-\dfrac{3r}{4}$
Toy	$m + M$	$\bar{x}$

Taking moments about AB:

$$(m + M)\bar{x} = \frac{3Mr}{8} - \frac{3mr}{4}$$

$$\Rightarrow \bar{x} = \frac{3(M - 2m)r}{8(M + m)}$$

b Let α = angle between OA and axis of cone

No equilibrium $\Rightarrow \bar{x} > r\tan\alpha$

$\tan\alpha = \dfrac{r}{3r} \Rightarrow \bar{x} > \frac{1}{3}r$

So $\dfrac{3(M - 2m)r}{8(M + m)} > \frac{1}{3}r$

$9(M - 2m) > 8(M + m) \Rightarrow M > 26m$

62 a $\bar{x} = \dfrac{\displaystyle\int_0^r rx^2\,dx}{\displaystyle\int_0^r rx\,dx} = \dfrac{\left[\frac{1}{3}rx^3\right]_0^r}{\left[\frac{1}{2}rx^2\right]_0^r} = \dfrac{\frac{1}{3}r^4}{\frac{1}{2}r^3} = \frac{2}{3}r$

b $\alpha = 72°$ (nearest degree)

63 a $\bar{x} = \dfrac{\frac{1}{2}\displaystyle\int_0^\pi \sin^2 x\,dx}{\displaystyle\int_0^\pi \sin x\,dx} = \dfrac{\frac{1}{4}\displaystyle\int_0^\pi 1 - \cos 2x\,dx}{\displaystyle\int_0^\pi \sin x\,dx}$

$$= \frac{\frac{1}{4}\left[x - \frac{1}{2}\sin 2x\right]_0^\pi}{[-\cos x]_0^\pi} = \frac{\frac{\pi}{4}}{2} = \frac{\pi}{8}$$

b $76°$ (nearest degree)

64 a $\dfrac{51a}{64}$ or $0.797a$ (3 s.f.)

b $\alpha = 70.6°$ (3 s.f.)

c $\beta = 38.7°$ (3 s.f.)

65 a

Shape	Mass	Mass ratio	Distance of CoM from O
Large hemisphere	$\frac{2}{3}\pi\rho(6a)^3$	27	$\frac{3}{8}\times 6a$
Small hemisphere	$\frac{2}{3}\pi\rho(2a)^3$	1	$\frac{3}{8}\times 2a$
Remainder	$\frac{2}{3}\pi\rho(6^3-2^2)a^3$	26	$\bar{x}$

Taking moments about O:

$26\bar{x}=27\times\left(\frac{3}{8}\times 6a\right)-1\times\left(\frac{3}{8}\times 2a\right)=60a$

$\Rightarrow \bar{x}=\dfrac{30a}{13}$

b

Shape	Mass	Mass ratio	Distance of CoM from O
Bowl B	$\frac{416}{3}\pi\rho a^3$	52	$\frac{30a}{13}$
Cylinder	$24\pi\rho a^3$	9	$9a$
Combined solid	$\frac{488}{3}\pi\rho a^3$	61	$\bar{y}$

Taking moments about O:

$61\bar{y}=52\times\dfrac{30a}{13}+9\times 9a=120a+81a=210a$

$\Rightarrow \bar{y}=\frac{201}{61}a$

c S will not topple.

66 a

Shape	Mass	Mass ratio	Distance of CoM from O
Hemisphere	$\frac{2}{3}\pi\rho a^3$	2	$\frac{3}{8}a$
Cone	$\frac{1}{3}\pi\rho a^3$	1	$\frac{a}{4}$
Remainder	$\frac{1}{3}\pi\rho a^3$	1	$\bar{x}$

Taking moments about O:

$2\times\frac{3}{8}a-1\times\frac{a}{4}=1\bar{x}\Rightarrow \bar{x}=\frac{a}{2}$

b Let N be the point of contact between the solid and the plane.

From $\triangle OGN \sin\alpha=\dfrac{\frac{a}{2}}{a}=\frac{1}{2}$

$\Rightarrow \alpha=\dfrac{\pi}{6}$

c $\dfrac{1}{\sqrt{3}}$

67 a

Shape	Mass	Mass ratio	Distance of CoM from O
Cone	$\frac{1}{3}\pi\rho(3r)^2h$	$3h$	$\frac{3}{4}h$
Cylinder	$\pi\rho(4r)^2r$	$16r$	$h+\frac{r}{2}$
Bollard	$\pi\rho(16r+3h)r^2$	$16r+3h$	$\bar{x}$

Taking moments about O:

$3h\times\dfrac{3h}{4}+16r\left(h+\dfrac{r}{2}\right)=(16r+3h)\bar{x}$

$\Rightarrow \dfrac{9h^2}{4}+16rh+8r^2=(16r+3h)\bar{x}$

$\Rightarrow \bar{x}=\dfrac{32r^2+64rh+9h^2}{4(16r+3h)}$

b $\alpha=74°$ (nearest degree)

68 a

Shape	Mass	Mass ratio	Position of CoM from O
Cylinder	$\pi\rho r^2 h$	3	$+\frac{h}{2}$
Cone	$\frac{1}{3}\pi\rho(2r)^2h$	4	$-\frac{h}{4}$
Tree	$\pi\rho r^2 h\left(1+\frac{4}{3}\right)$	7	$\bar{x}$

Taking moments about O:

$3\times\dfrac{h}{2}-4\times\dfrac{h}{4}=7\bar{x}\Rightarrow\dfrac{3h}{2}-h=7\bar{x}$

$\Rightarrow \bar{x}=\dfrac{h}{14}$

b $r=\frac{1}{4}h$

69 a

Shape	Mass	Distance of CoM from O
Hemisphere	$2M$	$h+\frac{3r}{8}$
Cylinder	$3M$	$\frac{h}{2}$
Combined solid	$5M$	$\bar{x}$

Taking moments about O:

$2M\left(h+\dfrac{3r}{8}\right)+3M\times\dfrac{h}{2}=5M\bar{x}$

$\Rightarrow \dfrac{7h}{2}+\dfrac{3r}{4}=5\bar{x}$

$\Rightarrow \bar{x}=\dfrac{14h+3r}{20}$

b $\frac{6}{7}r$

70 a $\dfrac{r}{2}$ below O

b $40.6°$

c The toy will not topple.

71 a $1.07\,\text{kg}$ (2 d.p.)   **b** $0.57\,\text{m}$ (2 d.p.)

72 a $3.41\,\text{kg}$ (2 d.p.)   **b** $0.58\,\text{m}$ (2 d.p.)

73 a e.g. the density of the rod increases as the distance from A increases.

b $\frac{11}{4}$

c $\frac{112}{9}\,\text{m}$

74 $4\sqrt{3}\,\text{m}$

Challenge

1 $e=\frac{1}{2}$

2 a Reaction of ring on ball $=\dfrac{mv^2}{R}$

Using $F=ma$ with frictional force:

$\dfrac{m\mu v^2}{R}=-m\dfrac{dv}{dt}$

$\dfrac{1}{v^2}\dfrac{dv}{dt}=\dfrac{-\mu}{R}$

$\dfrac{-1}{v}=\dfrac{-\mu t}{R}+c$

When $t=0$, $v=u\Rightarrow c=\dfrac{-1}{u}$

$\dfrac{-1}{v}=\dfrac{-\mu t}{R}-\dfrac{1}{u}\Rightarrow v=\dfrac{1}{\frac{\mu t}{R}+\frac{1}{u}}=\dfrac{uR}{R+u\mu t}$

b $\dfrac{e^{\frac{\pi}{2}}-1}{20}\,\text{s}$ or $0.191\,\text{s}$ (3 s.f.)

3 a $T_1=4Mg$, $T_2=T_3=0$ **b** $6.72°$

4 $4\,\text{cm}$

Online Full worked solutions are available in SolutionBank.

CHAPTER 4

Prior knowledge check

1 a $\dfrac{4}{3(2-3x)^2} + c$ **b** $\dfrac{4e^{3x}}{3} + c$ **c** $-\dfrac{\cos 5\pi x}{5\pi} + c$

2 $y = 3e^{\frac{x-1}{3(x+2)}}$

3 $\ln\frac{4}{3}$

Exercise 4A

1 $(16 - 12e^{-0.25t})\,\text{m s}^{-1}$

2 $v = t\sin t,\ x = \int t\sin t\,dt.$ Using integration by parts and the initial condition $x = 0$ at $t = 0$, we get $x = \sin t - t\cos t.$
At $t = \dfrac{\pi}{2}$, $x = -\dfrac{\pi}{2}\cos\dfrac{\pi}{2} + \sin\dfrac{\pi}{2} = 0 + 1 = 1.$

3 $2\ln 3\,\text{m}$

4 $11.5\,\text{m}$ (3 s.f.)

5 a $6\sqrt{2}\,\text{m s}^{-2}$ **b** $x = \frac{4}{3}\sin 3t$ **c** $t = \dfrac{\pi}{3}$

6 $v = \dfrac{3}{2} - \dfrac{3}{2 + t^2}$

7 a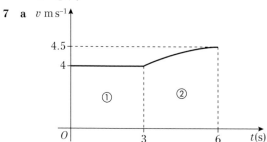

b $(27 - 3\ln 2)\,\text{m}$

8 a $4\,\text{m s}^{-1}$ **b** $(\pi - 2\sqrt{2})\,\text{m}$

9 a $v = 40 - 20\,e^{0.2t}\,\text{m s}^{-1}$ **b** $(200\ln 2 - 100)\,\text{m}$

10 a $c = 80,\ d = 1$ **b** $3200\ln\left(\dfrac{80+t}{80}\right)$

11 a $t = \ln 2.5,\ \ln 3$ **b** $\left(18 - \dfrac{15(\ln 3)^2}{2}\right)\text{m}$

12 a $t = 3$ **b** $(12 + 3\ln 12)\,\text{m}$

13 a $t = 1,\ t = \frac{3}{2}$ **b** $a = 25\,\text{m s}^{-2}$ **c** $s = \dfrac{3917}{54}\,\text{m}$
d $x = t^3 - \frac{5}{2}t^2 + 2t = t(t^2 - \frac{5}{2}t + 2)$
For $t > 0$, when $t^2 - \frac{5}{2}t + 2 = 0$
The discriminant of the quadratic equation is $\frac{25}{4} - (4 \times 2) < 0.$
The equation has no real roots. Therefore, P never returns to the origin for any $t > 0.$

14 a velocity (m s^{-1})
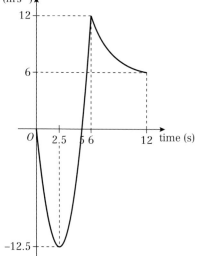

b $2.5 < t \le 6$ **c** $\left(\dfrac{142}{3} + 72\ln(2)\right)\text{m}$

Challenge

$\dfrac{dv}{dt} = \dfrac{60}{kt^2}$
Integrating gives:
$v = -\dfrac{60}{kt} + c$
Solving for k and c using $v = 0$ when $t = 2$, and $v = 9$ when $t = 5$ gives $v = 15 - \dfrac{30}{t}$. As $t \ge 2$, $\dfrac{30}{t} > 0$ so $v < 15$ and the speed of the car never reaches $15\,\text{m s}^{-1}$

Exercise 4B

1 $v^2 = \dfrac{x^2}{2} + 4x + 25$

2 $v = \pm\sqrt{(80 - 4x^2)}$

3 $\frac{1}{5}$

4 $16\,\text{m}$

5 a $\dfrac{6}{125}$
b $\pm 4\sqrt{14}\,\text{m s}^{-1}$ as the particle will pass through this position in both directions.

6 4

7 a $v^2 = 52 - 36\cos\dfrac{x}{3}$
b $2\sqrt{22}\,\text{m s}^{-1}$ $(\approx 9.38\,\text{m s}^{-1})$

8 $4.72\,\text{m s}^{-1}$ (3 s.f.), in the direction of x increasing.

9 a $1.95\,\text{m s}^{-1}$ (3 s.f.) **b** 26.8 (3 s.f.)

10 a $v = x + \dfrac{2}{x}$ **b** $2\sqrt{2}\,\text{m s}^{-1}$

11 $10\,\text{m}$

12 a $v = 3x^{\frac{2}{3}}$ **b** $x = (t+2)^3$

Challenge

$v^2 = \dfrac{1}{5}\left(25x - \dfrac{x^2}{2}\right) + \dfrac{163}{2}$

Exercise 4C

1 a $v = \ln(t+1)$ **b** $v = \ln(11)$

2 $0.137\,\text{s}$ (3 d.p.)

3 a $2.04\,\text{s}$ (3 s.f.) **b** $9.80\,\text{m}$ (3s.f.)

4 a $\dfrac{dv}{dt} = g - 2v$, so $\dfrac{dv}{(g - 2v)} = dt.$
Integrating both sides and using initial conditions:
$-\frac{1}{2}\ln(g - 2v) = t - \frac{1}{2}\ln g.$
$\Rightarrow 1 - \dfrac{2v}{g} = e^{-2t} \Rightarrow g(1 - e^{-2t}) = 2v.$
b $\dfrac{g}{4}(1 - e^{-4})\,\text{m}$

5 a $v = 12 - \dfrac{12}{e^{\frac{t}{4}}}$ **b** $12\,\text{m s}^{-1}$

6 $1.16\,\text{m}$ (3 s.f.)

7 $\frac{1}{2}\ln\left(\dfrac{u^2 + k}{k}\right)\text{m}$

8 a $\dfrac{dv}{dt} = -(a^2 + v^2) \Rightarrow \int\dfrac{dv}{a^2 + v^2} = -\int dt$
so $\dfrac{1}{a}\arctan\dfrac{v}{a} = -t + c$
Using $t = 0$, $v = U$ and $t = T$, $v = \frac{1}{2}U$ gives
$T = \dfrac{1}{a}\left[\arctan\dfrac{U}{a} - \arctan\dfrac{U}{2a}\right]$
b $\frac{1}{2}\ln\dfrac{4(a^2 + u^2)}{4a^2 + u^2}\,\text{m}$

9 $\dfrac{dv}{dt} = \dfrac{1600 - v^2}{64v} \Rightarrow \displaystyle\int \dfrac{64v}{1600 - v^2}\,dv = \displaystyle\int 1\,dt$

$\Rightarrow -32\ln(1600 - v^2) = t + c$

$T = -32\ln(1600 - 20^2) - (-32\ln(1600 - 10^2))$

$= -32\ln 1200 + 32\ln 1500 = 32\ln\dfrac{1500}{1200} = 32\ln\dfrac{5}{4}$

Mixed exercise 4

1 a $8(e^{0.5t} - 1)$ **b** $6\,\text{m s}^{-2}$

2 a $v = 18 - 10e^{-t^2}$ **b** $18\,\text{m s}^{-1}$

3 6

4 a $v = 10 - \dfrac{50}{2t + 5}$

b $(100 - 25\ln 5)\,\text{m} \approx 59.8\,\text{m}$

5 a $\dfrac{\pi}{2}\,\text{m s}^{-1}$

b From part **a**, $v = \dfrac{1}{4}\sin 2t + \dfrac{1}{2}t$

$x = \displaystyle\int v\,dt = -\dfrac{1}{8}\cos 2t + \dfrac{1}{4}t^2 + c$

Using inital conditions gives $c = \dfrac{1}{8}$, hence

$x = -\dfrac{1}{8}\cos 2t + \dfrac{1}{4}t^2 + \dfrac{1}{8}$

Substituting $t = \dfrac{\pi}{4}$ gives:

$x = -\dfrac{1}{8}\cos\dfrac{\pi}{2} + \dfrac{1}{4}\left(\dfrac{\pi}{4}\right)^2 + \dfrac{1}{8} = \dfrac{1}{64}(\pi^2 + 8)$

6 a $2.5\,\text{m s}^{-2}$ in the direction of x increasing.

b $8\,e^{-1}\,\text{m s}^{-2}$ in the direction of x decreasing.

c $\left(\dfrac{56}{3} - 8\,e^{-2}\right)\text{m} \approx 17.6\,\text{m}$ (3 s.f.)

7 a $v = 2t + \ln(t + 1)$ **b** $(2 + 3\ln 3)\,\text{m}$

8 a $T = 3\sqrt{10}\,\text{s}$

b velocity

c $\dfrac{7}{3} < t < 5$

d $59.2\,\text{m}$ (3 s.f.)

9 a $a = 6t - 20$ **b** $\dfrac{901}{27}\,\text{m}$

c The discriminant of the equation $x = t(t^2 - 10t + 32t) = 0$ is less than 0 for $t > 0$ and thus x can never be 0 for $t > 0$.

10 $\dfrac{1}{2kU}\,\text{s}$

11 a $t = \dfrac{2}{3}$ or $t = 2$ **b** $\dfrac{248}{27}\,\text{m}$

12 $\sqrt{\left(20 - \dfrac{12}{x}\right)}$

13 $x = 8$

14 $x = \dfrac{1}{5}$

15 a $v^2 = 16 + 6x - x^2$ **b** 5

16 a $v^2 = \dfrac{5x^3}{3} - \dfrac{x^4}{4} + \dfrac{2500}{3}$

b $\dfrac{50\sqrt{3}}{3}\,\text{m s}^{-1}$

17 a $6.04\,\text{m s}^{-1}$ (3 s.f.) **b** 2.56 (3 s.f.)

18 a $7\,\text{m s}^{-1}$ **b** $x = 7.56\,\text{m}$ (3 s.f.)

19 a $v = 2x + 3$ **b** $x = \dfrac{3}{2}(e^{2t} - 1)$

20 $v\dfrac{dv}{dx} = -k(U^2 + v^2) \Rightarrow \displaystyle\int \dfrac{v\,dv}{(U^2 + v^2)} = -\displaystyle\int k\,dx$

so $\dfrac{1}{2}\ln(U^2 + v^2) = -kx + c$

using distance $= 0$ when $v = U$ gives distance as $\dfrac{1}{2k}\ln\dfrac{8}{5}$

when $v = \dfrac{1}{2}U$

21 $209\,\text{m}$ (3 s.f.)

Challenge

$11\,776\,\text{km}$

CHAPTER 5

Prior knowledge check

1 a $11170\,\text{N}$ **b** $675\,\text{kJ}$

2 a $33.9\,\text{m s}^{-1}$ **b** $1.68\,\text{m}$

3 $13\dfrac{8}{9}\,\text{N}$

Exercise 5A

1 a $9.09\,\text{m s}^{-1}$ (3 s.f.) **b** $1.41\,\text{m s}^{-1}$ (3 s.f.)

c P first comes to rest when $t = \pi$.

d $14.2\,\text{m}$ (3 s.f.) **e** $OP = 20\,\text{m}$

2 a 10

b The van moves $10.6\,\text{m}$ in the first 4 seconds (3 s.f.)

3 a Maximum speed occurs when acceleration is zero, i.e. when force is zero. $\Rightarrow \dfrac{1}{6}(15 - x) = 0 \Rightarrow x = 15$

b $6.85\,\text{m s}^{-1}$

4 a $6.79\,\text{m s}^{-1}$ (3 s.f.) **b** $8.23\,\text{m s}^{-1}$ (3 s.f.)

c $8.10\,\text{N}$

5 $x = 0.677$ (3 s.f.)

6 a $0.25\dfrac{dv}{dt} = -\dfrac{8}{(t + 1)^2} \Rightarrow \displaystyle\int 0.25\,dv = -\displaystyle\int \dfrac{8}{(t + 1)^2}\,dt$

$\Rightarrow 0.25v = \dfrac{8}{(t + 1)} + c \Rightarrow v = \dfrac{32}{(t + 1)} + d$ where $d = 4c$

When $t = 0$, $v = 10$: $10 = \dfrac{32}{1} + d \Rightarrow d = -22$

$\Rightarrow v = 2\left(\dfrac{16}{(t + 1)} - 11\right)$

b $x = 32\ln 6 - 132$

7 $k = 66$

8 $\dfrac{1}{4}\ln 4\,\text{s} = 0.347\,\text{s}$ (3 d.p.)

9 $2\ln 2\,\text{m} = 1.39\,\text{m}$ (3 s.f.)

10 a $\ln 2.5\,\text{s} = 0.916\,\text{s}$ (3 d.p.)

b $(12 - 8\ln 2.5)\,\text{m} = 4.67\,\text{m}$ (3 s.f.)

11 $\dfrac{1}{2kg}\ln\left(\dfrac{\mu + ku^2}{\mu}\right)$

12 a $R(\rightarrow): -k(a^2 + v^2) = m\dfrac{dv}{dt} \Rightarrow \displaystyle\int 1\,dt = -\dfrac{m}{k}\displaystyle\int \dfrac{1}{a^2 + v^2}\,dv$

$\Rightarrow t = A - \dfrac{m}{ak}\arctan\left(\dfrac{v}{a}\right)$

When $t = 0$, $v = U$: $A = \dfrac{m}{ak}\arctan\left(\dfrac{U}{a}\right)$

When $t = T$, $v = \dfrac{U}{2}$: $T = \dfrac{m}{ak}\left(\arctan\left(\dfrac{U}{a}\right) - \arctan\left(\dfrac{U}{2a}\right)\right)$

b $\left(\dfrac{m}{2k}\right)\ln\left(\dfrac{4a^2 + 4U^2}{4a^2 + U^2}\right)$

13 a $12\left(1 - e^{-\frac{t}{4}}\right)$ **b** $12\,\text{m s}^{-1}$

14 a $m\dfrac{dv}{dt} = -mg - \dfrac{mgv}{k} \Rightarrow \dfrac{dv}{dt} = -g\left(1 + \dfrac{v}{k}\right)$

$\Rightarrow \displaystyle\int_0^T -g\,dt = \displaystyle\int_U^0 \dfrac{k}{k + v}\,dv \Rightarrow -g[t]_0^T = k[\ln(k + v)]_U^0$

$\Rightarrow -gT = k(\ln k - \ln(k + U)) \Rightarrow T = \dfrac{k}{g}\ln\left(\dfrac{k + U}{k}\right)$

Online Full worked solutions are available in SolutionBank.

b $H = \frac{k}{g}\left(U - k\ln\left(\frac{k+U}{k}\right)\right)$

15 a $v = \sqrt{\frac{1 - e^{-2gkx}}{k}}$ **b** $\sqrt{\frac{1}{k}}$

c This model has the particle rapidly approaching terminal velocity, within two metres of release the exponential term is of the order 10^{-19}.

Challenge

a Work done $= \int_a^b 3x^2 - x^{\frac{1}{3}}\,dx = \left[x^3 - \frac{3}{4}x^{\frac{4}{3}}\right]_a^b$
$$= b^3 - \frac{3}{4}b^{\frac{4}{3}} - a^3 + \frac{3}{4}a^{\frac{4}{3}}$$

Hence work done is independent of the initial velocity.

b 208 J (3 s.f.)

Exercise 5B

1 $F = \frac{k}{d^2}$, where d = distance from centre

distance $(x - R)$ above surface
$\Rightarrow$ distance x from centre $\Rightarrow F = \frac{k}{x^2}$

On surface $F = mg$, $x = R \Rightarrow mg = \frac{k}{x^2} \Rightarrow k = mgR^2$

$\therefore$ Magnitude of the gravitational force is $\frac{mgR^2}{x^2}$

2 For a particle of mass m, distance x from the centre of the earth.

$F = ma \Rightarrow \frac{k}{x^2} = mA$

On the surface of the earth, $x = R$, $A = g$
$\Rightarrow mg = \frac{k}{R^2} \Rightarrow k = mgR^2 \Rightarrow mA = \frac{mgR^2}{x^2} \Rightarrow A = \frac{gR^2}{x^2}$

3 $\sqrt{2gR}$

4 $\sqrt{\dfrac{U^2X + U^2R - 2gRX}{X + R}}$

5 $\sqrt{\dfrac{7gR}{5}}$

6 $2\sqrt{\dfrac{gR}{3}}$

7 a $F = \frac{k}{x^2}$ when $x = R$, $F = mg$ so $mg = \frac{k}{R^2} \Rightarrow k = mgR^2$

gravitational force on S $= \frac{mgR^2}{x^2}$

b speed $= \sqrt{\dfrac{7gR}{2}}$

Challenge

a 5.98×10^{24} kg   **b** 5500 kg m^{-3}

Exercise 5C

1 a $\frac{\pi}{2}$ s   **b** 1.83 m s^{-1} (3 s.f.)

2 a 1 m   **b** 1.36 m s^{-1} (3 s.f.)

3 a 5 m   **b** π s

4 $\frac{4}{3}$ m s^{-1}

5 17.9 m s^{-1} (3 s.f.)

6 a 1.26 m (3 s.f.)   **b** $x = 1.26\sin 4t$

7 a 0.133 m (3 s.f.)   **b** 0.0141 m (3 s.f.)

8 a 1.37 m (3 s.f.)   **b** 0.684 s (3 s.f.)

9 a 1.00 m s^{-1} (3 s.f.)   **b** 0.922 m s^{-1} (3 s.f.)

10 9.25 J (3 s.f.)

11 a 1.26 m s^{-1} (3 s.f.)   **b** $\frac{2}{3}$ s to fall 0.6 m

12 0.0738 s (3 s.f.)

13 a $x = 4\sin 2t \Rightarrow \dot{x} = 8\cos 2t \Rightarrow \ddot{x} = -16\sin 2t \Rightarrow \ddot{x} = 4x$
$\therefore$ S.H.M.

b $4, \frac{2\pi}{2} = \pi$ s   **c** 8 m s^{-1} **d** $\frac{\pi}{6}$ **e** $\frac{\pi}{12}$

14 a $x = 3\sin\left(4t + \frac{1}{2}\right) \Rightarrow \dot{x} = 12\cos\left(4t + \frac{1}{2}\right)$
$\Rightarrow \ddot{x} = -48\sin\left(4t + \frac{1}{2}\right) \Rightarrow \ddot{x} = 16x \therefore$ S.H.M.

b Amplitude = 3 m, Period $= \frac{\pi}{2}$ s

c $x = 1.44$ (3 s.f.)   **d** 0.660 (3 s.f.)

15 a 11.51 a.m. (nearest minute)

b 8.39 p.m. (nearest minute)

16 P takes 0.823 s to travel directly from B to A (3 s.f.)

Challenge

$\ddot{x} = -\omega^2 x$ $v^2 = \omega^2(a^2 - x^2)$

$v_1^2 = \omega^2(a^2 - x_1^2)$ (1)

$v_2^2 = \omega^2(a^2 - x_2^2)$ (2)

(2) $-$ (1): $v_2^2 - v_1^2 = \omega^2(a^2 - x_2^2) - \omega^2(a^2 - x_1^2)$

$v_2^2 - v_1^2 = \omega^2(a^2 - x_2^2 - a^2 + x_1^2)$

Rearranging gives $\omega^2 = \dfrac{v_2^2 - v_1^2}{x_1^2 - x_2^2}$ so $\omega = \left(\dfrac{v_2^2 - v_1^2}{x_1^2 - x_2^2}\right)^{\frac{1}{2}}$

$T = \dfrac{2\pi}{\omega} = 2\pi\left(\dfrac{x_1^2 - x_2^2}{v_2^2 - v_1^2}\right)^{\frac{1}{2}}$

Exercise 5D

1 a $F = ma \Rightarrow -T = 0.5\ddot{x}$
Hooke's law: $T = \frac{\lambda x}{l} = \frac{60x}{0.6} = 100x$
$\Rightarrow -100x = 0.5\ddot{x} \Rightarrow \ddot{x} = -200x \therefore$ S.H.M.

b $\frac{\pi}{10}\sqrt{2}$ s   0.3 m   **c** 4.24 m s^{-1} (3 s.f.)

2 a $F = ma \Rightarrow -T = 0.8\ddot{x}$
Hooke's law: $T = \frac{\lambda x}{l} = \frac{20x}{1.6} = \frac{25x}{2}$
$\Rightarrow -\frac{25x}{2} = 0.8\ddot{x} \Rightarrow \ddot{x} = -\frac{125x}{8} \therefore$ S.H.M.

b 3.21 s (3 s.f.)

3 a $F = ma \Rightarrow -T = 0.4\ddot{x}$
Hooke's law: $T = \frac{\lambda x}{l} = \frac{24x}{1.2} = 20x$
$\Rightarrow -20x = 0.4\ddot{x} \Rightarrow \ddot{x} = -50x \therefore$ S.H.M.

b 0.489 s (3 s.f.)   **c** 1.84 m (3 s.f.)

4 a $F = ma \Rightarrow -T = 0.8\ddot{x}$
Hooke's law: $T = \frac{\lambda x}{l} = \frac{80x}{1.2}$
$\Rightarrow -\frac{80x}{1.2} = 0.8\ddot{x} \Rightarrow \ddot{x} = -\frac{100x}{1.2} \therefore$ S.H.M.

5 a $x = 0.5\sin 10t$ **b** 50 m s^{-2}

6 a 0.5 m   **b** 2.11 m s^{-1} (3 s.f.)
c i 1.49 s   **ii** 0.3 m

7 a 0.351 s (3 s.f.)   **b** 2.56 J

8 a 2.45 m s^{-1} (3 s.f.)   **b** 1.25 (3 s.f.)

9 a $F = ma \Rightarrow T_B - T_A = 0.4\ddot{x}$
Hooke's law: $T = \frac{\lambda x}{l}$
AP: extension = $0.8 + x$
$\therefore T_A = \frac{12(0.8 + x)}{1.2} = 10(0.8 + x)$
BP: extension = $0.8 - x$
$\therefore T_B = \frac{12(0.8 - x)}{1.2} = 10(0.8 - x)$
$\therefore 10(0.8 - x) - 10(0.8 + x) = 0.4\ddot{x}$
$-20x = 0.4\ddot{x} \Rightarrow \ddot{x} = -50x \therefore$ S.H.M.

b 3.6 J

10 a $F = ma \Rightarrow T_B - T_A = m\ddot{x}$

Hooke's law: $T = \dfrac{\lambda x}{l}$

AP: extension $= 1.5l + x \Rightarrow T_A = \dfrac{3mg(1.5l + x)}{l}$

PB: extension $= 1.5l - x \Rightarrow T_B = \dfrac{3mg(1.5l - x)}{l}$

$\Rightarrow \dfrac{3mg(1.5l + x)}{l} - \dfrac{3mg(1.5l - x)}{l} = m\ddot{x}$

$-\dfrac{6mgx}{l} = m\ddot{x} \Rightarrow \ddot{x} = -\dfrac{6g}{l}x$ ∴ S.H.M.

b $2\pi\sqrt{\dfrac{l}{6g}}$ **c** $1.5l$

d $\sqrt{12gl}$ (or $2\sqrt{3gl}$)

11 a $F = ma \Rightarrow T_B - T_A = 0.5\ddot{x}$

Hooke's law: $T = \dfrac{\lambda x}{l}$

AP: extension $= 1 + x$ ∴ $T_A = \dfrac{15(1 + x)}{1} = 15(1 + x)$

BP: extension $= 1.5 - x$

∴ $T_B = \dfrac{15(1.5 - x)}{1.5} = 10(1.5 - x)$

∴ $10(1.5 - x) - 15(1 + x) = 0.5\ddot{x}$

$-25x = 0.5\ddot{x} \Rightarrow \ddot{x} = -50x$ ∴ S.H.M.

period $= \dfrac{2\pi}{\omega} = \dfrac{2\pi}{\sqrt{50}} = \dfrac{\pi}{5}\sqrt{2}$

b $1\,\text{m}$

Exercise 5E

1 a $1.64\,\text{m}$ (3 s.f.)

b $F = ma \Rightarrow 0.75g - T = 0.75\ddot{x}$

Hooke's law: $T = \dfrac{\lambda x}{l} = \dfrac{80(x + e)}{1.5}$

∴ $0.75g - \dfrac{80(x + e)}{1.5} = 0.75\ddot{x}$

from part **a**, $0.75g = \dfrac{80e}{1.5}$

∴ $0.75\ddot{x} = -\dfrac{80}{1.5}x \Rightarrow \ddot{x} = -\dfrac{80x}{1.5 \times 0.75}$ ∴ S.H.M.

c $0.745\,\text{s}$ (3 s.f.)

d $0.296\,\text{m}$ (3 s.f.)

2 a $0.049\,\text{m}$ (3 s.f.)

b $0.444\,\text{s}$ (3 s.f.)

c $2.83\,\text{m\,s}^{-1}$ (3 s.f.)

3 a $1.5\,\text{m\,s}^{-1}$

b In equilibrium:

R(↑): $T = 2mg$

Hooke's Law: $T = \dfrac{\lambda x}{l} = \dfrac{\lambda e}{1.5}$ ∴ $\dfrac{\lambda e}{1.5} = 2g$

When oscillating: $F = ma \Rightarrow 2g - T = 2\ddot{x}$

Hooke's Law: $T = \dfrac{\lambda(e + x)}{1.5}$

∴ $2g - \dfrac{\lambda(e + x)}{1.5} = 2\ddot{x}$

$\Rightarrow \dfrac{\lambda e}{1.5} - \dfrac{\lambda(e + x)}{1.5} = 2\ddot{x} \Rightarrow -\dfrac{\lambda x}{1.5} = 2\ddot{x}$

∴ $\ddot{x} = -\dfrac{\lambda}{3}x$ as $\lambda > 0$, this is S.H.M.

c 48 **d** $0.375\,\text{m}$

4 a $0.138\,\text{s}$ (3 s.f.) **b** $1.61\,\text{m\,s}^{-1}$

5 a $31.4\,\text{N}$ (3 s.f.)

b $F = ma \Rightarrow 0.4g - T = 0.4\ddot{x}$

Hooke's law: $T = \dfrac{\lambda x}{l} = \dfrac{31.36(x + 0.05)}{0.4}$

$0.4g - \dfrac{31.36(x + 0.05)}{0.4} = 0.4\ddot{x}$

$\ddot{x} = -\dfrac{31.36}{0.4^2}x$ ∴ S.H.M.

c $0.449\,\text{s}$, $0.07\,\text{m}$ **d** $0.98\,\text{m\,s}^{-1}$

e $0.156\,\text{s}$ to rise $11\,\text{cm}$ (3 s.f.)

6 a $0.416\,\text{s}$ (3 s.f.) **b** $0.0574\,\text{s}$ (3 s.f.)

7 a $0.221\,\text{m}$ **b** $0.517\,\text{s}$ (3 s.f.)

8 a $1.69\,\text{m}$ (3 s.f.)

b $T_A = \dfrac{15(e + x)}{0.6}$, $T_B = \dfrac{15(1.6 - (e + x))}{0.6}$

$F = ma$

$1.5g + \dfrac{15(1.6 - (e + x))}{0.6} - \dfrac{15(e + x)}{0.6} = 1.5\ddot{x}$

$1.5g + 40 - 25(e + x) - 25(e + x) = 1.5\ddot{x}$

$1.5g + 40 - 50e - 50x = 1.5\ddot{x}$

From part **a**, $50e = 1.5g + 40$

∴ $1.5\ddot{x} = -50x \Rightarrow \ddot{x} = -\dfrac{100}{3}x$ ∴ S.H.M.

c 0.398 (3 s.f.)

9 a $12.5\,\text{m\,s}^{-1}$ (3 s.f.) **b** $10.4\,\text{m}$ (3 s.f.)

c $1.56\,\text{s}$ (3 s.f.)

Challenge

Particle P: $T = 2\pi\sqrt{\dfrac{ml}{\lambda}} = 2\pi\sqrt{\dfrac{ml}{5mg}} = 2\pi\sqrt{\dfrac{l}{5g}}$

Particle Q: $3T = 2\pi\sqrt{\dfrac{l(m + km)}{5mg}} = 2\pi\sqrt{\dfrac{l(1 + k)}{5g}}$

$6\pi\sqrt{\dfrac{l}{5g}} = 2\pi\sqrt{\dfrac{l(1 + k)}{5g}}$ so $3\sqrt{l} = 2\sqrt{l} \times \sqrt{1 + k}$

squaring both sides $9 = 4 + 4k$ so $k = \dfrac{5}{4}$

Mixed exercise 5

1 a 108

b $11.8\,\text{m}$ (3 s.f.)

2 a $\dot{x} = \dfrac{-20}{3}(3t + 4)^{\frac{1}{2}} + \dfrac{76}{3}$

b $18.7\,\text{m}$ from O (3 s.f.)

3 $\dfrac{1}{3kU}$

4 $\left(\dfrac{g}{k}\right)^{\frac{1}{2}}(1 - \text{e}^{-2kD})^{\frac{1}{2}}$

5 $56.1\,\text{m}$ (3 s.f.)

6 $d = \dfrac{1}{10b}\ln(1 + bU^2)$

7 a R(↑): $-mg - \dfrac{mgv^2}{c^2} = ma \Rightarrow -g\left(\dfrac{c^2 + v^2}{c^2}\right) = v\dfrac{dv}{dx}$

$\Rightarrow \int g\,dx = -c^2\int\dfrac{v}{c^2 + v^2}dv \Rightarrow gx = A - \dfrac{c^2}{2}\ln(c^2 + v^2)$

At $x = 0$, $v = V$: $0 = A - \dfrac{c^2}{2}\ln(c^2 + V^2)$

$\Rightarrow A = \dfrac{c^2}{2}\ln(c^2 + V^2)$

Hence $gx = \dfrac{c^2}{2}\ln(c^2 + V^2) - \dfrac{c^2}{2}\ln(c^2 + v^2)$

$= \dfrac{c^2}{2}\ln\left(\dfrac{c^2 + V^2}{c^2 + v^2}\right)$

$\Rightarrow x = \dfrac{c^2}{2g}\ln\left(\dfrac{c^2 + V^2}{c^2 + v^2}\right)$

At the greatest height $v = 0$: $x = \dfrac{c^2}{2g}\ln\left(\dfrac{c^2 + V^2}{c^2}\right)$

$= \dfrac{c^2}{2g}\ln\left(1 + \dfrac{V^2}{c^2}\right)$

Online Full worked solutions are available in SolutionBank.

b $-g\dfrac{c^2 + v^2}{c^2} = \dfrac{dv}{dt}$

$\Rightarrow \displaystyle\int g\,dt = -c^2 \displaystyle\int \dfrac{1}{c^2 + v^2}\,dv$

$\Rightarrow gt = -c^2 \arctan\left(\dfrac{v}{c}\right) + C$

$\Rightarrow t = -\dfrac{c}{g}\arctan\left(\dfrac{v}{c}\right) + D$

When $t = 0$, $v = V$, so $D = \dfrac{c}{g}\arctan\left(\dfrac{V}{c}\right)$

So $t = \dfrac{c}{g}\arctan\left(\dfrac{V}{c}\right) - \dfrac{c}{g}\arctan\left(\dfrac{v}{c}\right)$

$v = 0 \Rightarrow t = \left(\dfrac{c}{g}\right)\arctan\left(\dfrac{V}{c}\right)$

8 Let the mass of the particle be m.
Let the resistance be kv^2, where k is a constant of proportionality.
If U is the speed for which the resistance is equal to the weight of the particle then

$kU^2 = mg \Rightarrow k = \dfrac{mg}{U^2}$. Hence the resistance is $\dfrac{mgv^2}{U^2}$

$R(\uparrow)$: $F = ma \Rightarrow -mg - \dfrac{mgv^2}{U^2} = ma \Rightarrow -\dfrac{g(U^2 + v^2)}{U^2} = \dfrac{dv}{dt}$ ✱

$\Rightarrow \displaystyle\int g\,dt = -U^2 \displaystyle\int \dfrac{1}{U^2 + v^2}\,dv \Rightarrow gt = A - U^2 \times \dfrac{1}{U}\arctan\left(\dfrac{v}{U}\right)$

When $t = 0$, $v = U$: $0 = A - U\arctan 1$

$\Rightarrow A = U\arctan 1 = \dfrac{\pi U}{4}$

Hence $gt = \dfrac{\pi U}{4} - U\arctan\left(\dfrac{v}{U}\right) \Rightarrow t = \dfrac{\pi U}{4g} - \dfrac{U}{g}\arctan\left(\dfrac{v}{U}\right)$

Let the time of ascent be T.

When $t = T$, $v = 0$: $T = \dfrac{\pi U}{4g} - \dfrac{U}{g}\arctan 0 = \dfrac{\pi U}{4g}$

b Equation ✱ can be written as $-\dfrac{g(U^2 + v^2)}{U^2} = v\dfrac{dv}{dx}$

$\Rightarrow \displaystyle\int g\,dx = -U^2 \displaystyle\int \dfrac{v}{U^2 + v^2}\,dx \Rightarrow gx = B - \dfrac{U^2}{2}\ln(U^2 + v^2)$

When $x = 0$, $v = U$: $0 = B - \dfrac{U^2}{2}\ln(2U^2)$

$\Rightarrow B = \dfrac{U^2}{2}\ln(2U^2)$

Hence $gx = \dfrac{U^2}{2}\ln(2U^2) - \dfrac{U^2}{2}\ln(U^2 + v^2)$

$\Rightarrow x = \dfrac{U^2}{2g}\ln\left(\dfrac{2U^2}{U^2 + v^2}\right)$

Let the total distance ascended be H.

When $h = H$, $v = 0$: $H = \dfrac{U^2}{2g}\ln\left(\dfrac{2U^2}{U^2}\right) = \dfrac{U^2}{2g}\ln 2$

9 a $R(\rightarrow)$: $-mk(V_0^2 + 2v^2) = ma \Rightarrow -k(V_0^2 + 2v^2) = v\dfrac{dv}{dx}$

$\Rightarrow \displaystyle\int k\,dx = -\displaystyle\int \dfrac{v}{V_0^2 + 2v^2}\,dv \Rightarrow kx = A - \tfrac{1}{4}\ln(V_0^2 + 2v^2)$

At $x = 0$, $v = V_0$: $0 = A - \tfrac{1}{4}\ln(V_0^2 + 2V_0^2)$

$\Rightarrow A = \tfrac{1}{4}\ln(3V_0^2)$

Hence $kx = \tfrac{1}{4}\ln(3V_0^2) - \tfrac{1}{4}\ln(V_0^2 + 2v^2)$

$\Rightarrow x = \dfrac{1}{4k}\ln\left(\dfrac{3V_0^2}{V_0^2 + 2v^2}\right)$

When $v = \tfrac{1}{2}V_0$: $x = \dfrac{1}{4k}\ln\left(\dfrac{3V_0^2}{V_0^2 + \tfrac{1}{2}V_0^2}\right)$

$= \dfrac{1}{4k}\ln\left(\dfrac{3V_0^2}{\tfrac{3}{2}V_0^2}\right) = \dfrac{\ln 2}{4k}$, $t = \dfrac{0.24}{kV_0}$

b 0.24 (2 d.p.)

10 a $P = Fv \Rightarrow F = \dfrac{P}{v}$

$R(\rightarrow)$: $F - \dfrac{mkv^2}{3} = ma \Rightarrow \dfrac{P}{v} - \dfrac{mkv^2}{3} = mv\dfrac{dv}{dx}$

$3P - mkv^3 = 3mv^2\dfrac{dv}{dx} \Rightarrow 3mv^2\dfrac{dv}{dx} = 3P - mkv^3$

b $\dfrac{1}{k}\ln\left(\dfrac{21P}{8(3P - mkv^3)}\right)$

11 a $F = \dfrac{k}{x^2}$ when $x = R$, $F = mg$

$\therefore \dfrac{k}{x^2} = mg$, $k = mgR^2$

b $\sqrt{\dfrac{8Rg}{5}}$ or $2\sqrt{\dfrac{2Rg}{5}}$

12 a $\dfrac{4\pi}{50}$ (or 0.251 (3 s.f.))

b $0.203\,\text{m s}^{-1}$ (3 s.f.)

c $0.318\,\text{m}$ (3 s.f.)

13 a $x = 2.5$

b $v^2 = 25x - 5x^2 + 32.75$

14 a $x = 3\sin\left(\dfrac{\pi}{4}t\right) \Rightarrow \dot{x} = \dfrac{3\pi}{4}\cos\left(\dfrac{\pi}{4}t\right)$

$\Rightarrow \ddot{x} = -3\left(\dfrac{\pi}{4}\right)^2 \sin\left(\dfrac{\pi}{4}t\right)$

$\Rightarrow \ddot{x} = -\left(\dfrac{\pi}{4}\right)^2 x \; \therefore$ S.H.M.

b Amplitude = $3\,$m
Period = $8\,$s

c $\dfrac{3\pi}{4}\,\text{m s}^{-1}$ (or $2.36\,\text{m s}^{-1}$ (3 s.f.))

d $0.405\,$s (3 s.f.)

15 a $1.54\,$s

b $0.116\,$m

c $1.03\,$s

16 a $F = ma \Rightarrow -T = 0.6\ddot{x}$

Hooke's law: $T = \dfrac{\lambda x}{l} = \dfrac{25x}{2.5} = 10x$

$\Rightarrow -10x = 0.6\ddot{x} \Rightarrow \ddot{x} = -\dfrac{10}{0.6}x \; \therefore$ S.H.M.

b Period is $1.54\,$s (3 s.f.)
Amplitude is $(4 - 2.5)\,\text{m} = 1.5\,$m

c P takes $0.468\,$s to move $2\,$m from B (3 s.f.)

17 a $F = ma \Rightarrow T_B - T_A = 0.4\ddot{x}$

Hooke's law: $T = \dfrac{\lambda x}{l}$

$T_A = \dfrac{2.5(0.4 + x)}{0.6}$, $T_B = \dfrac{2.5(0.4 - x)}{0.6}$

$\therefore \dfrac{2.5(0.4 - x)}{0.6} - \dfrac{2.5(0.4 + x)}{0.6} = 0.4\ddot{x}$

$\Rightarrow -5x = 0.4\ddot{x} \Rightarrow \ddot{x} = -\dfrac{5}{0.6 \times 0.4}x \; \therefore$ S.H.M.

b $1.38\,$s (3 s.f.)

c $0.229\,$s to reach D (3 s.f.)

18 a Spring AP extension: $2.4\,$m
Spring PB extension: $1.6\,$m

b $F = ma$

$T_B - T_A = 0.4\ddot{x}$

$T_A = \dfrac{20(2.4 + x)}{5}$

$T_B = \dfrac{18(1.6 - x)}{3}$

$6(1.6 - x) - 4(2.4 + x) = 0.4\ddot{x}$

$-10x = 0.4\ddot{x}$

$\ddot{x} = -25x$

$\therefore$ S.H.M.

c $4\,\text{m s}^{-1}$

19 a $\lambda = 3g$

b $0.5g - T = 0.5\ddot{x}$, $T = \dfrac{\lambda x}{l} = \dfrac{3g(0.2 + x)}{1.2}$

$\Rightarrow 5g - \dfrac{3g(0.2 + x)}{1.2} = 0.5\ddot{x} \therefore \ddot{x} = \dfrac{-3gx}{0.5 \times 1.2} = -5gx$

$\therefore$ S.H.M.

c $0.898\,\text{s}$ (3 s.f.)

d $2.01\,\text{m s}^{-1}$ (3 s.f.)

e $0.406\,\text{m}$ (3 s.f.)

20 a $F = ma \Rightarrow T_B - T_A = m\ddot{x}$

Hooke's law: $T = \dfrac{\lambda x}{l}$

$T_A = \dfrac{5mg(l - x)}{2l}$, $T_B = \dfrac{5mg(l + x)}{2l}$

$\therefore \dfrac{5mg(l - x)}{2l} - \dfrac{5mg(l + x)}{2l} = m\ddot{x}$

$\Rightarrow -\dfrac{5mgx}{l} = m\ddot{x} \Rightarrow \ddot{x} = -\dfrac{5g}{l}x \therefore$ S.H.M.

b $2\pi\sqrt{\dfrac{l}{5g}}$ **c** $\frac{1}{4}\sqrt{5gl}$

Challenge

$ma = -\dfrac{mMG}{(R + x)^2}$ so $a = -\dfrac{MG}{(R + x)^2}$

a $v\dfrac{dv}{dx} = -\dfrac{MG}{(R + x)^2}$ maximum height, H, is reached when $v = 0$

$\displaystyle\int_u^0 v\,dv = -\int_0^H \dfrac{MG}{(R + x)^2}\,dx \Rightarrow \left[\dfrac{v^2}{2}\right]_u^0 = \left[\dfrac{MG}{(R + x)}\right]_0^H$

$-\dfrac{u^2}{2} = \dfrac{MG}{(R + H)} - \dfrac{MG}{R} = MG\left(\dfrac{1}{R + H} - \dfrac{1}{R}\right) = MG\left(\dfrac{R - R - H}{R(R + H)}\right)$

$-\dfrac{u^2}{2} = \dfrac{MG}{(R + H)} - \dfrac{MG}{R} = MG\left(\dfrac{1}{R + H} - \dfrac{1}{R}\right) = MG\left(\dfrac{R - R - H}{R(R + H)}\right)$

$MG\left(\dfrac{H}{R(R + H)}\right) = \dfrac{u^2}{2} \Rightarrow 2MGH = u^2(R^2 + RH)$

$2MGH - RHu^2 = u^2R^2 \quad H(2MG - Ru^2) = u^2R^2$

$H = \dfrac{R^2u^2}{2MG - Ru^2} = \dfrac{Ru^2}{\dfrac{2MG}{R} - u^2}$

b $H \to \infty$ as $u^2 \to \dfrac{2MG}{R}$

Escape velocity $u = \sqrt{\dfrac{2MG}{R}} = \sqrt{\dfrac{2 \times 5.98 \times 10^{24} \times 6.7 \times 10^{11}}{6.4 \times 10^6}}$
$= 1.12 \times 10^4\,\text{m s}^{-1}$

Review exercise 2

1 $a = \dfrac{dv}{dt} = e^{2t}$

$v = \int e^{2t}\,dt = \frac{1}{2}e^{2t} + A$

When $t = 0, v = 0$

$0 = \frac{1}{2} + A \Rightarrow A = -\frac{1}{2}$

$v = \frac{1}{2}(e^{2t} - 1)$

2 a $v = 13 - 3e^{-\frac{1}{6}t}$ **b** $11.2\,\text{m s}^{-1}$ (3 s.f.)   **c** 13

3 a $v = 8 - 4\cos\frac{1}{2}t$ **b** $4(\pi - \sqrt{2})\,\text{m}$

4 a $v = 2e^{-2t} - 1$ **b** $\frac{1}{2}(1 - \ln 2)\,\text{m}$

5 a $a = \dfrac{dv}{dt} = 3(t + 4)^{-\frac{1}{2}}$

$v = -3\int(t + 4)^{-\frac{1}{2}}\,dt = A - 6(t + 4)^{\frac{1}{2}}$

When $t = 0, v = 18$

$18 = A - 6 \times 2 \Rightarrow A = 30$

$v = 30 - 6\sqrt{t + 4}$

b $162\,\text{m}$

6 a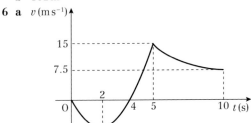

b $2 < t < 5$

c $\displaystyle\int_0^4 3t(t - 4)\,dt = [t^3 - 12t]_0^4 = -32$

$\displaystyle\int_4^5 3t(t - 4)\,dt = [t^3 - 12t]_4^5 = 7$

So distance travelled in the interval is $32 + 7 = 39\,\text{m}$

d 6.98 (3 s.f.)

7 a $v = (p + qt)^{-1}$

$a = -q(p + qt)^{-2} = -qv^2$

b $p = \frac{1}{20}, q = \frac{3}{1600}$

c $x = \dfrac{1600}{3}\ln\left(1 + \dfrac{3}{80}t\right)$

8 $a = \dfrac{d}{dx}\left(\frac{1}{2}v^2\right) = 4x$

$\frac{1}{2}v^2 = \int 4x\,dx = 2x^2 + A$

$v^2 = 4x^2 + B$

At $x = 2, v = 4$

$16 = 16 + B \Rightarrow B = 0$

$v^2 = 4x^2$

9 $a = \dfrac{d}{dx}\left(\frac{1}{2}v^2\right) = 1 - 4x^{-2}$

$\frac{1}{2}v^2 = \int(1 - 4x^{-2})\,dx$

$= x + \dfrac{4}{x} + A$

$v^2 = 2x + \dfrac{8}{x} + B$, where $B = 2A$

At $x = 1, v = 3\sqrt{2}$

$18 = 2 + 8 + B \Rightarrow B = 8$

$v^2 = 2x + \dfrac{8}{x} + 8$

At $x = \frac{3}{2}$

$v^2 = 2 \times \frac{3}{2} + 8 \times \frac{2}{3} + 8 = \dfrac{49}{3}$

10 $8.76\,\text{m s}^{-1}$ (3 s.f.)

11 a $v^2 = 4k^2\left(1 - \dfrac{2}{x + 1}\right)$

b $v = 2k\sqrt{1 - \dfrac{2}{x + 1}}$

As x is positive, $\dfrac{1}{1 + x}$ is positive and $1 - \dfrac{1}{x + 1} < 1$

So $v < 2k$

12 a At the maximum value of v, $a = 0$

$a = \frac{1}{12}(30 - x) = 0$

$x = 30$

b $v^2 = 5x - \dfrac{x^2}{12} + 25$

13 a $0.9\,\text{m}$

b $a = p - qx$

At $x = 0, a = 20$

$20 = p - 0 \Rightarrow p = 20$

$a = 20 - qx$

At $x = 2, a = 12$

$12 = 10 - 2q \Rightarrow q = 4$

c $1\,\text{m}$

Online Full worked solutions are available in SolutionBank.

14 a $v^2 = 60 - \dfrac{72}{2x+1}$

b $0.1\,\text{m}$

15 a $a = \dfrac{\text{d}}{\text{d}x}\left(\tfrac{1}{2}v^2\right) = \dfrac{k}{2x^2} + \dfrac{k}{4c^2}$

$\tfrac{1}{2}v^2 = \int\left(\dfrac{k}{2x^2} + \dfrac{k}{4c^2}\right)\text{d}x = -\dfrac{k}{2x} + \dfrac{kx}{4c^2} + A$

$v^2 = -\dfrac{k}{x} + \dfrac{kx}{2c^2} + B$

At $x = 2a$, $v = -\sqrt{\dfrac{k}{c}}$

$\dfrac{k}{c} = -\dfrac{k}{2c} + \dfrac{2kc}{2c^2} + B \Rightarrow B = \dfrac{k}{2c}$

At $x = c$

$v^2 = -\dfrac{k}{c} + \dfrac{kc}{2c^2} + \dfrac{k}{2c} = 0$

The particle comes to instantaneous rest where $x = c$.

b $a = \dfrac{\text{d}}{\text{d}x}\left(\tfrac{1}{2}v^2\right) = \dfrac{k}{2x^2} - \dfrac{k}{4c^2}$

$\tfrac{1}{2}v^2 = \int\left(\dfrac{k}{2x^2} - \dfrac{k}{4c^2}\right)\text{d}x = -\dfrac{k}{2x} - \dfrac{kx}{4c^2} + C$

$v^2 = -\dfrac{k}{x} - \dfrac{kx}{2c^2} + D$

At $x = c$, $v = 0$

$0 = -\dfrac{k}{c} - \dfrac{kc}{2c^2} + D \Rightarrow D = \dfrac{3k}{2c}$

Hence

$v^2 = -\dfrac{k}{x} - \dfrac{kx}{2c^2} + \dfrac{3k}{2c} = \dfrac{-2kc^2 - kx^2 + 3kcx}{2c^2x}$

$= -\dfrac{k}{2c^2x}(x - c)(x - 2c)$

When $v = 0$, $(x - c)(x - 2c) = 0$
$x = c, 2c$
After leaving $B\,(x = c)$, the particle next comes to rest at $A\,(x = 2c)$.

16 a $T = \dfrac{5}{6}$ **b** $OA = 10\ln 2\,\text{m}$

17 a $k\ln\left(\dfrac{k}{k+U}\right) + U\,\text{m}$ **b** $\ln\left(\dfrac{k+U}{k}\right)\,\text{s}$

18 a $\dfrac{13(e^{\frac{13t}{50}} - 1)}{(e^{\frac{13t}{50}} + 1)}\,\text{m s}^{-1}$

b As $\dfrac{(e^x - 1)}{(e^x + 1)} < 1$, $v < 13\,\text{m s}^{-1}$

so speed cannot exceed $13\,\text{m s}^{-1}$

19 a $\dfrac{30(e^{\frac{t}{3}} - 1)}{e^{\frac{t}{3}}}\,\text{m s}^{-1}$ **b** $30\,\text{m s}^{-1}$

20 a $T = \dfrac{1}{a}\left(\arctan\dfrac{20}{a} - \arctan\dfrac{12}{a}\right)$

b $\dfrac{1}{2}\ln\left(\dfrac{a^2 + 400}{a^2 + 144}\right)\text{m}$

21 53.6 (3 s.f.)

22 a $v^2 = 20 - 16e^{-0.1x}$

b $x = 10\ln 4$

c For all x, $e^{-0.1x} > 0$
So $v^2 = 20 - 16e^{-0.1x} < 20$
Hence $v < \sqrt{20}$

23 a $v^2 = 2x^2 - x^3 + 144$ **b** $12\,\text{m s}^{-1}$

24 a $t = -2\ln\left(\tfrac{2}{5}\right)$ or $2\ln\left(\tfrac{5}{2}\right)$ **b** $(4 + 20e^{-1})\,\text{m}$

25 a $F = ma$

$\dfrac{48000}{(t + 2)^2} = 800a$

$a = 60(t + 2)^{-2}$

$v = \int 60(t + 2)^{-2}\,\text{d}t$

$= A - \dfrac{60}{t + 2}$

When $t = 0$ $v = 0$

$0 = A - \dfrac{60}{2} \Rightarrow A = 30$

$v = 30 - \dfrac{60}{t + 2}$

As $t \to \infty$, $\dfrac{60}{t + 2} \to 0$, $v \to 30$

As t increases, the car approached a limiting speed of $30\,\text{m s}^{-2}$

b $(180 - 120\ln 2)\,\text{m}$

26 a 6 **b** $17\,\text{m}$

27 a $\dfrac{g(e^{3t} - 1)}{3e^{3t}}\,\text{m s}^{-1}$ **b** $\dfrac{g}{9}(5 + e^{-6})\,\text{m}$

28 $gk^2\ln\left(\dfrac{gk}{gk + U}\right) + Uk$

29 a As $F \propto \dfrac{1}{x^2}$, $F \propto -\dfrac{k}{x^2}$

At $x = R$, $F = -mg$

$-mg = -\dfrac{k}{R^2} \Rightarrow k = mgR^2$

$F = -\dfrac{mgR^2}{x^2}$

b $\sqrt{\dfrac{gR}{6}}$

30 a $F = ma$

$-\dfrac{cm}{x^2} = ma$

$a = -cx^{-2}$

$\tfrac{1}{2}v^2 = -\int cx^{-2}\,\text{d}x = \dfrac{c}{x} + A$

$v^2 = \dfrac{2c}{x} + B$

When $x = R$, $v = U$

$U^2 = \dfrac{2c}{R} + B \Rightarrow B = U^2 - \dfrac{2c}{R}$

$v^2 = \dfrac{2c}{x} + U^2 - \dfrac{2c}{R} = U^2 + 2c\left(\dfrac{1}{x} - \dfrac{1}{R}\right)$

b $\tfrac{1}{2}RU^2$

31 a $F = ma$

$-\dfrac{k}{x^2} = ma$

$a = -\dfrac{k}{ma^2}$

At $x = R$, $a = -g$

$-g = -\dfrac{k}{mR^2}$

$k = mgR^2$

$-\dfrac{mgR^2}{x^2} = ma$

$a = v\dfrac{\text{d}v}{\text{d}x} = -\dfrac{gR^2}{x^2}$

b $X = \dfrac{2gR^2}{2gR - U^2}$

32 $\dfrac{2}{3}$

33 a $19.7\,\text{N}$ (3 s.f.)   **b** $5.44\,\text{m s}^{-1}$ (3 s.f.)

34 a $1.3\,\text{m}$ **b** $2.6\,\text{m s}^{-1}$

   c $5.2\,\text{m s}^{-2}$ **d** $0.79\,\text{s}$ (2 d.p.)

35 a The period of motion is $6\,$s.

$$T = \frac{2\pi}{\omega}$$

$$\omega = \frac{\pi}{3}$$

Measuring the time, t seconds, from an instant when C is at Q and the displacement from the centre of the oscillation, O: $x = a\cos\omega t$

The amplitude is $2l$ and $\omega = \frac{\pi}{3}$

After 0.75 s, C is at P:

$$x = 2l\cos\left(\frac{\pi}{3} \times 0.75\right) = 2l\cos\frac{\omega}{4} = \sqrt{2}\,l$$

$$b = a - x = 2l - \sqrt{2}\,l = (2 - \sqrt{2})l$$

b $v = \dfrac{\sqrt{2}\,\pi l}{3}$

c $0.28\,$s (2 d.p.)

36 a At C: $v^2 = \omega^2(a^2 - x^2)$

$$0 = \omega^2(a^2 - 1.2^2) \Rightarrow a = 1.2$$

At A: $v^2 = \omega^2(a^2 - x^2)$

$$\left(\tfrac{3}{10}\sqrt{3}\right)^2 = \omega^2(1.2^2 - 0.6^2)$$

$$\omega^2 = \frac{27}{108} = \frac{1}{4} \Rightarrow \omega = \frac{1}{2}$$

Checking $a = 1.2$ and $\omega = \frac{1}{2}$ at B

$$v^2 = \omega^2(a^2 - x^2)$$

$$= \tfrac{1}{4}(1.2^2 - 0.8^2) = \tfrac{1}{5}$$

$$v = \tfrac{1}{5}\sqrt{5}$$

b At O, $x = 0$:

$$v^2 = \tfrac{1}{4}(1.2^2 - 0^2) = 0.36$$

$$v = 0.6\,\text{ms}^{-2}$$

c $0.15\,\text{ms}^{-2}$

d $0.412\,$s (3 s.f.)

37 a $\frac{\pi}{4}\,\text{mh}^{-1}$

b 4 hours

38 a Let the piston be modelled by the particle P.

Let O be the point where $AO = 0.6\,$m

When P is at a general point in its motion.

Let $OP = x$ metres and the force of the spring on P be T newtons.

Hooke's Law:

$$T = \frac{\lambda x}{l} = \frac{48x}{0.6} = 80x$$

$R(\rightarrow)$: $F = ma$

$$-T = 0.2\ddot{x}$$

$$-80x = 0.2\ddot{x}$$

$$\ddot{x} = -400x = -20^2 x$$

Comparing with the standard formula for simple harmonic motion, this is simple harmonic motion with $\omega = 20$.

$$T = \frac{2\pi}{20} = \frac{\pi}{10}$$

b $6\,\text{ms}^{-1}$

c $\frac{\pi}{15}\,$s

39 a Let E be the point where $OE = 0.6\,$m

When P is at a general point in its motion, let $EP = x$ metres and the force of the spring on P be T newtons.

Hooke's Law:

$$T = \frac{\lambda x}{l} = \frac{12x}{0.6} = 20x$$

$R(\rightarrow)$: $F = ma$

$$-T = 0.8\ddot{x}$$

$$-20x = 0.8\ddot{x}$$

$$\ddot{x} = -25x = -5^2 x$$

Comparing with the standard formula for simple harmonic motion, this is simple harmonic motion with $\omega = 5$.

$$T = \frac{2\pi}{\omega} = \frac{2\pi}{5}$$

b $6.25\,\text{ms}^{-2}$

c $0.68\,\text{ms}^{-1}$ (2 s.f.)

d As it passes through C, P is moving away from O towards B.

40 a When P is at the point X, $AX = x$ metres and the force of the spring be T newtons.

Hooke's Law:

$$T = \frac{\lambda x}{l} = \frac{21.6x}{2} = 10.8x$$

$R(\rightarrow)$: $F = ma$

$$-T = 0.3\ddot{x}$$

$$-10.8x = 0.3\ddot{x}$$

$$\ddot{x} = -36x = -6^2 x$$

Comparing with the standard formula for simple harmonic motion, this is simple harmonic motion with $\omega = 6$.

$$T = \frac{2\pi}{\omega} = \frac{2\pi}{6} = \frac{\pi}{3}$$

b $9\,\text{ms}^{-1}$

c $\frac{\pi}{18}\,$s

d $1.16\,$m (3 s.f.)

41 a $4mg$

b Hooke's Law:

$$T = \frac{\lambda e}{l} = \frac{4mg\left(\frac{1}{4}l + x\right)}{l} = mg + \frac{4mgx}{l} \qquad (1)$$

Newton's second law:

$R(\downarrow)$: $F = ma$

$$mg - t = m\frac{d^2 x}{dt^2} \qquad (2)$$

Substitute (1) into (2):

$$mg - \left(mg + \frac{4mgx}{l}\right) = m\frac{d^2 x}{dt^2}$$

$$-\frac{4mgx}{l} = m\frac{d^2 x}{dt^2}$$

$$\frac{d^2 x}{dt^2} = -\frac{4gx}{l}$$

c $\frac{1}{2}\sqrt{3gl}$

d First P moves freely under gravity until it returns to B. Then it moves with simple harmonic motion about O.

42 a $\frac{9a}{2}$

b When P is at a general point X, let $OX = x$.

At this point the extension of the string is $0.5a + x$.

Hooke's Law:

$$T = \frac{\lambda e}{l} = \frac{8mg\left(\frac{1}{2}a + x\right)}{a} = mg + \frac{2mgx}{a} \qquad (1)$$

Newton's second law:

$R(\downarrow)$: $F = ma$

$$mg - t = m\frac{d^2 x}{dt^2} \qquad (2)$$

Substitute (1) into (2):

$$mg - \left(mg + \frac{2mgx}{a}\right) = m\frac{d^2 x}{dt^2}$$

$$-\frac{2mgx}{a} = m\frac{d^2 x}{dt^2}$$

$$\frac{d^2 x}{dt^2} = -\frac{2gx}{a}$$

Comparing with the standard formula for simple harmonic motion, this is simple harmonic motion

with $\omega = \sqrt{\dfrac{2g}{a}}$

$$T = \frac{2\pi}{\omega} = 2\pi\sqrt{\frac{a}{2g}} = \pi\sqrt{\frac{2a}{g}}$$

c $\frac{1}{2\sqrt{2}}a$

Online Full worked solutions are available in SolutionBank.

d As $a > \frac{1}{2}a$, the string will become slack during its motion. The subsequent motion of P will be partly under gravity, partly simple harmonic motion.

Challenge

1 a $\dfrac{dv}{dt} = -e^{2kv}$

$$\int_u^0 e^{-2kv}\, dv = -\int_0^t dt$$

$$-\frac{1}{2k}\left[e^{-2kv}\right]_u^0 = -t$$

$$t = \frac{1}{2k}\left(1 - e^{-2ku}\right)$$

$$= \frac{1}{2k}\left(\frac{e^{2ku} - 1}{e^{2ku}}\right)$$

b $\left(\dfrac{1}{4k^2} - \dfrac{1 + 2ku}{4k^2 e^{2ku}}\right)$ m

2 $a = 8x\dfrac{dx}{dt}$

Using $a = v\dfrac{dv}{dx}$ and $v = \dfrac{dx}{dt}$

$v\dfrac{dv}{dx} = 8xv$

$\dfrac{dv}{dx} = 8x$

$\int dv = \int 8x\, dx$

$v = 4x^2 + c$

$t = 0,\ x = 0,\ v = -k:\ -k = c$

$\dfrac{dx}{dt} = 4x^2 - k$

Displacement has maximum when $\dfrac{dx}{dt} = 0$

$4x^2 - k = 0 \Rightarrow x = \dfrac{\sqrt{k}}{2}$

So maximum displacement is $\frac{1}{2}\sqrt{k}$

3 $\dfrac{1}{2gk}\ln\left(\dfrac{4 + 4kU^2}{4 + kU^2}\right)$

Exam-style practice: AS level

1 a $v = \dfrac{12e^{\frac{t}{2}} - 1}{e^{\frac{t}{2}} + 1}$

b $(e^{\frac{t}{2}} - 1) < (e^{\frac{t}{2}} + 1) \Rightarrow \dfrac{(e^{\frac{t}{2}} - 1)}{(e^{\frac{t}{2}} + 1)} < 1$

so $v = \dfrac{12(e^{\frac{t}{2}} - 1)}{(e^{\frac{t}{2}} + 1)} < 12\,\text{m s}^{-1}$

2 a $R(\rightarrow):\ T\cos 60° = m \times 2a\cos 60° \times \dfrac{kg}{4a} \Rightarrow T = \dfrac{mkg}{2}$

b $mg\left(1 - \dfrac{\sqrt{3}\,k}{4}\right)$ **c** $k < \dfrac{4}{\sqrt{3}}$

d $R(\rightarrow):\ T\cos\theta = m \times 2a\cos\theta \times \dfrac{3g}{a}$

$\Rightarrow T = 6mg$

$R(\uparrow):\ T\sin\theta = mg \Rightarrow 6mg\sin\theta = mg$

$\Rightarrow \sin\theta = \frac{1}{6}$

$QX = 2a\sin\theta = \dfrac{a}{3}$

$QO = 2a\sin 60° = a\sqrt{3}$

$\therefore QX:QO = \frac{1}{3}:\sqrt{3} = 1:3\sqrt{3}$

3 a i $1.1a$ **ii** $3.25a$

b a

c $20.5°$

d Weight acts through midpoint.

Exam-style practice: A level

1 a i 12.5 hours   **ii** 5 m

b $2.46\,\text{m h}^{-1}$

c 7.05 hours (3 s.f.)

2 a i $\dfrac{11a}{12}$ **ii** $\dfrac{5a}{12}$

b $24°$

3 a $\frac{1}{4}\ln\left(\dfrac{k^2 + 4U^2}{k^2 + U^2}\right)$ m

b $\displaystyle\int_U^{2U} \dfrac{1}{k^2 + v^2}\, dv = -\int_0^t 2\, dt$

$\left[\dfrac{1}{k}\arctan\dfrac{v}{k}\right]_U^{2U} = -[2t]_0^t$

$\dfrac{1}{k}\arctan\dfrac{2U}{k} - \dfrac{1}{k}\arctan\dfrac{U}{k} = -2t$

So $t = \dfrac{1}{2k}\left(\arctan\dfrac{U}{k} - \arctan\dfrac{2U}{k}\right)$

4 a $47.4\,\text{m s}^{-1}$

b $23.4 < v < 77.4$

5 a $V = \pi\displaystyle\int_0^{\frac{\pi}{4}}\cos 2x\, dx = \pi\left[\frac{1}{2}\sin 2x\right]_0^{\frac{\pi}{4}} = \pi\left(\frac{1}{2} - 0\right) = \dfrac{\pi}{2}$

b $\dfrac{\pi}{4} - \dfrac{1}{2}$

c $\alpha = 74°$ (2 s.f.)

6 a At P, K.E. $= \frac{3}{2}ag$ and P.E. $= mg(a - a\cos 60°) = \frac{1}{2}mg$

so total energy is $2amg$.

When the string makes angle θ with OQ,

total energy $= \frac{1}{2}mv^2 + amg(1 - \cos\theta)$

$\Rightarrow \qquad v^2 = 2ag(1 + \cos\theta)$

$T - mg\cos\theta = \dfrac{mv^2}{r}$

$\Rightarrow T = mg\cos\theta + 2mg(1 + \cos\theta)$

$\qquad = mg(2 + 3\cos\theta)$

String goes slack when $T = 0$, so $2 + 3\cos\theta = 0$

$\Rightarrow \cos\theta = -\frac{2}{3}$

$\Rightarrow \qquad \theta = -131.8°$

This corresponds to an angle of $48.2°$ with the upward vertical.

P does not make complete circles, as the string goes slack before P reaches the top of the circle.

Index